Optical Interconnection

Foundations and Applications

For a complete listing of the *Artech House Optoelectronics Library*, turn to the back of this book.

Optical Interconnection

Foundations and Applications

Christopher Tocci and H. John Caulfield
Editors

Artech House
Boston • London

Library of Congress Cataloging-in-Publication Data

Includes bibliographical references and index.
ISBN 0-89006-632-9
1. Optical interconnects. 2. Electrooptical devices. I. Tocci, Christopher.
II. Caulfield, H. J. (Henry John), 1936–
TA1660.O67   1994   94-5943
621.39'1–dc20   CIP

A catalogue record for this book is available from the British Library

© 1994 ARTECH HOUSE, INC.
685 Canton Street
Norwood, MA 02062

All rights reserved. Printed and bound in the United States of America. No part of this book may be reproduced or utilized in any form or by any means, electronic or mechanical, including photocopying, recording, or by any information storage and retrieval system, without permission in writing from the publisher.

International Standard Book Number: 0-89006-632-9
Library of Congress Catalog Card Number: 94-5943

10 9 8 7 6 5 4 3 2 1

*To my wife, Meche, my parents, Audrey and Anthony Tocci,
my grandparents, Mabel and Olin Burr,
and finally, the inimitable—P.J.*

C. S. T.

To George, Claude, Mary, and all of their descendents

H. J. C.

Contents

Foreword   xi

Preface   xv

Section I   Motivational Aspects: An Introduction to Optical Interconnection

Chapter 1   An Introduction to Optical Interconnection   3
- 1.1   Information Processing's Dependence Upon Connectivity   3
- 1.2   The Evolution of Optical Interconnection Technology   4
- 1.3   More Contemporary Issues   5
  - 1.3.1   At the Device Level   5
  - 1.3.2   At the System Level   6
- 1.4   Organization of the Book   7
- References   8

Chapter 2   Formalism of the Current Interconnect Bottleneck   9
- 2.1   Introduction   9
  - 2.1.1   Interconnect Bottleneck in Conventional Microelectronic Systems   9
  - 2.1.2   Advantages of Optical Interconnects   13
- 2.2   Power-Speed Tradeoff Analysis   14
  - 2.2.1   Chip-Level Interconnects   15
  - 2.2.2   MCM-Level Interconnects   35
- 2.3   Connection Density Analysis   42
  - 2.3.1   Introduction   42
  - 2.3.2   Lower Boundaries on Interconnect Area   43
  - 2.3.3   Interconnect Area Upper Boundaries for Specific Optical Architectures   50
  - 2.3.4   Discussion   55
  - 2.3.5   Conclusions   59

		References	60
Section II		**Current Optical Interconnection Approaches and Limitations**	
Chapter 3		Bulk Waveguide Routing—Fiber Optics	65
3.1	Introduction		65
	3.1.1	Intrashelf Communication	69
	3.1.2	Intershelf Communication	75
	3.1.3	System Backplane Requirements	76
3.2	Fiber Embedded Substrate Technology		77
	3.2.1	Rigid Substrates	78
	3.2.2	Flexible Substrates	82
3.3	Technology Capabilities and Limitations		84
	3.3.1	Fiber Array Connector Capabilities	85
	3.3.2	Interconnection Density and Fiber Packaging Issues	89
	3.3.3	Passive and Active (Rare Earth-Doped) Fiber Backplane Networks	90
3.4	Environmental Ruggedness Issues		96
3.5	Summary		98
		References	99
Chapter 4		Free-Space Routing	103
4.1	Introduction		103
4.2	Free-Space Optical Routing Components and Architectures		110
4.3	Network Implementations		117
4.4	Spatial Light Modulators and Processor-Related Issues		128
4.5	Laser Optical Power Supplies and System Diagnostic Tools		135
4.6	Summary		137
		References	138
		Further Reading	139
Chapter 5		Integrated Optical Waveguide Routing—Micro-optics	141
5.1	Introduction		141
5.2	Optical Waveguides		143
	5.2.1	Introduction	143
	5.2.2	Properties of Guided Waves	144
	5.2.3	Slab Waveguides	151
	5.2.4	Channel Waveguides	160
	5.2.5	Planar Medium	166
	5.2.6	Waveguide Fabrication	167
5.3	Waveguide Couplers		177
	5.3.1	Coupling of Laser Diode Beam into Fiber	178
	5.3.2	Fiber-Waveguide Coupling	180
	5.3.3	Other Transversal Coupling Schemes	185
5.4	Waveguide Lenses		185

5.5	Active Guided Wave Devices	193
	5.5.1 Electrooptic Waveguide Devices	195
	5.5.2 Acoustooptic Modulators	211
	5.5.3 Magnetooptic Modulators	218
5.6	Conclusions	220
	References	222

Chapter 6 Optoelectronic Interconnects and Electronic Packaging — 227

6.1	Introduction	227
6.2	Design and Implementation Considerations	228
6.3	Material and Processing Requirements	236
6.4	Electronics for Integration	237
6.5	Integrated Optoelectronic Detectors and Receivers	237
6.6	Integrated Optoelectronic Transmitters	240
6.7	Packaging	244
	References	252

Section III Applications of Optical Interconnects

Chapter 7 Polymer-Based Photonic Integrated Circuits and Their Applications in Optical Interconnects and Signal Processing — 255

7.1	Introduction	255
7.2	Formation of Polymer Microstructure Waveguides	256
7.3	Formation of Linear and Curved Channel Waveguide Array	260
7.4	Formation of Waveguide Holograms by Local Sensitization with Applications to WDM	264
7.5	χ^2 Electrooptic Polymer	268
7.6	Rare Earth Ion-Doped (REI) Polymer Waveguide Amplifier	272
7.7	Polymer-Based Electrooptic Waveguide Modulator	279
7.8	Fanout Density and Optical Interconnection	281
7.9	Reliability Test	284
7.10	Further Applications	287
	7.10.1 Nonlinear All Optical Switch	287
	7.10.2 Systolic Arrays for Optical Computing	288
	7.10.3 Position Sensor	289
7.11	Conclusions	290
	References	291

Chapter 8 Optical Interconnection for use in Hostile Environments — 295

8.1	Introduction	295
8.2	Optical Interconnection Applications in Hostile Environments—A Brief Overview	296
	8.2.1 Backplane Interconnection Applications	296
	8.2.2 Smart Skins and Structures	297
	8.2.3 Additional Candidates for Embedding	298

8.3	Case Example 1: Extraction of Cryogenically Generated Signals	299
	8.3.1 Problems with Proposed Integrated Optical Cryogenic Coupler	314
	8.3.2 One Solution to the Leaked Channel Light	314
	8.3.3 Inefficient Handling of Unused Channel Optical Power Solution: Recycle Light	315
8.4	Case Example 2: MCM High Bandwidth Interconnections	316
	References	318

Chapter 9 An Approach for Implementing Reconfigurable Optical Interconnection Networks for Massively Parallel Computing — 319

9.1	A 3-D Integrated Optical Computer Architecture	320
9.2	Large Guided-Wave Networks	324
9.3	Network Analysis	326
	9.3.1 Crossbar Networks	328
	9.3.2 n-Stage Network	331
	9.3.3 Multistage Interconnection Networks	332
	9.3.4 Duobanyan Network	336
	9.3.5 Benes Network	338
	9.3.6 Active and Passive Splitter/Combiner Networks	340
9.4	Comparison of Networks	344
9.5	A 3-D Integrated Optical Crossbar Network	347
9.6	System Operation	351
9.7	Conclusion	353
	References	354

Appendix	Theoretical Analysis for Cryogenic Coupler's Usable Dynamic Range (Per Channel)	357
List of Acronyms		367
About the Authors		371
Index		377

Foreword

Interconnection is one of the most important concepts in optics and photonics with regard to information technology. The first and most-established aspect of optical interconnection is its application to the high-speed, large-capacity transmission of information. While long-distance communication is already a reality, researchers are also exploring applications on a smaller scale: at the board-to-board, chip-to-chip and on-chip level. The second aspect, which I find more interesting, is that interconnection is involved in almost all aspects of optical information-processing systems. This stems from one of the advantages optics enjoys over electronics: its ability to convey information at large bandwidths in free space with less crosstalk. Even simple imaging by a single lens can be taken as a multichannel, point-to-point interconnection from an object plane to an image plane. This book fully discusses optical interconnection, emphasizing the "reality" of the approach (as described in the Preface). I believe the book will provide every reader with information insights for future developments in optics/photonics and information technology.

Satoshi Ishihar
Japan
November 1993

The demand for increased-bandwidth capacity for telecommunications and computer networks is leading to visions of high-capacity fiber-optic "information superhighways." Beyond using optical elements only for transmission, however, there has been a growing realization that increasing demands for system speed and capacity are making it attractive to consider optical and photonic technology in signal-processing, for example, the "intelligence" of communications and computing systems [1].

Ten years ago, participants at topical meetings on optical computing and photonic switching were discussing visions of all-optical computers and transparent optical telecommunications networks—the assumption being that all switching, logic, and signal-processing operations would be performed with optical components. More recently, it has been acknowledged that it appears unlikely that optics will ever replace electronics for those applications where electronics is able to perform satisfactorily. However, the combination of optical and electronic components appears capable of dramatically enhancing and extending the capabilities of current-day electronic systems.

Over the last ten years, photonics technology has made remarkable advances, and commercial applications are growing rapidly [2]. Small microlasers have been demonstrated with threshold currents well below 1 mA. The low power dissipation of these lasers increases their integrability with other optical or electronic components. There have also been major strides in materials growth and in device design and fabrication that facilitate the integration of optical components to the point where one can now design complex circuits in which there is an intimate intermixture of photonic and electronic components. Such circuits will permit the system designer the freedom of using optics where this technology offers advantages and electronics where this is the desirable technology, without major sacrifices of power or complexity in the optics/electronics conversions.

Beyond long distance transmission, there are two areas where optical technology would appear to have some clear advantages over existing technology. One is in the area of high-speed operations, where optical devices exhibiting switching times of less than 1/10 of a piesecond have been demonstrated in the laboratory [3]. The other area, optical interconnection, is the subject of this volume.

Two recent studies are relevant. In May 1993, the IEEE sponsored a workshop on "Interconnections within High-Speed Digital Systems." The central focus of this workshop was a design problem assigned to groups of participants [4]. They were asked to define a family of hardware components and modules to serve as basic building blocks for high-speed interconnections. Any technology could be used for the solution. Over the near term (from now to about five years out), most groups opted for electronic interconnection solutions. In the longer term (ten years out), however, optical interconnections started to dominate.

In September 1993, AT&T and Corning published the results of a survey on potential applications of free space digital optics [5]. They surveyed a spectrum of communications and computing researchers, government and industry representatives, and potential users. The overwhelming consensus was that free-space digital optics will: lower energy required for chip-to-chip communications; allow massive pinouts per chip or high bandwidth interconnections; decrease signal skew problems; and enable new connection-intensive architectures. There was less consensus on the time frame, but most respondents believed there would be significant impact in less than ten years.

In this volume, Tocci and Caulfield have invited a number of experts to provide perspectives on the state of the art for optical interconnections and their visions of future development. It provides a fascinating overview of a technology that may prove to be crucial to the future "information age."

P. W. E. Smith
University of Toronto
December 1993

REFERENCES

[1] P. W. E. Smith, "Optics in Telecommunications: Beyond Transmission," in "International Trends in Optics", J. Goodman, ed. Academic Press, New York (1991).
[2] "The Light Fantastic," *Business Week*, May 10, 1993.
[3] S. R. Friberg, A. M. Weiner, Y. Silberberg, B. G. Sfez, and P. W. Smith, "Femtosecond Switching in a Dual-Core-Fiber Nonlinear Coupler," *Optics Letters*. Vol. 133, pp. 904–906 (1988).
[4] *IEEE LEOS Newsletter*, Oct 1993, pp. 7–16.
[5] "Free-Space Digital Optics: Commercial Applications Survey Results," Braxton Associates, Boston, MA, Aug. 1993.

Preface

This book is intended not as a text on optical interconnection technology but as a discussion of optical interconnection philosophy with verifiable technological thinking and examples. This is a key point, because many books on emerging fields have the tendency to be little more than a compilation of not-so-current symposia or journal reports with some editorial verbiage thrown in to ease digestion. It is the explicit intention of the editors that this text be useful for a number of years after you purchase it because the philosophies proposed here will only mature with time, not become lost or greatly reduced in meaning due to technological evolution.

Many textbooks have addressed optics for communication, signal processing, and computing. However, there were no textbooks dedicated to how optical designs handle the efficient routing and coupling of photonic information. This book attempts to address this issue explicitly through a number of different authors with very different backgrounds and approaches. The only underlying principles that the individual authors were required to adhere to were: explain your optical interconnection approach (philosophy) and support your approach with the "ility" checks of reproducibility, manufacturability, testability, and reliability (san-"ity"). The "ility" checks will automatically enforce "reality" on each design approach to optical interconnection. These constraints assure the reader that each method examined in the text is not frocked with the same "snake oil" mentality that seems to show up in many of the research papers.

The book's other primary purpose is an attempt to stem the "overselling" of general optical processing without regard to the optical connectivity issues. By adopting this mentality, the reader can truly begin to evaluate the research journals, trade periodicals, and tidal wave of optical processing ideas in terms of their approach from the aforementioned "ilities," hence, efficacy. Without a fairly rigorous examination of real optical interconnection issues, the required baseline for this connectivity form will grow in a haphazard fashion. Further, each optical computer or processor idea developed will put all its emphasis on the computing or processing effort with only a cursory study or understanding on the assumed, but absolutely important, issue of efficient photonic connectivity.

The editors feel, as do the individual contributors, that optical interconnection is defined by technology, architectural concepts, and all associated manufacturing processes which define the "reality" of the approach. This baselining by practical examination is paramount for future optical computing and processing ideas and, ultimately, products.

The editors would like to thank the contributors for their effort and insights into this fairly well misunderstood topic. Thanks to them, perhaps, the readers will not make the same errors in the technical assessment of optical processing and computing machinery as many technical journals, periodicals, and even conferences have done. The editors have certainly gained a tremendous insight into this fundamentally important topic.

Acknowledgments

At this time, we would like to acknowledge the following software companies whose donated products helped create the textual, mathematical, and graphical conversion formats required for the creation of this book. Halcyon Software of 1590 La Pradera Drive, Campbell, California, 95008, provided DoDOT 4.0 Graphics Toolbox. Pacific Micro of 201 San Antonio Circle, C250, Mountain View, California, 94040, provided Mac-In-Dos Common-Link for the PC. Design Science of 4028 Broadway, Long Beach, California, 90803, provided MathType 3.0 equation editor. Wolfram Research, Inc. of 100 Trade Center Drive, Champaign, Illinois, 61820, provided Mathematica 2.1. Inset Systems of 71 Commerce Drive, Brookfield, Connecticut, 06804-3405, provided HiJaak for Windows 1.0.

The editors would also like to thank Mr. Mark Walsh of Artech House for accepting the idea for this book from an unknown writer and struggling optiker; Ms. Lisa Tomasello (Acquisitions Assistant at Artech) for keeping the editors honest and on time (in a pleasant way); and Mr. Kim Field for greatly assisting the electronic generation of this text. This has been an incredible effort over two years with eight of the best names in the field of optical interconnection.

Dr. Tocci would also like to thank the management at Baird Corporation, in particular Dr. Robert J. Krupa and Mr. Charlie Ross, for the acceptance of the many phone calls and unexplained behavior associated with this editor during the many months of editing.

Christopher S. Tocci
H. John Caulfield
August 1993

Section I
Motivational Aspects: An Introduction to Optical Interconnection

Chapter 1
An Introduction to Optical Interconnection
H. John Caulfield and Christopher S. Tocci

As switching components become smaller and less expensive, we begin to notice that most of our costs are in wires, most of our space is filled with wires, and most of our time is spent transmitting from one end of the wire to the other [1].

This text will address issues of optical interconnection with respect to the following questions. What are the electrical versus optical interconnection issues? Can we throw the electrical wiring away? If so, when? If not, why not? (The ground rules of optical interconnection.) Are optical fibers and waveguides just fancy wires or something more powerful? (The promise of something better with optical interconnection.) Can we throw away the basic idea of guiding or confining the message carriers in wires or fibers and go to unconfined "free space"? (The diversity of optical interconnection.) What are the advantages, costs, and tradeoffs within and among these options? (The reality of optical interconnection.)

1.1 INFORMATION PROCESSING'S DEPENDENCE UPON CONNECTIVITY

While this book is not directly concerned with processing information, the subject cannot be completely ignored when considering interconnection of signals. After all, interconnection, whether by software or dedicated hardware, defines what process is going to take place as well as how the process is going to take place. Therefore, interconnection is processing by virtue of intent of what information goes where, when, and how.

A complicated hardware processor is often an orderly collection of simple connections among simple processing units whose timing, position, and variability are controlled by an intention implemented by software. The central processing unit (CPU) or other hardware processor appears to do much more than this, but it appears to do more by performing hundreds or thousands of simple routings very fast. This fundamental observation is the basis of information—its perception and its processing. In simple terms, information exists (i.e., reality) and is assimilated into the perceiver (e.g., man, plant, machine) by the perceiver's ability to process or reformulate the information into a compatible form for consumption and ultimate usage. The reformulation is taking one form of energy (i.e., information) and converting it into another by repositioning the constituent parts of the original information. Computers do not create information. There is no magic. Computers rearrange and often simplify—by partially destroying—the input information. This repositioning is an intentional reconnection of the fundamental attributes of the input. Consequently, the ability of a system to optimally acquire the appropriate reconnection map is crucial to the perceiver's speed and depth of assimilation or understanding. It is in this domain that man's intention of information handling is reflected in the directed "wiring" of any machine.

1.2 THE EVOLUTION OF OPTICAL INTERCONNECTION TECHNOLOGY

From an ancient historical perspective, optical interconnection was conceived and used long before radio telecommunications came along. A simple and direct method was required for communicating with one's comrades during military maneuvers. This was especially true during the heat of battle where verbal communications were badly compromised by the noise of battle. Light sources were cheap and readily available (e.g., fires, candles, and the sun). Controlling light was equally easy and many materials could be used to control light direction (e.g., a mirror or shiny metallic plate) and concentration (e.g., a simple lens). From a modern sense of interconnection, most applications use optical signaling techniques for the same reasons that our earlier ancestors did. Optical communication is directionally quiet, generally providing line-of-sight security. It is also largely immune to interference, which makes it reliable. The fact that optical communication has a much higher information bandwidth than its purely electronic counterparts was not a motivator for its use until the last 40 years. The development of the modern fiber optic cable was a way to circumvent the bandwidth limitations associated with long telecommunication links. However, just as the printed circuit board replaced large amounts of bulky hook-up wire in electronic equipment, bulky fiber optic cabling will evolve into integrated optical "printed" circuits.

It is interesting to notice the evolutionary analogy between electrical with optical interconnection. Initially, electrical circuits and subsystems were interconnected with hook-up wiring. During the early 1960s, printed circuits began to emerge, greatly reducing electronic interconnection while enhancing performance. As the printed circuit evolved, circuitry became smaller, and many functions were integrated into a single package called an integrated circuit (IC). These smaller electronic circuits had increased information handling, lower power dissipation, and smaller footprints. Further, with the advent of mass foundry production techniques in silicon, germanium, and other III-V and II-VI compounds, these "superchips" became extremely cost effective at producing a given function as compared to unit cost.

This same reasoning has already started evolving in optics. The electrical hook-up wire was replaced with fiber optics, which exhibited much higher bandwidth without the associated electrical crosstalk and "hash" noise mechanisms that cause signal degradation. This was especially important to the telecommunications industry. Both analog and digital data links were economically feasible in the mid to late 1960s. Further refinement of packaging and optoelectronic components and integration techniques allowed multi-GHz bandwidths and tens of kilometers of distance between repeaters. These improvements have now suggested, as have many badly restricted computational high-throughput machine architectures, that integrating the optoelectronic device into IC packaging and replacing electrical wire with optical interconnection is the next evolution of signal handling or transference. Remember that guided optical information does not interact with neighboring optical signals; hence, signal or channel densities may increase several orders of magnitude. In a three-dimensional interconnection, this could easily result in an overall signal throughput increase of six to seven orders of magnitude in any given system.

Integrated optical interconnection has a big help in its birth because much of what is understood has been inherited from the microelectronic industry. Such integration techniques give optical interconnection technology the immediate advantage and maturity of scalability and compatibility with existing ICs. However, optical interconnection is not without it brand of design pain either. Energy coupling becomes paramount in determining the efficacy of an interconnection. For instance, mode coupling between two converging, or diverging, couplers must be geometrically matched, or there is the characteristic "reflection" of power. RF designers will empathize immediately to the geometric considerations. However, geometric mode matching at optical wavelengths is much tougher than at RF, microwave, or millimeter electrical wavelengths. Compound this last statement of efficient power coupling with the greatly increased bit error rate (BER) requirements of future computing and telecommunication networks, and one gets the sense of what is at stake in the upcoming decades of information handling and processing.

1.3 MORE CONTEMPORARY ISSUES

Sections 1.3.1 and 1.3.2 will briefly clarify the two fundamental problem areas associated with electrical interconnection. It is the purpose of this text to address these problem areas in some detail in the light (no pun intended) of optical interconnection.

1.3.1 At the Device Level

Ironically, the issue of increased information bandwidth became important in the very large scale integrated (VLSI) design arena. In a seminal paper, the issues of the die feature size and interconnection density were the fundamental problem areas for electrical interconnection [2]. These limitations were interconnection bandwidth associated with lumped resistor-capacitor (RC) time constants among logic gates and interconnection scalability of VLSI devices with decreasing geometries.

The information bandwidth limitation is reasoned that when device features are reduced in scale by a, the effective interconnection capacitance is reduced by a, but the interconnection resistance is increased by a. The net RC time constant remains the same; hence, there is no increase in signal bandwidth with decreasing device geometry. Clearly, after working hard to compact more circuitry onto a chip, it is frustrating that this same scale reduction does little to nothing for improving information throughput.

The scalability of interconnection problem arises when more complex VLSI chips are decreased in size and interconnected to other increasingly complex VLSI chips within the same package. If we want to interconnect N VLSI chips with M interconnections, then Rent's Rule indicates that $M = N^{2/3}$. This is an empirical relationship based on statistically independent VLSI chip functionality. Further, Rent's Rule assumes logic gates and not memory elements because memory requires fewer connections. However, when larger VLSI subsystems are created on larger substrate areas, Rent's Rule's exponent may be dropped to 1/2. Nevertheless, with the megagate designs of today, and tens-of-megagates in the next four to five years, the interconnection issue will dominate most system and architectural design considerations.

Another scalability problem associated with electrical interconnection is the electromigration effect. As stated earlier, if designs are scaled down by α, cross-sectional areas of conductors are reduced by α^2 while currents are reduced by α. This creates an increase in current density of α, hence stressing the conduction lines into failure by internal electronic bombardment of the lattice structure of the conductor.

1.3.2 At the System Level

In the early days of computing, parallel processing was considered the direction that computing would evolve. Unfortunately, the number of wires grew so large that interconnection became a major problem. John Von Neumann, who had such a profound impact on so many fields in so few years, invented a way around this interconnection problem. His approach used only a few wires from the CPU to memory and vice versa. Each item had an address, so any item could be read into or out of memory without a one-to-one connection. This brilliant insight made modern computers possible.

To gain speed, most modern computers use multiple processors in parallel to attack complex problems. For these situations, Von Neumann's invention is no longer well suited and slows down operations considerably. Ironically, this situation has become known as the Von Neumann bottleneck. Computer scientists of the modern era often forget that parallel processors preceded modern serial processors.

Almost every other aspect of the computer has changed in recent years. At one time, the processing elements were large, slow, and expensive, while interconnections (e.g., wire) were relatively small, fast and inexpensive. Today, the situation is reversed. Processing units are small, fast, and inexpensive; so the size, speed, and cost of interconnections now dominate computing. Computers are described less by their processors than by their interconnections (e.g., cosmic cubes, connection machines, and butterfly machines). In short, interconnections are the heart of modern multiprocessors. Further, many current multiprocessor architectures are composed of constituent subprocessor arrays which, in themselves, create another interconnection challenge.

What we ultimately seek are messages (i.e., information packets) that use their own headers and built-in rules to find their own optimum routes through complex networks. This compact and self-contained message format is exactly what modern communication systems use today. Hence, one might come to the realization that smart communication and multiprocessor systems might have very similar architectural topologies. This is indeed happening. Less intelligence is built into the hardware or architecture and more into the message package as to how, when, and where the information is to be processed by a simple processor.

It is to this mentality that the interconnection processes become much more important than what processors are used. The information packets contain enough information so that little more than a somewhat simple or "dumb" processor architecture can anneal or adapt the appropriate connection map to implement the optimal solution with some given constraints.

1.4 ORGANIZATION OF THE BOOK

The text is structured into three parts describing the motivation, current approaches, and specific applications associated with optical interconnection.

Section I (Chapters 1–2) introduces the reader to the underlying reasons why optical interconnection has become a necessity for future computer and processing systems. The speed and density issues associated with current VLSI technology are approaching their theoretical limit in two fundamental ways. First, the smallest electronic geometries are around the .2 micron dimension, which affects fabrication yields and creates immense electromagnetic crosstalk problems as clocking speeds increase. Consequently, performance compromises between interconnection density and speed have to be made to keep the overall system throughput high. The second reason for conversion to optical connectivity arises from switching energy requirements. This energy requirement is the amount of energy required for a logic gate to change its state. For insulated gate VLSI electronic structures (e.g., GaAs, MOS, CMOS, and InP.), the following equation gives the power required to change the state of a gate within a certain period of time.

$$P_G = \frac{1}{2}C(\Delta V)^2 f$$

where P_G denotes the power dissipated at the gate (watts); C denotes the gate capacitance (farads); ΔV denotes the minimum voltage difference between logic states (volts); and f denotes the operating frequency (Hz or second^{-1}).

Imagine a chip, such as the Intel i586 or P5 CPU, that has a three-million gate count and an internal clock of 66 MHz. Substituting into the above equation with some typical VLSI values gives a chip heat dissipation per gate of: $C = 0.2$ pF, $\Delta V = 1.7$V, $f = 66$ MHz, and $P_G = 19.1$ μW.

Hence, there is a maximum possible heat dissipation of 57W for the entire VLSI package. Obviously, this chip is headed for a certain meltdown unless most of the gates are not functioning most of the time. Further, the input gate capacitance, C, is not truly indicative of all GaAs input capacitances present. Any Miller effect or compound active device topology could increase this value by 5 to 10 times with an accompanying increase in total device heating. Clearly this chip, like all others of increasing density and speed, cannot take full advantage of its internal architecture and/or clocking capability.

This brief example is a prime cornerstone in the power consideration of computational or signal processing operations. In Chapter 2, Michael Feldman will show more detailed constraints associated with electronic implementations of high-throughput architectures and compare them with the optical approach.

Section II (Chapters 3–6) begins the discussions associated with the different optical approaches for interconnection. In Chapter 3, Ronald Nordin explains the use of optical fiber as embedded optical conductors in PCB and MLB fabrications. In Chapter 5, Robert Shih and Tomasz Jannson will describe the emergence of integrated optical waveguides for routing and switching of optical information and will introduce the hybridized designs of interconnection that merge conventional electronic circuits with electro-optical devices and constitute the optoelectronic interconnect and its impact on device packaging. This is described in more detail by Lynn Hutcheson in Chapter 6. In Chapter 4, Rick Morrison will discuss the pragmatics of free-space optical connection and introduce the reader to binary or diffractive optics and its increasing role in optical signal routing.

Section III (Chapters 7–9) gives the reader some practical application areas that have been investigated using some forms of optical interconnection. This section was included to assure the reader that such interconnection modality definitely has merit and, in some cases, represents the only way certain signals can be handled, either because of bandwidth, density, or environmental reasons.

REFERENCES

[1] Hillis, W.D., *The Connection Machine*, MIT Press, Cambridge, Massachusetts, 1985.
[2] Goodman, J. W., F. I. Leonberger, S. Y. Kung and R. A. Athale, "Optical Interconnections for VLSI Systems," *IEEE Proc.*, Vol. 72, 1981, pp. 850–866.

Chapter 2
Formalism of the Current Interconnect Bottleneck

Michael Feldman

2.1 INTRODUCTION

Recently, attention has been given to using optical technology to improve connection of devices on a single chip or wafer and between chips and modules [1–4,6]. For circuits requiring complex interconnection, optical interconnections will occupy less area than their electronic counterparts [7].

2.1.1 Interconnect Bottleneck in Conventional Microelectronic Systems

The performance of many existing computational systems based on microelectronic integrated circuits is limited by the performance of communication links between devices, rather than by the devices themselves [1–4]. Interconnects are often responsible for the majority of the area, power, and time delay. This is particularly evident for high-speed systems requiring connection of widely separated devices.

The interconnect bottleneck occurs at all levels of electronic packaging. In this chapter, three levels of packaging will be discussed:

- Level one: chip-Level (intrachip) interconnects are connections between devices on a single very large scale integrated (VLSI) circuit or chip.
- Level two: multichip Module (MCM) level (intra-MCM or chip-to-chip) interconnects are connections between VLSI chips within a single MCM.
- Level three: board-Level (inter-MCM) interconnections are connections between MCMs on a single printed circuit board.

The electronic interconnect bottleneck is severe at each packaging level. Although the bottleneck tightens as one progresses to higher and higher levels of packaging, the impact on performance at each level is largely determined by the particular application. For example, many systems are designed to concentrate large amounts of processing power on individual chips and minimize the chip-to-chip communication links. For such systems, the level-one connection bottleneck is typically the most severe. For other more globally interconnected systems, the level-two or level-three bottleneck may be the most critical factor limiting performance.

2.1.1.1 Level One: Chip-Level Interconnects

The communication bottleneck at the chip level stems in part from the large capacitance and resistance associated with long interconnect lines. Both the capacitance and resistance of a VLSI interconnect increase at a rate proportional to the connection line length, L. Therefore, the switching energy of the interconnect (proportional to the capacitance) grows with L, and the interconnect rise time (proportional to the product of the resistance and capacitance) grows with L^2.

The communication bottleneck will become more severe and widespread as VLSI technology advances with further increases in device densities and switching speeds [1,3,4]. The dramatic increases in VLSI performance during the last few decades were produced by scaling down device dimensions combined with increasing the total chip area. While scaling down the chip minimum feature size improves the performance of individual devices, it degrades the performance of the interconnects. When all circuit dimensions are reduced (scaled) by a factor of S ($S > 1$), the gate delay of each transistor is reduced by a factor of $1/S$ and the switching energy by $1/S^2$ [1]. However, the RC delay of a fixed length connection link increases by a factor of S^2 [1,3]. As the total chip size increases, longer connection lengths are required, further worsening the interconnect bottleneck.

The area of high-density, VLSI circuits is often determined by the interconnects. This problem is caused primarily by the planar nature of VLSI connections. Typical VLSI interconnects are limited to two levels of low-resistivity metal interconnects. Incorporation of more levels of interconnects, an active research area, would alleviate this problem only temporarily. Abstract VLSI computational models that allow for an arbitrary number of interconnect layers but restrict the number of layers to a constant value, independent of the chip area, have been used to determine asymptotic growth rates of VLSI connection networks as a function of the number of processing nodes, N, in the network [5,6]. These models have shown that while the area of the processing circuitry grows at a rate proportional to N, the area of the interconnects of many popular networks grows at a rate proportional to N^2. Thus, for large values of N, the interconnect area will dominate the area of the chip.

VLSI connections are subject to several additional detrimental effects, including signal crosstalk and electromigration. To conserve area, it is desirable to place interconnect lines as close as possible. This can lead to crosstalk due to signal interference. Crosstalk effects become more severe with feature size scaling and place additional limitations on interconnect bandwidth. Electromigration is the physical displacement of atoms in a conductor as a result of the passage of a high-current density, which eventually leads to the deterioration of the interconnect. Electromigration effects also worsen with integrated circuit (IC) scaling [3].

2.1.1.2 Level Two: MCM-Level Interconnects

Multichip modules were introduced to reduce the large difference in performance of chip-level interconnects and chip-to-chip, board-level interconnects. Chip-to-chip, board-level connections are used when single-chip carriers are employed rather than MCMs. A single-chip carrier is a package that contains I/O ports (typically pins or surface-mount technology pads) on the outside. These I/O ports are attached to corresponding locations on a PC board. The chip is placed in the interior of the package. The chip I/O pads may be flip-chip bonded or wire bonded to the SCC substrate. For a typical 1×1–cm chip, the SCC may be 2- to 3-inches long. Due to the large size of the SCC package, the distance between chips is typically quite large when SCCs are employed.

MCMs allow multiple VLSI chips to be placed in a single package. The chips are typically attached to the MCM substrate with flip-chip bonding or tape automated bonding (TAB). Up to 100 1×1–cm chips can be placed within a 4×4–inch MCM package.

MCMs allow for a level of communication intermediate between chip-level and board-level connections. Due to the close proximity of the chips, time delay and resistive effects are not as severe as for board-level interconnects. In addition, the electrical connections between an MCM and a chip typically contain much less capacitance and inductance than for a board-level interconnect. Finally, the connection density and speed of level-two connections is typically much higher than board-level connections. For the most part, this is due to developing MCM technological capabilities that are outpacing PC board technology, partly because of the smaller size of MCM substrates.

There are three principal differences between level-two and level-one interconnects.

1. Level-two interconnects contain bonding pads needed to make connections between the chip and the MCM substrate. The large area of chip bonding pads (and their associated circuitry) reduces the connection density. The large capacitance and inductance tends to increase the interconnect power dissipation and reduce the interconnect bandwidth.
2. Level-two signal line dimensions are typically much larger than chip-level interconnects. The smallest level-two connections are approximately 5- to 8-μm wide with a pitch of 15- to 25-μm. (Typical density and speed of state-of-the-art packaging-level interconnects are given in Table 2.1.) Much lower density connections are typically employed to reduce the line resistance, especially for long distance and high-speed connections.
3. Transmission line effects tend to dominate the interconnect performance, especially for long links. Transmission line effects become significant when the interconnect

Table 2.1
IC Process Parameters Expressed as Functions of Circuit Minimum Feature Size, $\tilde{\lambda}$ (in μm), for First-Level Aluminum Lines

Parameter	Symbol	Value	Units
Minimum line width	W_{MIN}	$1.5\,\tilde{\lambda}$	μm
Gate input capacitance	C_{IN}	$170\,\tilde{\lambda}\,\epsilon_{OX}$	fF
Output capacitance	C_{OA}	$0.72\tilde{\lambda}^2$	fF
Contact output capacitance	C_{OB}	$3.1\tilde{\lambda}^2$	fF
Line capacitance	C_{LA}	$\epsilon_{OX}/(0.35 W_{MIN})$	fF/μm^2
Fringing line capacitance	C_{LB}	~ 0.061	ff/μm
Line resistance	R_L	$\rho/(0.09\tilde{\lambda})$	Ω/sq
Power supply voltage	V	$2.9\,\tilde{\lambda}^{1/2}$	V
Inverter saturation current	I_O	0.26	mA

Note: ϵ_{OX} = the permittivity of silicon dioxide = $3.9 \times 8.85 \times 10^{-3}$ fF/μm; ρ = the resistivity of aluminium = 0.0274Ω-μm.

distance is larger than about 1/10th of the electrical signal wavelength. Transmission line effects can result in the attenuation of electrical signals through the skin effect. They can also result in multiple reflections if not properly terminated.

For these reasons, VLSI technological advances are concerned with packing more devices onto a single chip and thereby avoiding chip-to-chip interconnects.

2.1.1.3 Level Three. Board-Level Interconnects

For MCM-to-MCM interconnects, the interconnect bottleneck tightens even further. Both the connection density and bandwidth capabilities are typically reduced by more than an order of magnitude. MCMs can be attached to a PC board by pin grid array and pin-in-hole (PIH) technology, surface-mount technology, or flip-chip technology. Pin spacing is typically 25–100 mils. The associated capacitances and inductances typically limit the data rate on the PC board to <500 Mb/s. The line pitch on a PC board is typically 100–300 μm.

To alleviate the VLSI communication bottleneck, chip designers have employed the principle of locality [2]. That is, circuits are designed such that connections as short as possible are used. Mesh- and ring-connected systolic processor arrays are examples of this design philosophy. These networks have proven useful in solving problems such as matrix-matrix and matrix-vector multiplication. However, they are fundamentally limited to solving problems in a time that grows at least as fast as the square root of the number of PEs. Increased global communication is required to solve problems faster, which requires longer connection lengths.

2.1.2 Advantages of Optical Interconnects

Optical interconnects can improve communication in microelectronic systems by greatly alleviating most of the previously mentioned problems [3,7–10]. The use of optical interconnects to replace particular electrical connection links with optical signal paths can result in the following advantages: increased communication speed, reduced power consumption, reduced area, reduced crosstalk, increased reliability, increased fault tolerance, reduced cost, enhanced circuit testing capabilities, and provision of a path toward dynamic interconnects.

Most of these advantages stem from differences between fundamental properties of photons and electrons. While electrons are charged fermions, subject to strong mutual interactions and strong reactions to other charged particles, photons are neutral bosons, virtually unaffected by mutual interactions and Coulomb forces. Thus, unlike electrons, multiple beams of photons can cross paths without significant interference.

Similarly, photons can propagate through transparent materials without appreciable attenuation or power dissipation. Thus, neglecting speed of light delays, the speed of an optical link is limited only by the switching speed and capacitance of the transmitters and detectors. (For a 10-cm connection length, and a 50° hologram deflection angle, the speed-of-light delay is 0.5 ns.) Hence, the speed and power requirements of an optical interconnect are independent of the connection length. Because electrical VLSI connections have a switching energy directly proportional to the line length and an RC delay that grows quadratically with line length, for long enough communication links, optical connections will dissipate less power and provide faster data rate communication. This will be discussed in detail in Section 2.2.

The increased interconnect density capabilities of optical communication systems stem from the noninteracting nature of photons, allowing multiple light beams to cross paths without mutual effects. This property allows holographic interconnects to achieve a three-dimensional connection density with only two-dimensional optical elements. It will be shown in Section 2.3 that optical interconnects can achieve a higher interconnect density for many popular multiprocessor connection networks than fundamental limits allow for planar VLSI connections. It will also be shown that only fully three-dimensional VLSI interconnects could compete with the interconnect density capacity of an optical system. However, because the optical interconnect system consists only of a fixed number of planar elements, the cost of the system is proportional only to its planar area, not its volume.

This high-density interconnect capacity enables an optically interconnected system to occupy a smaller area than an equivalent fully electrical VLSI system. This may reduce the cost of optical interconnects below that of electrical connections. The relative insensitivity of holograms to small defects may further reduce the cost of an optically interconnected VLSI system. If CGHs are employed, fabrication can be performed with standard VLSI processing procedures. However, if requirements on the alignment of an optical system are severe, this may increase the cost of the system. Alignment restrictions can be more severe than VLSI alignment tasks when two planar elements need to be aligned with high precision (≤ 5 μm) while separated by a large distance (≥ 1 cm).

There are several additional advantages obtained from the use of optical interconnects. Reduced crosstalk results from the lack of interaction among photons. The reliability of optical connection systems may be higher than VLSI interconnects because optical elements are not subject to critical failure mechanisms, such as electromigration, although the reliability of many proposed optoelectronic transmitters is still largely unknown. Fault tolerance can be achieved by designing and integrating holographic components after fabrication and testing of an MCM. The hologram can then be designed to interconnect only the working devices. Testing capabilities are enhanced by providing access to the interior of the chip through the use of integrated on-chip photodetectors.

2.2 POWER-SPEED TRADEOFF ANALYSIS

The ultimate goal of almost every technological advance in microelectronics computational systems involves either performance or cost. Performance is generally measured in terms of the number of computations per second and computation accuracy. Interconnects affect the computational speed of speed through the maximum data rate (or bit rate). The data rate of an interconnect is defined as the number of binary digital signals (bits) that can be reliably transmitted per second over the interconnect. Note that the data rate is different from the interconnect bandwidth, because bandwidth generally refers to sinusoidal signals. Thus, for a given interconnect, the bandwidth will generally be significantly higher than the maximum data rate.

The maximum interconnect data rate is limited by the interconnect rise time, defined as the time needed for the receiving gate input voltage to rise from 10% to 90% of its final value. The data rate and rise time are related by $D_R \leq 1/2t$, where D_R is the data rate and t is the interconnect rise time.

Limits on the power dissipation of computational systems are generally determined by system cost goals. Cost (and convenience) goals typically determine the type of cooling mechanism employed. For example, in conventional workstations, forced-air cooling is used with fans that are run at speeds sufficiently low so as to not distract users. In some high-performance workstations, water-cooling is used. The CRAY Y/MP super computer uses more elaborate (and much more expensive) refrigeration systems with low-temperature liquids. The cooling system that is chosen determines the amount of power that can be dissipated without causing any devices to exceed a particular critical temperature. The critical temperature is chosen to ensure a particular reliability.

The power dissipation and data rate of an interconnect are intricately related by $P = E \times D_R$, where P is the power dissipation and E is the interconnect switching energy. Switching energy is defined as the total energy required to switch the state of a receiving inverter from one state to the other state and back again. Note that for a given technology interconnect, as the employed data rate increases, the power dissipation increases. Thus, the actual data rate used is typically determined by power dissipation limits as determined by cooling and heat-sinking mechanisms that are dictated by cost limitations.

In the next two sections, expressions for the switching energy and the rise time of optical and electrical level-one interconnects are derived. Expressions for level-two interconnects are given in Section 2.2.2. We neglect the switching energy and rise time associated with the driver of the laser diode and light propagation delay. The validity of these assumptions is discussed in Section 2.2.1.8.

2.2.1 Chip-Level Interconnects

2.2.1.1 Chip-Level Optical Interconnect Model

A system for performing optical level-one interconnects is shown in Figure 2.1. Electronic signal lines are replaced with integrated optical signal transmitters, detectors, and a hologram. The optical transmitters can be GaAs lasers, light modulators, or LEDs. Note that for level-one interconnects, we assume that the lasers are integrated onto the VLSI chip. Several methods of creating hybrid systems combining GaAs and silicon are under development [8,9]. In Section 2.2.2, we will describe a model in which the optoelectronic transmitters are located on chips that are distinct from chips containing processing circuitry. We will first analyze a system consisting of a semiconductor laser transmitting a signal to a single photodetector (fanout of 1).

The optical detector circuits, (Figure 2.2), are CMOS compatible optical gates [5] consisting of a photodiode and a load transistor. The optical gate is required to drive a standard CMOS inverter gate, also shown in Figure 2.2. Although the use of a sophisticated amplifier circuit following the detector would decrease optical power requirements, it is impractical to fabricate complex amplifiers near every photodetector in a high-density interconnection scheme.

The switching energy of the optical interconnect illustrated in Figures 2.1 and 2.2 contains two components: E_1, the electrical energy supplied by the power supply of the

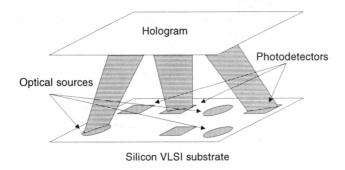

Figure 2.1 A system for performing optical level -1interconnects.

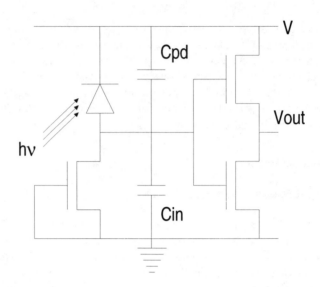

Figure 2.2 An optical detector circuit.

detector circuit; and E_2, the electrical energy required to drive the optical emitter. E_1 is given by

$$E_1 = (2 \times 2 \times 0.5)(C_{PD} + C_{IN})V^2 \tag{2.1}$$

where C_{PD} is the photodiode capacitance, C_{IN} is the input capacitance of a minimum size CMOS gate, and V is the power supply voltage. E_1 includes the energy dissipated in the photodiode and the load transistor during the charging of the receiving gate and the energy dissipated in the load transistor while the gate is discharging. The first factor of 2 in (2.1) is due to our definition of switching energy, which accounts for both the charging and discharging of the gate. The second factor of 2 results from the presence of the resistive load during the charging of the gate.

The second component, E_2, can be determined by noting that the photocurrent, I_p, generated within the detector is given by

$$I_p = \frac{2q(P_L - P_{th})\eta}{h\nu} \tag{2.2}$$

where h is Plank's constant, ν is the optical frequency, q is the electronic charge, P_{th} is the electrical power required to bias the laser at threshold, and P_L is the average total electrical power required to drive the laser. The total optical link efficiency, η, is defined as

$$\eta = \eta_L \eta_H \eta_D \tag{2.3}$$

where η_L is the external differential efficiency of the laser (incremental optical power out divided by incremental electrical power in) and η_H is the efficiency of the hologram. The photodetector quantum efficiency, η_D, is approximately equal to $(1 - e^{-\alpha d})$ (where α is the absorption coefficient and d is the thickness of the detector active region) if surface reflections and near-surface absorption is neglected. The switching energy, E, of an interconnect is related to the average power dissipation, P, by

$$E = 2\tau P \tag{2.4}$$

where τ is the interconnect rise time, defined as the time needed for the receiving gate input voltage to rise from 10% to 90% of its final value. For an optical interconnect, $\tau \approx 2R(C_{PD} + C_{IN})$, where R is the resistance of the load transistor. Setting R to a value of V/I_P allows the signal to rise to the power supply voltage, and using (2.2) and (2.4) yields the following expression for E_2:

$$E_2 = 2V(C_{PD} + C_{IN})\frac{h\nu}{\eta q} + 2\tau P_{TH} \tag{2.5}$$

Thus the total switching energy of an optical interconnect, E_0, given by the sum of E_1 and E_2 is

$$E_0 = 2V(C_{PD} + C_{IN})\left(\frac{h\nu}{\eta q} + V\right) + 2\tau P_{TH} \tag{2.6}$$

Because in many cases $\frac{h\nu}{\eta q} \gg V$, (2.6) simplifies to

$$E_0 \approx 2V(C_{PD} + C_{IN})\frac{h\nu}{\eta q} + 2\tau P_{TH} \tag{2.7}$$

Note that for small rise times, the switching energy is directly proportional to the detector circuit capacitance, $(C_{PD} + C_{IN})$. For long rise times, the switching energy is proportional to τ.

2.2.1.2 Chip-Level Electrical Interconnect Model

In this section, a model for CMOS VLSI interconnections is developed for a driving gate sending a signal to a single receiving gate (a fanout of 1). The model [10] is based on the circuit diagram in Figure 2.3. The use of additional "repeater" inverter gates placed periodically along the line to restore the signal rise time is not accounted for in this model. Also, the switching energy and delay time of the additional inverter stages required to

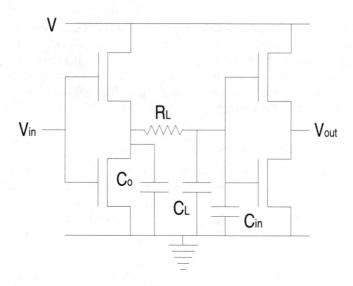

Figure 2.3 Electrical interconnection of two CMOS gates.

drive large line-driving gates are neglected. (Correspondingly, the energy and time delay associated with the drivers of the optical sources was neglected in the previous section.) R_L is the resistance per square of the transmission line. The three capacitances in the figure are the input capacitance of the receiving gate, C_{IN}; the output capacitance of the CMOS driving gate, C_O; and the line capacitance, C_L. C_{IN} includes the gate capacitances of the two transistors composing the receiving gate. (The Miller capacitance is neglected.) The output capacitance, C_O, can be divided into two capacitances as follows:

$$C_O = MC_{OA} + C_{OB} \qquad M \geq 1 \tag{2.8}$$

where M is the ratio of the gate width of the driving inverter to the gate width of the minimum size inverter allowed by the process design rules. C_{OA} is the junction capacitance of a minimum size inverter gate. C_{OB} is the additional capacitance due to the area needed to form a contact to the drain regions.

The line capacitance, C_L, is given by

$$C_L = LWC_{LA} + LC_{LB} \tag{2.9}$$

where L and W are the length and width of the line, C_{LA} is the parallel plate line capacitance, and C_{LB} is the fringing capacitance. The total capacitance of an electrical interconnect line, C_T, is given by

$$C_T = C_O + C_L + C_{IN} \tag{2.10}$$

and the switching energy of an electrical interconnect is given by

$$E_E = C_T V^2 = (MC_{OA} + C_{OB} + C_{IN} + C_{LA}LW + C_{LB}L)V^2 \tag{2.11}$$

The interconnect rise time can also be estimated from the model of Figure 2.3.

By approximating the dynamic resistance of a CMOS inverter as $V/2I_O$, where I_O is the maximum current that can be supplied by a minimum size CMOS inverter gate (which can be determined from SPICE [11] simulations), the total transmission line rise time can be calculated as

$$\tau \approx R_L C_{LA} L^2 + R_L C_{LB} L^2/W + 2C_{IN} R_L L/W + [V/(MI_O)](C_T) \tag{2.12}$$

The terms are found by multiplying each capacitance with the sum of the resistances that occur before it on the transmission line [12]. The first two terms are due to the distributed RC time delay of the transmission line itself. The third term is the RC time delay associated with the line charging the receiving gate. The fourth term is due to the driving gate charging all three capacitances.

A computer simulation program was developed to determine the validity of this interconnection model. Employing the SPICE circuit simulation program as a subroutine, the computer program can determine rise time and switching energy from user inputs of line length and SPICE process parameters. The program was used to model polysilicon interconnect lines for the MOS implementation service (MOSIS) [13] 3-μm CMOS process. The simulation results agree to within 15% of the analytic estimations over a wide range of signal line properties.

For fixed switching energy and line length, M and W should be chosen to minimize the interconnect rise time. For example, Figure 2.4 shows a plot of delay time versus M and W using (2.11) and (2.12) and using a switching energy of 11pJ and a line length of 1.5 mm for polysilicon lines fabricated by the MOSIS 3-μm CMOS process. Note that because the switching energy is held constant, M and W are related by (2.11), and thus there is only one independent variable. The figure clearly indicates that the delay time is minimized for a particular value of M. This optimum value of M is denoted by M_{OPT}, and the corresponding value of linewidth by W_{OPT}.

In general, W_{OPT} and M_{OPT} are the values of W and M, respectively, that minimize the rise time for given values of switching energy and line length. Thus, W_{OPT} and M_{OPT} can be calculated by employing La Grange multipliers; the equation to be minimized is (2.12) and the equation of constraint is (2.11). This results in

$$M_{OPT} = \frac{(E_E/V^2) - (C_{OB} + C_{IN} + C_{LB}L)}{C_{OA} + C_{LA}L\sqrt{B}} \tag{2.13}$$

$$W_{OPT} = \frac{(E_E/V^2) - (C_{OB} + C_{IN} + C_{LB}L)}{(C_{OA}/\sqrt{B}) + C_{LA}L} \tag{2.14}$$

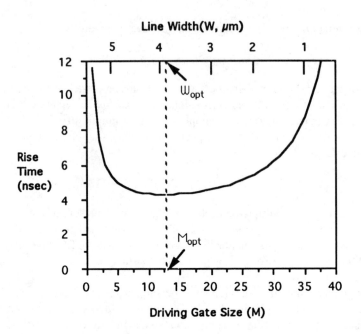

Figure 2.4 Interconnect delay time as a function of driving gate size and line width for constant energy loss: $E = 11$ pJ, $L = 1.5$ mm for a polysilicon interconnect line.

where $B = (R_L C_{LB} L + 2 R_L C_{IN}) I_0 V C_{OA} / (E_E C_{LA})$. These two equations are valid if they yield values of M and W that are physically realizable; that is, if $W_{OPT} \geq 1$ and $W_{OPT} \geq W_{MIN}$, where W_{MIN} is the minimum line width allowed by the process design rules. Equations (2.13) and (2.14) allow the optimization of line width and driving gate dimensions in the design of electrical interconnects.

2.2.1.3 Comparison of Electrical and Optical Level-One Interconnects for a Fanout of One, Based on 3-μm CMOS Design Rules

The switching energies of optical and electrical interconnects can be compared by examining the differences between (2.11) describing the switching energy of an electrical interconnect, E_E, and (2.6) describing the switching energy, E_O, of an optical interconnect with a semiconductor laser transmitter. To illustrate the behavior of these equations, both E_E and E_O are plotted as functions of rise time, τ, in Figure 2.5.

The optical interconnect switching energy curve is a plot of (2.6) for a 3.6-μm thick, 10-μm square detector size, yielding a photodiode capacitance of 5.3 fF (including sidewall capacitances). The detector quantum efficiency, approximately equal to $1 - \exp(-\alpha d)$, is approximately 30% for an assumed semiconductor laser wavelength of 0.8 μm ($\alpha \approx 1000$ cm^{-1}). Also, a total optical link efficiency, η, of 9.0% (e.g., laser

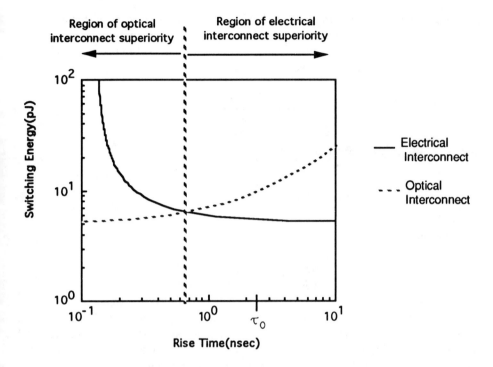

Figure 2.5 Switching energy versus rise time for a 1.0-mm aluminum line and for a 9.0% efficient optical interconnect with a 1-mW laser diode threshold power.

external efficiency = 40%, hologram efficiency = 75%, and detector efficiency = 30%) and a laser threshold power, P_{TH}, of 1.0 mW is assumed. (Values of $V = 5V$ and $C_{IN} = 17$ fF are based on 3-μm MOSIS CMOS specifications.) For small rise times (less than 0.3 ns in this case), the second term in (2.6) is negligible, and the first term yields an optical switching energy of ≈ 5 pJ, which is independent of τ. For long rise times, the second term dominates, and E_O is proportional to τ.

The electrical switching energy curve is a plot of (2.11), where again 3-μm MOSIS CMOS parameters are employed, this time for a 1-mm first-level aluminum line. The values of M and W in this equation were chosen by one of two different methods. We define τ_O as the rise time given by (2.12) with $M = 1$ and $W = W_{MIN}$. For rise times longer than τ_O, M and W are each set to their minimum values. In this region, data rates are slow enough that there is no need to increase M or W to reduce resistance, and hence, E_E can be kept at its minimum value. For rise times smaller than τ_O, however, M and/or W must be increased to reduce the rise time. In this region, values of M and W are given by (2.13) and (2.14) to minimize E_E for each value of τ. (If one of the equations yields a physically unrealizable result, the corresponding parameter is set to its minimum value and the other parameter value is determined from (2.11).) Because E_O is an increasing function of rise time and E_E is a decreasing function, E_O will be less than E_E for high data rate applications.

The ratio of the energy dissipated in an optical interconnect to that dissipated in an electrical interconnect can be obtained from (2.7) and (2.11):

$$\frac{E_O}{E_E} = 2\frac{h\nu}{qV}\frac{(C_{PD} + C_{IN})}{\eta C_T} + \frac{2\tau P_{TH}}{C_T V^2} \qquad (2.15)$$

If $\tau \ll \tau_2$, where τ_2 is defined as

$$\tau_2 = \frac{h\nu(C_{PD} + C_{IN})V}{q\eta P_{TH}} \qquad (2.16)$$

(2.15) reduces to

$$\frac{E_O}{E_E} = 2\frac{h\nu}{qV}\frac{(C_{PD} + C_{IN})}{C_T}\frac{1}{\eta} \qquad (2.17)$$

The second factor in (2.17) is simply the ratio of energies of optical and electrical fundamental particles. The third factor is the ratio of the capacitances associated with each type of interconnect. Note that the capacitance associated with electrical interconnects grows with increasing line length, whereas that associated with optical interconnects remains constant. This is illustrated in Figure 2.6, where E_E is plotted versus τ for several values of line length, L. Thus, in general, given any parameter values, for a large enough value of L or a small enough value of τ, E_O will be less than E_E.

The energy-versus-time plot in Figure 2.6 was converted to a power-versus-speed plot in Figure 2.7 by using the following relations:

$$1/2\tau = \text{maximum data transmission rate} \qquad (2.18a)$$

$$\text{data rate} = \frac{\text{total power dissipation}}{\text{switching energy}} \qquad (2.18b)$$

From Figures 2.6 and 2.7 (or (2.18b)), it can be seen that when the switching energy of an optical interconnect is less than that of a corresponding electrical interconnect, the optical interconnect is able to transmit data at higher rates than the corresponding electrical interconnect when subjected to the same power dissipation limits. Conversely, the optical interconnect is able to dissipate less power if data are transmitted at the same rate.

For a given rise time, we define the line length for which $E_O/E_E = 1$ as the break-even line length, denoted L_{be}. Line-length L_{be} is plotted versus τ in Figure 2.8, using the same values of η and process parameters as were used for Figure 2.6. Note that $L_{be}(\tau)$ is

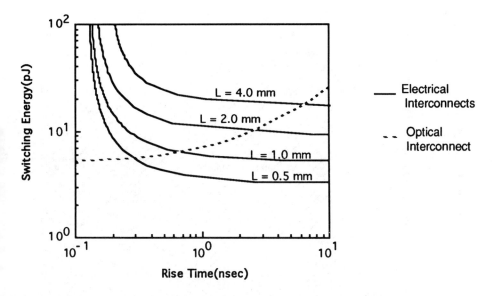

Figure 2.6 Switching energy versus rise time for aluminum lines for four different lengths and for a 9.0% efficient optical interconnect with a 1-mW laser diode threshold power.

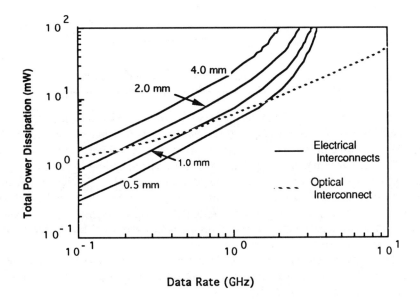

Figure 2.7 Total power dissipation per interconnect versus maximum data transmission rate for the same interconnect parameters as in Figure 2.6.

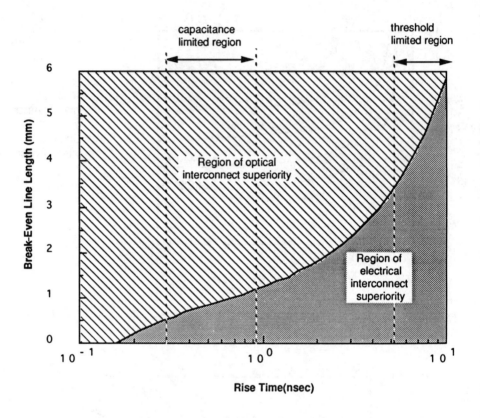

Figure 2.8 Break-even line length versus rise time for an aluminum interconnect line (3-μm minimum feature size) and a 9.0% efficient optical system, with 1.0-mW threshold power lasers.

given by the value of L for which an E_E curve crosses the E_O curve in Figure 2.6. For interconnect lengths longer than L_{be}, optical interconnects will have a smaller switching energy. For lengths shorter than L_{be}, electrical interconnects will have a smaller switching energy. Thus, for interconnect lengths and rise times corresponding to points lying in the shaded area labeled "region of optical interconnect superiority" in Figure 2.8, optical interconnects will have a smaller switching energy than the corresponding electrical interconnects. Similarly, the "region of electrical interconnect superiority" in Figure 2.8 indicates values of line length and rise time for which electrical interconnects have a smaller switching energy.

Although the dependence of L_{be} on τ is rather complicated in general, there are three regions where the curve obeys rather simple equations. For long rise times, the second term in (2.15) is dominant, and thus

$$L_{be}^{th} \approx \frac{2\tau P_{th}}{\{V^2(W_{MIN}C_{LA} + C_{LB})\}} \tag{2.19}$$

We define this length as the threshold limited break-even line length, L_{be}^{th}, and the corresponding values of τ for which $L_{be} \approx L_{be}^{th}$ as the threshold-limited region. The threshold-limited region, indicated in Figure 2.8, occurs for $\tau > \approx 2\tau_2$, where τ_2 was defined by (2.16). Note that L_{be}^{th} is equal to the energy required to bias a laser diode at threshold, divided by the energy per unit length required to charge an electrical interconnect line capacitance.

On the other hand, for $\tau \ll \tau_2$, the second term in (2.15) can be neglected, and hence,

$$L_{be} \approx \left(2\frac{h\nu}{qV}\frac{C_{PD} + C_{IN}}{\eta} - MC_{OA} - C_{OB} - C_{IN}\right)/(WC_{LA} + C_{LB}) \quad (2.20a)$$

If τ is near τ_1, which is defined as the rise time given by (2.12) with $M = 1$, $W = W_{MIN}$, and $L = L_{be}$, then M and W can be replaced with their minimum values, and (2.20a) becomes

$$L_{be}^C \approx \left(2\frac{h\nu}{qV}\frac{C_{PD} + C_{IN}}{\eta} - C_{OA} - C_{OB} - C_{IN}\right)/(W_{MIN}C_{LA} + C_{LB}) \quad (2.20b)$$

This length is defined as the capacitance-limited break-even line length, L_{be}^C. If the capacitances associated with the transmitting and receiving gates of the electrical interconnect are neglected, then L_{be}^C is proportional to the photodiode circuit capacitance divided by the electrical interconnect line capacitance per unit length. The capacitance-limited region, defined as the values of τ for which $L_{be} \sim L_{be}^C$, occurs for $\approx \tau_1/10 < \tau < \approx \tau_2/2$. This region is also indicated in Figure 2.8.

A third region of the break-even line length versus τ plot can be defined for values of τ for which the break-even line length is limited primarily by the RC delay of an electrical interconnect. From (2.12), one can show that

$$\tau > R_L C_{LA} L^2 + V C_{OA}/I_O \quad (2.21a)$$

and hence

$$L_{be}^{RC} < \sqrt{\frac{\tau - (VC_{OA}/I_O)}{R_L C_{LA}}} \quad (2.21b)$$

This boundary, defined as the RC-limited break-even line length, is an upper boundary on the break-even line length that holds for all τ. Because for the case illustrated in Figure 2.8 (3 μm IC minimum feature size, 9.0% efficient optical system) $L_{be}^{RC} \gg L_{be}$, an RC-limited region does not exist for this case. However, it will be shown in Section 2.2.1.5

that L_{be}^{RC} decreases with IC dimension scaling, resulting in an RC-limited region for small values of τ.

2.2.1.4 Fanout Considerations

Digital systems often require fanout, the sending of an output signal from a single gate to the inputs of several receiving gates. The rise time of an electrical transmission line performing F-fold fanout depends on the manner in which the F gates are distributed along the interconnect line. Figure 2.9 illustrates three types of fanout that are representative of the majority of cases.

The first type, illustrated in Figure 2.9(a), occurs when all of the receiving gates are located very close to each other on the IC. For an optical interconnect to perform this type of fanout, there would be no advantage in placing more than one photodetector by the receiving gates as this would increase the switching energy without reducing line capacitance or RC delay. Thus the optical system would provide a one-to-one connection, and the fanout would be performed electrically as indicated in Figure 2.9(b). For both electrical and optical systems, better performance can be obtained by placing an additional inverter immediately before the receiving gates. In this case, the break-even line length is identical to the case of a one-to-one gate connection.

A second type of fanout, denoted remote fanout, is illustrated in Figure 2.9(c). Here the receiving gates are not located in the same vicinity and hence require distinct interconnect lines. For simplicity, we assume that the F lines providing the F-fold fanout are each equal in length. Electrical interconnect remote fanout is modeled by replacing C_{IN} and C_L in (2.10) with FC_{IN} and FC_L, respectively. Optical interconnect remote fanout can be accomplished by allowing the hologram to divide the incoming beam into F output beams, focusing each output beam onto a distinct detector as illustrated in Figure 2.9(d). This can be accounted for by replacing $(C_{PD} + C_{IN})$ in (2.7) with $F(C_{PD} + C_{IN})$. Note that this results in no change in L_{be}^{RC}, and has only a small effect on L_{be}^{RC}, because the fanout factor, F, increases the values of terms in both the numerator and denominator of (2.20b). However, the threshold limited break-even line length becomes

$$L_{be}^{th} \approx 2\tau P_{th}/[V^2 F(W_{MIN} C_{LA} + C_{LB})] \qquad (2.22)$$

As F increases, L_{be}^{th} decreases linearly. Also, the rise time, which determines the transition from the capacitance-limited region, τ_2, to the threshold-limited region, becomes

$$\tau_2 = V(h\nu/q)(C_{PD} + C_{IN})F/(P_{th}\eta) \qquad (2.23)$$

As the fanout increases, the threshold-limited region is shifted to longer rise times and the width of the capacitance-limited region is extended. These effects are illustrated in

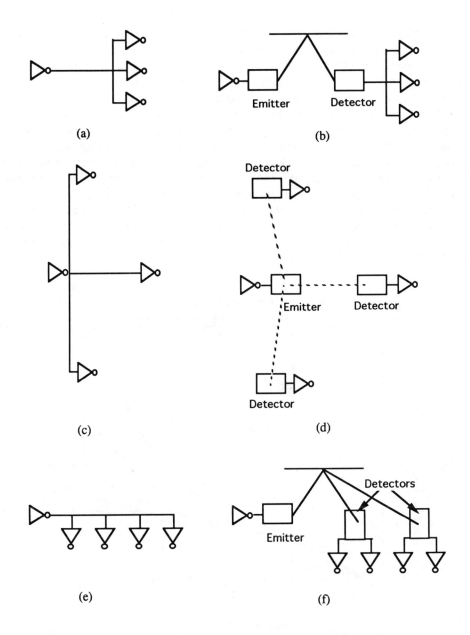

Figure 2.9 Fanouts for electrical interconnects (a,c,e) and optical interconnects (b,d,f).

Figure 2.10 where L_{be} is plotted versus τ for 5 different values of remote fanout. Note that for $F = 8$, the capacitance-limited region extends from $\tau \approx 0.4$ ns to $\tau \approx 5$ ns.

The third type of fanout occurs when the receiving gates are evenly distributed along the interconnect line as illustrated in Figure 2.9(e). In this case there is little or no reduction in line capacitance when an optical link replaces an electrical one (for a high density of gates). If optical interconnects were to be implemented by placing a photodetector next to each gate, the additional capacitance of this detector would be larger than the small amount of interconnect line that is removed. The advantage obtained in this case is only a reduction in RC delay. Depending on the delay times required, it is advisable to break the interconnect line into segments as indicated in Figure 2.9(f). In this way, fanout is performed partially by the hologram and partially by the electronics.

Note that this situation is quite different from that of an electrical interconnect with repeater inverter gates placed at each detector location. Although RC delay can be reduced in the electrical interconnect case, additional delay is introduced due to the product of the line resistance and the input capacitance of the repeaters.

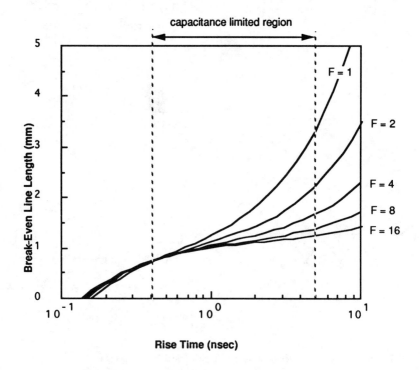

Figure 2.10 Break-even line length versus rise time for five different values of remote fanout. The efficiency of the optical system is 9.0%, the laser threshold power is 1 mW. The minimum feature size of the IC is 3 μm. The capacitance limited region is indicated for a fanout of 8.

Despite these advantages, an optical interconnect system utilizing a semiconductor laser as the signal transmitter is limited in fanout capabilities due to the limited light power that can be produced by a single laser. Denoting this maximum light power as P_{MAX} yields

$$P_{MAX} = (P_L - P_{th})\eta_L \qquad (2.24a)$$

with P_L, P_{th}, and η_L as determined in Section 2.2.1.1. Using (2.2) and (3.3) gives

$$P_{MAX} = V(C_{PD} + C_{IN})h\nu F/(\eta_H \eta_D q_\tau) \qquad (2.24b)$$

Thus, the maximum fanout is given by

$$F < \eta_H \eta_D q_\tau P_{MAX}/[V(C_{PD} + C_{IN})h\nu] \qquad (2.24c)$$

For $P_{MAX} = 50$ mW [15,8] and $\tau = 1$ ns, the maximum fanout is ≈ 67. Note that high-power laser diodes typically have threshold powers ≥ 1 mW [15,8]. The effects of larger laser threshold powers on break-even line length are discussed in Section 2.2.1.6.

2.2.1.5 Effects of Scaling Electronic Circuit Dimensions

Break-even line length has been calculated as a function of delay time based on typical parameters for a 3-μm CMOS process. For this process, it was found that L_{be} is <2.0 mm, for $\tau < 2$ ns, for aluminum first-layer lines. To determine how L_{be} changes with current and future semiconductor device processes, a scaling rule is applied. The scaling rule chosen is that of [3] in which all dimensions scale linearly, and the power supply voltage scales quasistatically. Denoting the minimum feature size of the process (the size of the minimum transistor gate length) as $\tilde{\lambda}$ gives the expressions for the process parameters listed in Table 2.1. The form of each expression was obtained from the scaling rules. The constants were chosen so that the parameters agree with those of the 3-μm MOSIS process for $\tilde{\lambda} = 3$-μm. We assume that the optical wavelength and the detector capacitance, C_{PD}, remain constant. The fringing capacitance was calculated by modeling the line as a parallel plate with half cylinders on each side [16].

In Figure 2.11, L_{be} is plotted versus τ for four different values of $\tilde{\lambda}$ for an aluminum interconnect line. The behavior of L_{be} with decreasing $\tilde{\lambda}$ can be described by examining the behavior of the L_{be}-versus-τ curve in the three regions described by (2.20b), (2.21b) and (2.22). The RC-limited line length can be written in terms of $\tilde{\lambda}$ as follows:

$$L_{be}^{RC} = 0.22\,\tilde{\lambda}\{\tau - (8\text{ psec}/\mu\text{m}^{5/2})\tilde{\lambda}^{5/2}\}^{1/2}(\rho\epsilon_{OX})^{-1/2} \qquad (2.25)$$

$$\approx 0.22\tilde{\lambda}\{\tau/(\rho\epsilon_{OX})\}^{1/2}$$

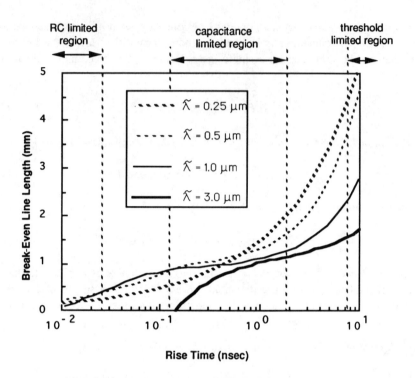

Figure 2.11 Break-even line length versus rise time for a 9.0% efficient optical system with 1-mW threshold power lasers and an aluminum interconnect line, both for a fanout of 8. The minimum feature size of the IC varies from 3 μm to 0.25 μm. The RC limited, capacitance limited, and threshold limited regions are indicated for a 1-μm IC minimum feature size.

for $\tilde{\lambda}^{5/2} \ll \tau/8$ psec, $\tilde{\lambda}$ in μm. For small $\tilde{\lambda}$, L_{be}^{RC} is directly proportional to $\tilde{\lambda}$. Thus, as feature sizes reduce, the RC-limited break-even line length decreases. The RC-limited region occurs for values of τ for which L_{be} approaches L_{be}^{RC}. This occurs for large values of M and W or for $\tau \ll \tau_1$. The RC-, capacitance-, and threshold-limited regions are indicated in Figure 2.11 for $\tilde{\lambda} = 1\mu m$.

The situation is just the opposite in the capacitance-limited and threshold-limited regions. From (2.20b) and Table 2.1, it can be seen that for small $\tilde{\lambda}$, the capacitance-limited line length approaches

$$L_{be}^{C} \to \frac{h\nu}{q} \frac{1}{V} \frac{1}{h} \frac{C_{PD}}{W_{MIN}C_{LA} + C_{LB}} \quad (2.26)$$

In this expression, all values are independent of $\tilde{\lambda}$, except the power supply voltage, V, which varies with $\tilde{\lambda}^{1/2}$. Thus, L_{be}^{C} is proportional to $\tilde{\lambda}^{-1/2}$ (for small $\tilde{\lambda}$), and as feature sizes scale, capacitance-limited break-even line length increases. However, L_{be}^C cannot increase indefinitely. Because the minimum supply voltage, V, is given by $2kT/q$ [17], the maximum value of L_{be}^C is given by

$$(L_{be}^C)_{MAX} = \frac{h\nu}{kT} \frac{1}{\eta} \frac{C_{PD}}{W_{MIN}C_{LA} + C_{LB}} \quad (2.27)$$

This is an upper boundary on break-even line length for arbitrarily small minimum features (and for $\tau \ll \tau_2$). This boundary is proportional to $h\nu/kT$, the ratio of a single quanta of light energy to a single quanta of thermal energy. It is also proportional to the ratio of the photodiode capacitance to the line capacitance per unit length.

The threshold limited break-even line length region ($\tau \gtrsim \tau_2$) is affected in two ways. First, the transition rise time, τ_2, decreases with decreasing $\tilde{\lambda}$. For small $\tilde{\lambda}$, τ_2 scales with $\sqrt{\tilde{\lambda}}$, and thus L_{be} approaches L_{be}^{th} for smaller values of τ. Second, L_{be}^{th} scales with $1/\tilde{\lambda}$ (from (2.22) and Table 2.1), resulting in larger break-even line lengths in the threshold-limited region with feature size scaling.

2.2.1.6 Effects of Improved Optical Link Parameters

Up to this point, optical interconnects have been evaluated based on a laser threshold power of 1 mW and an optical system efficiency of 9.0%. Although GaAlAs lasers have been fabricated with threshold currents as low as 1 mA [18] ($P_{th} \approx 1.5$ mW), the reported threshold currents of lasers fabricated as hybrids on top of silicon have been much larger [8,9]. Note that the laser threshold power affects the break-even line length only in the threshold-limited region ($\tau \gtrsim 2\tau_2$), where $L_{be}^{th} \propto P_{th}$. However, as the threshold power increases, τ_2 decreases (as shown in (2.23)). This behavior is illustrated in Figure 2.12, where L_{be} is plotted versus τ for several values of P_{th} for a fanout of 8- and 1.0-μm minimum feature size. For large values of P_{th} ($P_{th} \geq 100$ mW in this case), $L_{be}^{th} > L_{be}^{RC}$, the L_{be}-versus-τ curve is virtually equivalent to a plot of L_{be}^{RC} versus because τ.

Recall that the optical system efficiency, η, was defined in (2.3) as the product of the laser differential conversion efficiency, the hologram diffraction efficiency, and the detector quantum efficiency. Although from (2.21b) and (2.22), L_{be}^C and L_{be}^{th} are independent of η, it is evident from (2.20b) that L_{be} is inversely proportional to η. This dependence of L_{be} on η is illustrated in Figure 2.13, where L_{be} is plotted versus τ, for $\tilde{\lambda} = 1.0$ μm and $F = 8$, for five different values of η. Again, because for $\eta = 0.1\%$, L_{be}^C is very large, this particular curve is virtually equivalent to a plot of L_{be}^{RC} versus τ. As η increases, the energy-limited line length decreases. Figures 2.12 and 2.13 illustrate that as optical communication technology improves, the break-even line length will decrease dramatically. On the other hand, L_{be} is relatively insensitive to microelectronic technology improvements through minimum feature size scaling as indicated by Figure 2.11.

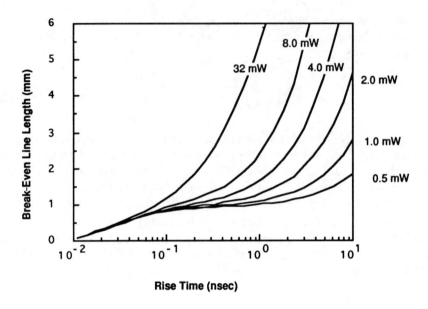

Figure 2.12 Break-even line length versus rise time for varying threshold powers of the laser transmitters. Optical system efficiency is 9.0%, IC minimum feature size is 1.0 μm, and the fanout is 8.

2.2.1.7 The Use of Light Modulators as Optical Signal Transmitters

As mentioned earlier, modulators are an alternative to semiconductor lasers as optical signal transmitters [19–21]. Figure 2.14 illustrates an optical interconnect system with light modulators as optical transmitters. (Systems using reflective modulators have also been proposed [20]). A dc-biased laser illuminates the back side of a VLSI chip that has a transparent substrate (e.g., silicon on sapphire). Modulators and detectors are integrated with the silicon circuitry on the transparent substrate. Modulators are attractive because they may be easier to integrate on a VLSI chip and they may dissipate less on-chip power because the electrical-to-optical power conversion occurs off chip. Although several silicon based light modulator technologies have been developed [19,22], these modulators have thus far exhibited limited switching speeds (<2 MHz). Methods for combining GaAs based MQW modulators with silicon are under development.

In this section, the switching energy and delay time of an optical interconnect employing a light modulator as the optical signal transmitter are calculated. The switching energy calculation consists only of on-chip energy dissipation; energy dissipated in the off-chip laser is neglected.

An optical link with a light modulator as the signal transmitter can be modeled by replacing η, η_L, and P_L with $\acute{\eta}$, η_M, and P_M, respectively, in (2.2) and (2.3); where $\acute{\eta}$, η_M, and P_M are defined as follows: $\acute{\eta} = \eta_M \eta_H \eta_D$, P_M is the optical power input to a modulator,

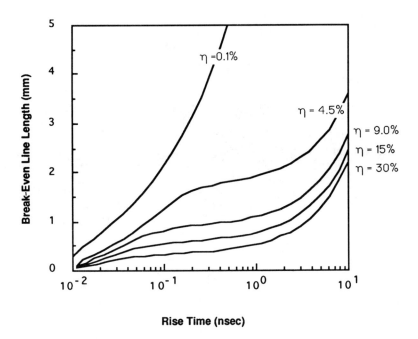

Figure 2.13 Break-even line length versus rise time for varying optical system efficiency. Minimum feature size is 1.0 μm, and the fanout is 8.

and η_M, the optical modulator efficiency, is equal to the optical power emitted by a modulator divided by P_M. Thus (2.5) becomes

$$E_2 = 4V(C_{PD} + C_{IN})h\nu/(\acute{\eta}q) \tag{2.28}$$

This is the optical energy required of an external source for illumination of each modulator. (The form of this expression is a factor of two larger than that of (2.5) because the modulator requires source illumination even when it is transmitting no light.) Electrical energy dissipated in each detector circuit is $2(C_{PD} + C_{IN})V^2$, as described in Section II. Additional electrical power is required to switch the state of a modulator. This power is given by $C_M V_M^2/(2\tau)$, where C_M is the capacitance of the modulator, and V_M is the modulator driving voltage supply. This results in a total interconnect switching energy of

$$E_O = 2VF(C_{PD} + C_{IN})\{2h\nu/(\acute{\eta}q) + V\} + C_M V_M^2 \tag{2.29}$$

for an optical link with an on-chip light modulator transmitter and a hologram performing a F-fold fanout.

For a multiple quantum well modulator, typical values are $\eta_M = 60\%$, and the modulator switching energy, $C_M V_M^2 = 2pJ$ (for a 225 μm² device) [23]. In Figure 2.15,

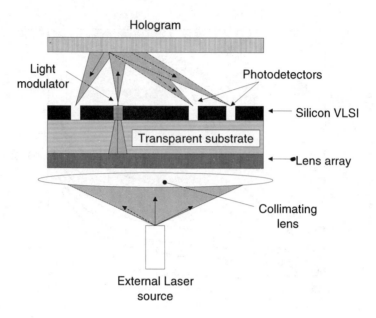

Figure 2.14 Optical interconnect system with modulators as optical signal transmitters.

break-even line length is plotted versus rise time for $\hat{\eta} = 36\%$ (e.g., $\eta_M = 60\%$, $\eta_H = 75\%$, $\eta_D = 80\%$) for several values of remote fanout. Because increasing F affects only one of the terms in (2.29), break-even line length decreases with increasing fanout when F is small. For large values of F, the first term in (2.29) dominates, and the break-even line length is relatively insensitive to changes in fanout. Note that the break-even line length for modulators as optical interconnect signal transmitters is smaller than for laser diodes (Fig. 2.10), especially for large rise times.

2.2.1.8 Summary of Chip-Level Interconnect Comparison

Switching energy and delay time have been expressed as functions of interconnect line length and IC process parameters for both the optical link of Figures 2.1 and 2.2 and the electrical link of Figure 2.3. Although the analysis was performed for free-space intrachip optical interconnects, fiber optic or optical guided wave interconnects could be accounted for by replacing η_H in (2.3) with the link efficiency of the guided wave link, including coupling losses and attenuation.

Several approximations are incorporated into the expressions developed here. For example, for electrical interconnects, the switching energy and delay time of the additional inverters required to drive large line-driving gates is neglected. A similar approximation is made for optical interconnects where the energy and delay associated with the driving circuitry of the lasers is neglected. Both of these approximations become invalid only for small rise times or large fanouts when large amounts of electrical power are required, assuming laser diodes with small threshold powers are used. For such cases, both optical and electrical interconnects would suffer from approximately equal amounts of additional energy dissipation and delay time. Other assumptions, including the neglect of the speed of light delays for optical interconnects, the neglect of transmission-line effects, and the lack of accounting for the use of repeaters for electrical interconnects are valid for the range of delay times and line lengths considered here.

The expressions for switching energy and rise time were used to plot break-even line length as a function of rise time for a 3-μm CMOS process (Fig. 2.8). For interconnect lengths longer than the break-even line length, optical interconnects have a smaller ratio of power-dissipation to data-transmission rate than have corresponding electrical interconnects. Although for small rise times the switching energy for optical interconnects is approximately constant, the switching energy for electrical interconnects increases as the rise time decreases because small rise times require large capacitive line-driving gates. Therefore, the break-even line length decreases with increasing data transmission rates as indicated in Figure 2.8. The break-even line length is <2 mm for rise times less than ~2.5 ns. As the remote signal fanout (Figs. 2.9(c,d)) is increased, Figure 2.10 indicates that the break-even line length decreases, especially for rise times greater than ~1 ns. The effect of IC dimension scaling on the break-even line length was found by applying a scaling rule to the parameters in the previously developed equations. Results plotted in Figure 2.11 indicate that for a conservative estimate of total optical link efficiency of 9.0%, a laser diode threshold power of 1.0 mW, and a signal fanout of 8, the break-even line length remains below ~2 mm for rise times $\leq$2 ns and IC minimum feature sizes >0.25 μm. Figure 2.12 indicates the large sensitivity of the break-even line length to the laser diode threshold power. Figure 2.15 shows that the use of light modulators rather than laser diodes as optical signal transmitters can result in smaller break-even line lengths (<1 mm for a 3-μm IC minimum feature size), especially for rise times longer than ~1 ns. These results suggest that optical interconnects may be advantageous for intrachip communication in large-area VLSI circuits or wafer scale integrated circuits, especially when high data rates and/or large fanouts are required.

2.2.2 MCM-Level Interconnects

2.2.2.1 Optical Interconnect Model

The principal difference between the MCM-level and chip-level models for optical interconnects is that for MCM-level interconnects we assume that the optical signal transmitter may be located on a different chip than the driving transistor. We will still assume that the detector is integrated onto the same chip as the receiving gate. The reasoning for these assumptions is as follows.

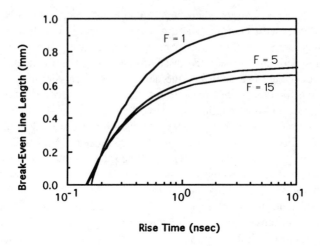

Figure 2.15 Break-even line length versus rise time for implementation of optical interconnects with light modulators. Fanout varies from 1 to 15. Optical link efficiency is 45%. Minimum feature size is 3 µm.

1. The performance of optoelectronic transmitters integrated onto VLSI chips is currently several orders of magnitude below that of similar transmitters located on specially designed distinct chips.
2. The performance of integrated detectors is comparable to discrete detectors to within an order of magnitude.
3. Placing the lasers off-chip allows the use of laser chips with few or no transistors, corresponding to the highest performance, state-of-the-art lasers.

However, placing the detectors off-chip would require integrating complex amplifiers containing many transistors onto the detector chip needed to drive the bonding pads. This will greatly increase the power dissipation, especially for large fanout connections and will require the use of VLSI-compatible detectors. Note that the principal disadvantage of this approach is that current VLSI chips cannot be used.

The MCM-level optical interconnect model is illustrated in Figure 2.16. Note that it is very similar to the previous model (Fig. 2.2) with the addition of a bonding pad and a short lumped RC line. With this module the switching energy of an optoelectronic connection, E_O, is given by

$$E_O = 2VC_{PD}F\{h\nu/(q\eta)\} + 2\tau P_{th} + V^2(2C_B + C_L L_1) \qquad (2.30)$$

where V is the power supply voltage, C_{PD} is the photodetector circuit capacitance, F is the fanout, ν is the optical frequency, q is the electronic charge, η is the total optical system efficiency (the product of detector quantum efficiency, hologram efficiency, and

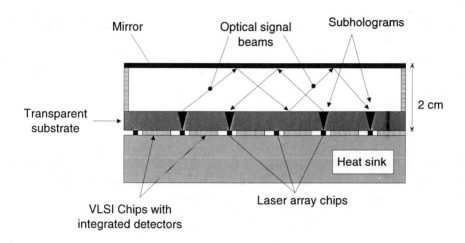

Figure 2.16 Optical system configuration with a double-pass hologram.

laser differential efficiency), τ is the electrical signal rise time, P_{th} is the laser diode threshold power, C_B is the bonding pad capacitance, C_L is the line capacitance per unit length, and L_1 is the distance from the driving gate to the laser. The second term in (2.30) gives the energy required to bias the laser diode at its threshold voltage. The first term gives the additional electrical energy needed to drive the laser and detector circuits to charge F photodetector capacitances to a voltage of V. The last term gives the power required to drive the bonding pads and the metal line between pads.

2.2.2.2 Electrical Interconnect Models

For MCM-level interconnects, the model for electrical interconnects changes much more dramatically than for optical interconnects. First, as for optical interconnects, the bonding pad capacitances must be included. This can be done rather easily by modifying (2.11) to yield

$$E_E = V^2\{C_B(F + 1) + FC_LL\} \qquad (2.31)$$

where L is the connection length.

Second, the value of C_L is larger for MCM-level connections than for chip-level interconnects. C_L in (2.31) is the capacitance per unit length of the MCM substrate microstrip lines, which is significantly larger than the capacitance per unit length of the intrachip metal lines. C_L is calculated from the following equation:

$$C_L = \epsilon_{OX}\left\{1.15\left[\frac{W}{T}\right] + 2.80\left[\frac{D}{T}\right]^{0.22}\right\} \qquad (2.32)$$

where W is the width of the line, T is the thickness of the line, and d is the spacing between the line and the ground plane as illustrated in Figure 2.17.

The first term in (2.32) is the parallel-plate line capacitance, and the second term is the fringing capacitance. Typical values for the ratios in (2.32) are $D/T \sim 1$ and $W/T \sim 2$, yielding,

$$C \approx 5.1\epsilon_{OX} \tag{2.33}$$

The switching energy of the electrical link on the MCM substrate is (2.31). Note that in (2.31), the switching energy associated with the on-chip connections, including the pad drivers, is neglected.

The third difference is that for long interconnect lengths and/or high-speed signals, transmission line effects become significant. Transmission line effects can cause attenua-

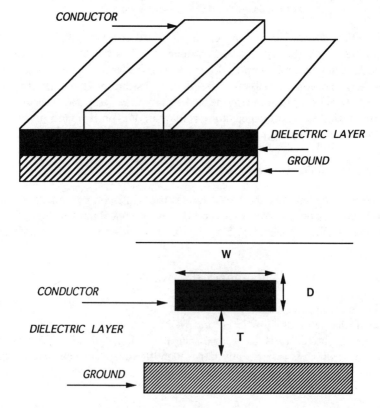

Figure 2.17 Crossection of a microstrip transmission line (top); conductor microstrip transmission line (bottom).

tion due to skin effects, degradation of signal integrity due to reflections, and increased power dissipation occurring in terminating resistors.

Despite these differences, (2.31) is still valid as a lower boundary on the switching energy, even at high speeds, if the entire electrical connection link is charged to the power supply voltage, V. In practical systems, this can only occur if the line length is much smaller than the wavelength of the electrical signal. In addition, this region in which the entire line is charged to the supply voltage is fundamentally bounded by

$$L \leq \lambda_e/2 \tag{2.34}$$

where λ_e is the wavelength of the electrical signal. When L is equal to this boundary (one-half of the wavelength), the entire line can be charged only to the same voltage only if the signal were an ideal square wave pulse (infinitely small transition times). Nevertheless, (2.31) is an approximate lower boundary as long as condition (2.34) is satisfied.

When the line length is longer than one-half the wavelength of the electrical signal, the pulsed transmission of signals along terminated transmission lines can be employed. In this case, the power dissipation for electrical interconnects may be reduced for long interconnect lines supporting high-frequency signals below that indicated by (2.31).

To model this case, we use an idealized model of electrical transmission lines. The idealized model is shown in Figure 2.18. The microstrip is modeled as a distributed LC line. Line resistance is neglected. The line is assumed to be perfectly terminated with a matched terminating impedance, Z_M. The capacitance of the NAND receiving gates is neglected, but the capacitance of the bonding pads is included.

The switching energy for the transmission line model, E_T, is given by

$$E_T = V^2 \{C_B(F + 1) + F/(D_R Z_O)\} \tag{2.35}$$

where Z_O is the characteristic impedance for TEM transmission. The first term in (2.35) is the energy required to charge the bonding pads in Figure 2.18. The second term is the energy dissipated by a single pulse in the purely resistive terminating impedance, Z_O. This impedance is given by

$$Z_O = \sqrt{L/C} \tag{2.36}$$

The inductance L, for a microstrip is typically calculated from the relation

$$c = 1/\sqrt{LC_{VAC}} \tag{2.37}$$

where C_{VAC} is the line capacitance per unit length when the dielectric layer in Figure 2.17 is replaced with a vacuum. The resulting inductance calculated from this equation is

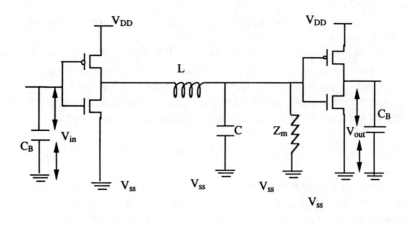

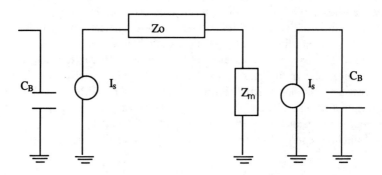

Figure 2.18 Schematic diagram (top) and model (bottom) of a perfectly matched transmissions line.

assumed to be unchanged when the real dielectric layer is inserted. Substituting (2.37) into (2.36) yields

$$Z_O = \sqrt{\epsilon_{ox}/cC_L} \tag{2.38}$$

Substituting (2.38) into (2.35) yields

$$E_{TL} = (2C_B V^2 + TcC_L)/\sqrt{\epsilon_{ox}} \tag{2.39}$$

Thus, a lower boundary on the power dissipation for electrical transmission lines is given by

$$P_{TL} = (2C_B V^2/T + cC_L)/\sqrt{\epsilon_{ox}} \tag{2.40}$$

2.2.2.3 Comparison Between Optical and Electrical Interconnects For MCM-Level Connections

The power dissipation for optical and electrical interconnects is compared in Figure 2.18(a,b). The optical interconnect curves in Figure 2.19 are based on (2.30). The lines labelled "optical interconnects (laser off-chip)" are obtained from (2.23) with $L_1 = 5$ mm and the value of C_B indicated. The value of C_B of 0.4 pF corresponds to the value employed in a prototype optical interconnect system developed at the University of North Carolina at Charlotte. The value of C_B of 0.14 pF is the value of C_B to be used in a second-generation system currently under development. The line labeled "optical interconnects (lasers on-chip)" is obtained by setting C_B and L_1 to 0 in (2.4) and corresponds to having the lasers integrated onto the silicon ICs.

The electrical interconnect plots in Figure 2.19 are obtained from two different models. When the line length is shorter than one-half of the signal wavelength, the lumped RC model, given by (2.31), is employed. When the line length is longer than one-half the signal wavelength, the ideal lossless perfectly matched transmission line model, given by (2.40), is employed. (Note that (2.30) and (2.39) yield equal values when the line

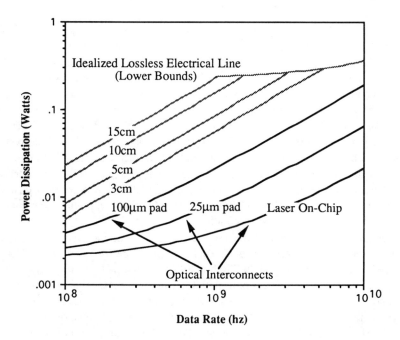

Figure 2.19 Power dissipation versus data rate for optical interconnects and lossless electrical interconnects. Note that the electrical interconnect depends on connection length but the optical interconnect does not.

length is equal to one-half the signal wavelength.) Thus, the electrical interconnect lines represent lower boundaries on the power dissipation for electrical connections.

Figure 2.19, values of V = 3V and $\epsilon_{OX} = 3.5\epsilon_0$ are used. Although $3V$ is a reasonable voltage for high-performance CMOS VLSI chips, many researchers have proposed using smaller voltage swings to reduce the power dissipation. Although this method does reduce the power dissipated in the terminating resistors, it requires highly dissipative amplifiers.

Note from Figure 2.19 that for data rates up to 20 GHz, the power dissipation for optically interconnected links can be much less than lower boundaries on the power dissipation of electrical links. At very high frequencies, the only difference between the optical and electrical interconnect curves is differences that result from the reduced number of bonding pads that can be used for optical interconnects. Also note that for frequencies below 1 to 2 GHz, there is only a small advantage in employing integrated on-chip optoelectronic transmitters.

2.3 CONNECTION DENSITY ANALYSIS

2.3.1 Introduction

A fine-grain parallel processing computer system consists of a large array of simple processing elements (PE). For many classes of problems, such a system can potentially have dramatically higher computational throughputs than conventional single-processor machines (e.g., [24–26]). However, many high-performance processor arrays require complex inter-PE communication networks involving many nonlocal interconnects or connections between chips or modules. The interconnections in VLSI circuit implementations of such processor arrays are responsible for a significant portion of the circuit delay, silicon area, and power dissipation. Selective replacement of some conventional connections with free-space optical connections [1] can yield circuit designs that require less area, consume less power, and operate at higher speeds. The conditions under which optical interconnects are attractive from the standpoint of circuit speed and power [27] have already been described. In this chapter, we examine the area requirements of optical interconnections for multiprocessor computers.

An optically interconnected processor array consists of individual electronic PEs that are interconnected by optical beams. Each PE would be identical to an electrically connected PE, except instead of including bonding pads and pad-and-link drivers, the PE would require one or more optoelectronic signal transmitter (a laser [28] or light modulator [29–31]) and one or more photodetectors [5]. A holographic optical element (HOE) placed above the PE plane serves to interconnect the transmitters and detectors in the desired pattern [32,33]. The PEs themselves may all reside on the same chip or wafer or on an ensemble of VLSI circuit chips, all bonded to a common substrate. The holographic interconnects are implemented in the same manner for both cases. That is, unlike in VLSI implementations, the interconnection mechanism is the same for both intrachip and chip-to-chip interconnects. We refer to such a processor array as a VLSIO (very large scale integrated optoelectronic) computer.

The purpose of this chapter is to compare VLSIO machines to conventional VLSI processor arrays in terms of interconnect density capabilities. In particular, for a computer containing N PEs, we calculate the growth rate of the area devoted to computation, A, as an asymptotic function of N (neglecting constant factors) for both optical and electrical interconnects. For electrical interconnects, the area devoted to computation, A_E, corresponds to the silicon area, and thus, the manufacturing cost is proportional to A_E. For optical interconnects, we assume all the PEs reside on one wafer. Interconnects are performed by a constant number (independent of N) of thin optical elements placed above the wafer. The area devoted to computation, A_O, consists of the silicon area plus the area of each optical element. Thus, again, cost is proportional to A_O. For both optical and electrical interconnects, the total area consists of three components: the area of the interconnects, A_I, the area of the PEs, and the area of the nexi, or the I/O ports of the PEs. The total area asymptotic growth rate, $A_E(N)$ or $A_O(N)$, is determined by the fastest of the growth rates of these three components.

Section 2.3.2 describes lower boundaries on the area devoted to computation for electrical interconnects, A_E, and for optical interconnects, A_O. In Section 2.3.3, upper boundaries on the hologram area for optical interconnects, A_I, are derived based on the use of computer generated holograms (CGH). These boundaries are then used in Section 2.3.4 to compare the growth rates of cost and cost per performance for VLSI and VLSIO processor array implementations.

2.3.2 Lower Boundaries on Interconnect Area

This section determines lower boundaries on the area devoted to interconnects needed to connect PEs of a processor array in specific interconnection network patterns. In Section 2.3.2.1, we first compare lower boundaries on interconnect area (neglecting PE and nexus area) that are based on basic properties of the electrical and optoelectronic technologies. These lower boundaries, involving only a few basic assumptions, are relatively fundamental. In Section 2.3.2.2, we introduce assumptions that allow us to account for the finite area of the PEs and nexi and again compare area lower boundaries. This second approach provides a more realistic area growth model, but, due to the additional assumptions, it is a somewhat less fundamental model than the former.

The following three functions, defined formally in [36] will be used throughout this chapter: $\Omega\{f(N)\}$ grows at a rate at least as great as $f(N)$; $O\{f(N)\}$ grows at a rate at most as great as $f(N)$; and $\Theta\{f(N)\}$ grows at a rate proportional to $f(N)$.

2.3.2.1 Fundamental Lower Boundaries

In this section, lower boundaries on interconnect area are obtained by modeling the PE array as a graph. The vertices of the graph, corresponding to the PEs themselves, are laid out in a two-dimensional plane, and the area of the PEs is neglected. The edges are determined by the interconnection network. For electrical interconnects, edges are implemented as wires (of constant width) laid out in a fixed number of two-dimensional planes (as in [34,35]). For optical interconnects, the edges are implemented by optical beams that are guided by optical elements as in Figure 2.20. Thus, the difference in area of lower boundaries for the two technologies arises from the essentially two-dimensional planar implementation of VLSI and the three-dimensional capabilities of optics.

For either interconnection technology, a boundary can be drawn that divides the processor array so that at least $(N/2) - 1$ of the vertices lie on either side of the boundary [35]. The minimum bisection width of an interconnection network [34], $\omega(N)$, can be defined as the number of simultaneous communication links (the number of links that can be implemented in one unit of time) crossing such a boundary, minimized with respect to all possible boundaries. For example, a mesh interconnection network has a minimum bisection width of $\Theta(N^{1/2})$, while a hypercube has $\omega = \Theta(N)$.

If the circuit is interconnected electronically, Thompson [34] has shown that the circuit area, A_E, is bounded from below by

$$A_E = \Omega(\omega^2) \qquad (2.41)$$

If the circuit is interconnected optically, consider a plane located directly below the optical connection elements as in Figure 2.20.

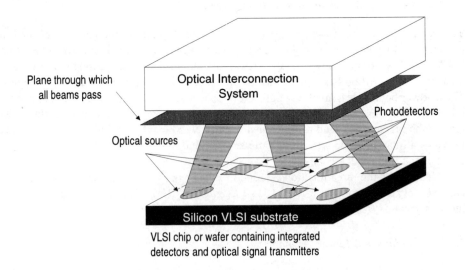

Figure 2.20 A general VLSIO circuit.

Note that A_O grows at a rate at least as great as the area of this plane. Because all light beams pass through this plane, an amount of information equal to $\Phi(\phi)$ may be simultaneously transferred across this plane. And thus, from Gabor's theorem [7,36],

$$A_O = \Omega(\omega) \tag{2.42}$$

Note that this boundary, based on information theory, applies to thick as well as thin HOEs.

2.3.2.2 Interconnect Area Lower Boundaries for Finite-Size PEs

The limits above arise from the area required to connect zero-area nodes of a two-dimensional planar graph with paths in the same plane (electrical case) or paths in a three-dimensional volume (optical interconnect case). However, somewhat different lower boundaries on the area of a processor array are obtained by accounting for the finite area of the PEs and the I/O ports of the PEs, as in the following model. We will first describe the entire model and then justify the assumptions involved.

Complexity Model for Interconnect Area Growth

Nexus Area. In this model, we again assume that the circuit consists of N PEs laid out on a two-dimensional plane. Each PE has a single nexus [34] or interconnect I/O port associated with it. The nexus consists only of the signal receivers and transmitters associated with the corresponding PE. The circuitry required for transmitting and receiving signals to or from the nexus is considered part of the corresponding PE (unlike in [34]). We assume that the area of a single nexus is $\Theta(F)$, where F is the degree of the PE (i.e., the number of directly connected PEs). For simplicity, we assume the degree of each PE in a network is the same.

We distinguish between two types of interconnect networks. In a dedicated link network, each PE has F signal transmitters (one for each connected PE) and $\Theta(F)$ signal receivers. Thus each PE is capable of simultaneously transmitting a different output signal to F different PEs. In a shared link network, each PE has $\Theta(F)$ receivers but only one transmitter with a fanout of F. The single-output signal from a PE is broadcast to all directly connected PEs.

PE Area. We will also assume that the area growth of a single PE (A_{PE}) is the same for all PEs in a given network. For shared link networks, we assume that the area of a PE is $\Omega(F)$. We distinguish between two types of PEs for dedicated links. A PE capable of transmitting only one output signal at a time is denoted a $F{:}1$ PE and has area $\Omega(F)$. A PE capable of transmitting F distinct signals simultaneously is denoted a $F{:}F$ PE and has area $\Omega(F^2)$.

The functions of the PEs can be described more formally as follows. An F:1 PE is capable of performing the Boolean function, f, given by

$$O = f(S; I_1, I_2, \ldots, I_F) \qquad (2.43)$$

where O is the single output and S is the internal state of the processor, and the I_j's are the inputs. An F:F PE is capable of simultaneously performing F Boolean functions f'_i given by

$$O_i = f'_i(S; I_1, I_2, \ldots, I_F) \quad i = 1, 2, \ldots, F \qquad (2.44)$$

where the O_is are the F outputs of the PE. (Thus, an F:F PE is actually a circuit that has a parallelism of at least F). The PEs in a shared link network are assumed to be F:1 PEs. A PE in a dedicated link network can be either an F:1 PE with an additional selector circuit used to address one of its F transmitters or an F:F PE.

Electrical Interconnects. For electrical interconnect networks, we allow wires on a two-dimensional grid in the PE plane to interconnect the nexi as in [34,35]. A wire is not allowed to cross over a nexus or PE unless the wire is an input or output to that PE.

Optical Interconnects. For VLSIO implementations, the connections between transmitters and detectors are performed by optical beams that pass through thin optical elements (HOEs or classical optical components). The only properties of the optical beams included in the model are: the limitation on simultaneous information transfer described in Section 2.3.2.1 and diffraction limited divergence properties. That is, if a beam is confined to a circular area of diameter, d, normal to the direction of propagation, then after propagating a distance, z, the beam will occupy an area at least as large as a circle with diameter D, where

$$D \geq 1.2\pi z/(d\lambda) \qquad (2.45)$$

Because we are neglecting constant factors, we need only the relationship

$$D = \Omega(z/d) \qquad (2.46)$$

Because the optical components are all assumed to be thin, the wave amplitude distribution at the exit pupil of an element is equal to the product of the incident wave amplitude distribution and the transmittance function of the element. Therefore, transmitters requiring different connection patterns must illuminate different optical element regions. We also assume that the angle by which any optical component can deflect a beam is limited by a fixed maximum value, Φ, where $\Phi < 90°$.

Justification of Model Assumptions

Nexus Area. It is clear that the area of a nexus is $O(F)$ because both shared and dedicated links require at most F receivers. However, it is not clear that the area is $\Omega(F)$ because the communication algorithms on many networks are designed such that a single PE is capable of receiving at most m signals at a time, with $m < F$ (in particular $m=1$). In this case, each PE may need fewer than F receivers, with each receiver performing a logical "wired" OR operation among several input signals. As in other VLSI models [35], we assume that a single receiver is capable of performing only a fixed number of simultaneous logical wired OR operations or, equivalently, has a fixed maximum fan-in of k. The justification (for both electrical and optical interconnects) is that each signal transmitter connected to a receiver transmits a small but finite amount of power, P_O, when in a logical "0" state. Because these powers are summed together at the receiver, the receiver will interpret the output as a logical 1 if the fan-in exceeds a value of k, where $k \times P_O \geq$ the receiver threshold power. Thus, the number of receivers for any node of degree F, is $\Omega(F/K)$, and the area is $\Omega(F)$, because we are ignoring constant factors.

PE Area. It can be shown from Ullman's or Thompson's models [34,35] that electrical fan-in of F input signals in time log F required for shared links takes $\theta(F)$ area (assuming nexi receivers can be located in the interior of the PE). Thus, $A_{PE} = \Omega(F)$ is justified. For F:1 PEs in dedicated link networks, an electrical fan-in of this type is needed as well as a selector circuit. Because both functions take $A_{PE} = \theta(F)$, again $A_{PE} = \Omega(F)$ is justified. For F:F PEs, we note that many basic functions of the form given by (2.44), such as sorting and general permutations of the inputs, have information content $\Omega(F)$. Thus, from [35], $A_{PE}T^2 = \Omega(F^2)$, and for constant time, $A_{PE} = \Omega(F^2)$.

Electrical Interconnects. The model presented here for electrical interconnects is similar to many previous VLSI complexity models [34,35] with one additional assumption. We assume that wires may not intersect PEs, while most traditional VLSI models allow for wires to cross through nexi between I/O ports or through processors between circuit elements. Our additional assumption ensures that the area of an individual PE grows at the same rate as a PE in the corresponding optically interconnected network. Although using the traditional VLSI model approach will result in smaller circuit areas for some networks, it does so only by increasing the area of the individual PEs and, therefore, increases the latency of each PE. For many networks, this additional assumption results in little or no effect on electrical interconnect area lower boundaries. The lower boundaries that are affected by this assumption are discussed in Section 2.3.

Optical Interconnects. The model accounts for basic properties of free-space propagation of light, assuming Raman-Nath region diffraction [37] from the holograms. The value of Φ, the maximum hologram deflection angle, is a technological limit determined by such factors as hologram resolution and system alignment tolerance. For current technology, CGHs are limited by $\Phi \approx 50°$ *[38]. We do not account for effects of noise and crosstalk.*

Resulting Lower Boundaries

Optical Interconnect Lower Boundaries. Lower boundaries on the interconnect area for optical interconnects can be determined from the complexity of the nexi. Because $\Omega(F)$

receivers per nexus are required, the number of signal beams capable of simultaneously passing through the plane of Figure 2.20 is $\Omega(NF)$, and thus, the total interconnect area (based on the degrees of freedom needed in the hologram plane [36,39]) is

$$A_I = \Omega(NF) \tag{2.47}$$

Because ω, the minimum bisection width is less than or equal to NF, this boundary is at least as strong as the boundary given by (2.42).

Electrical Interconnect Lower Boundaries. While the strongest optical interconnect lower boundary is determined by the nexus complexity, the strongest electrical interconnect lower boundary is determined by the minimum bisection width (2.41). However, note that, for shared link networks, $\omega \leq N$, because each PE is capable of sending at most one distinct signal across any boundary. This may lead to rather weak lower boundaries on A_E for shared link networks. However, due to the assumption prohibiting wires from intersecting PEs, it can be shown with geometric arguments that $A_E = \Omega(\omega'F)$, where ω' is the minimum bisection width of the corresponding dedicated link network. Thus, in general, for shared link networks,

$$A_E = \Omega\{\max(\omega'F,\ \omega^2)\} \tag{2.48}$$

Comparison of Lower Boundaries

The lower boundaries determined from equations derived in this section are given in Table 2.2 for four popular interconnect networks. A fully connected network with dedicated links, shown in Figure 2.21(a), allows any PE to directly transmit a signal to any other

Table 2.2
Lower Bounds on Order of Area Dependence on Number of Processing Elements

Architecture	Link	Communication Complexity (C)	Lower Bounds (Optics) (A_I)	Complexity	Area (NA_{PE})	Lower Bounds (Electronics) (A_E)
Crossbar	Shared	N	N^2	$F{:}1$	N^2	$N^3(N^2\log^2 N)$
	Dedicated	N^2	N^2	$F{:}1$	N^2	$N^4(N^3)$
				$F{:}F$	N^3	N^4
Hypercube	Shared	N	$N/\log N$	$F{:}1$	$N/\log N$	N^2
	Dedicated	N	$N/\log N$	$F{:}1$	$N/\log N$	N^2
				$F{:}F$	$N/\log^2 N$	
Shuffle exchange	Either	$N/\log N$	N	$F{:}1$ or $F{:}F$	N	$N^2/\log^2 N$
Mesh	Either	$N^{1/2}$	N	$F{:}1$ or $F{:}F$	N	N

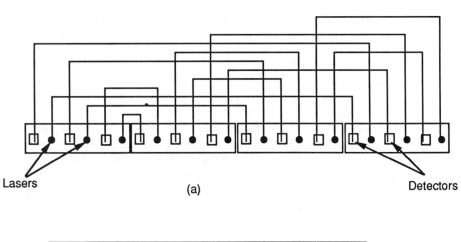
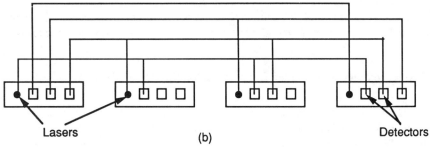
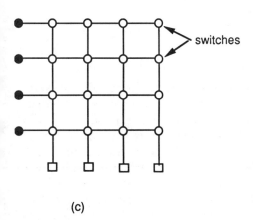

Figure 2.21 Crossbar interconnection schemes: (a) a dedicated-link crossbar interconnection network; (b) a shared-link crossbar network; (c) a crossbar switch network.

PE. In a shared link crossbar, shown in Figure 2.21(b), message passing or time multiplexing schemes are typically used to allow a receiving processor to determine which messages are destined for itself. A related interconnection network is the crossbar switch of Figure 2.21(c). The cross-point switches may be simple toggle switches controlled by an external processor or complicated processing units capable of message passing and parallel routing. This connection network can be evaluated in our model by considering each cross-point switch as a PE in a mesh connected arrangement.

The first column of Table 2.2 lists the networks in order of decreasing minimum bisection width. The fourth column contains lower boundaries on the HOE area for optical implementations of each architecture as determined from (2.47). Note that A_O is equal to the maximum of A_I (column 4) and NA_{PE} (column 6). Lower boundaries on A_E were determined from (2.41) for dedicated link networks and from (2.48) for shared link networks. There are two networks in Table 2.2 affected by the assumption for electrical interconnects prohibiting wires to intersect PEs. The estimated lower boundaries on the area growth of electrical interconnect circuits without this assumption are listed in parentheses in Table 2.2. A comparison of columns four and seven shows that, in general, the higher the minimum bisection width, ω, the greater the potential advantage for optical interconnects.

2.3.3 Interconnect Area Upper Boundaries for Specific Optical Architectures

In this section, we examine particular optical interconnect architectures in order to calculate upper boundaries on the order of the interconnect area growth rate for each interconnection network. These growth rates will be compared to the lower boundary for the electrical interconnections discussed above. The optical interconnect model described in Section 2.3.2.2 is used to obtain these upper boundaries.

2.3.3.1 Double-Pass HOE Architecture

A double-pass HOE architecture (Fig. 2.22) allows a higher density of space-variant interconnects than a single reflective CGH architecture [38]. For this architecture, the HOE is divided into facets or subholograms. Each transmitter illuminates a distinct transmitter subhologram located directly above it. For a shared link interconnect network, the transmitter subhologram then divides the beam into F beams and focuses each beam onto the appropriate detector subhologram after reflection off the planar mirror. Each detector subhologram acts as a single lens and focuses a single beam onto a detector located directly below it. Because the double-pass HOE architecture does not rely on any space-invariant properties of the communication links, this architecture can be used to implement any of the interconnection networks.

Section 2.3.2.2 showed that the dimensions of a beam illuminating a subhologram are given by [37]

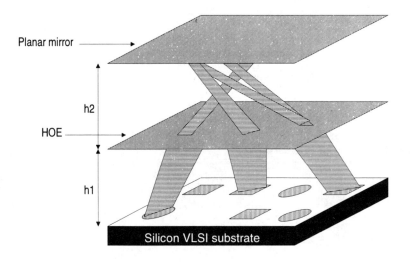

Figure 2.22 Refractive system utilizing a multielement double-pass HOE.

$$w_{Parallel} = \theta\{h_2/(\sqrt{A_s}\cos^3\phi)\}$$
$$w_{Perpendicular} = \theta\{h_2/(\sqrt{A_s}\cos\phi)\}$$
(2.49)

where ϕ is the beam deflection angle, A_s is the area of the transmitter subhologram, and $w_{Parallel}$ and $w_{Perpendicular}$ are the beam dimensions parallel and perpendicular, respectively, to the direction of the beam tilt. These equations can be derived directly from (2.46). The area of a detector subhologram, A_D, is

$$A_D = \theta\{\lambda^2(h_2^2 + L^2)^2/(h_2^2 A_s)\} \tag{2.50}$$

Because the maximum value of ϕ is limited to Φ, $h_2 = O(L_{MAX})$ where L_{MAX} is the maximum distance between source and detector subholograms. Substituting L_{MAX} for both h_2 and L in (2.50) gives

$$A_D = O(L_{MAX}^2/A_S) \tag{2.51}$$

For shared link networks,

$$A_l = NA_S + NFA_D \tag{2.52}$$

To simplify this expression we introduce the parameter, L', defined as the number of PEs traversed by the longest common link. For a shared link network,

$$L' = \theta\left\{\frac{L_{MAX}}{\sqrt{A_S FA_D}}\right\} \qquad (2.53)$$

Setting $A_S = \theta u(FA_D)$ [37] and combining (2.51), (2.52), and (2.53) gives,

$$A_I = O(NFL'^2) \qquad (2.54)$$

Similarly, for dedicated link networks,

$$A_I = NFA_S + NFA_D \qquad (2.55)$$

$$L' = \theta\left\{\frac{L_{MAX}}{\sqrt{FA_S FA_D}}\right\} \qquad (2.56)$$

Setting $A_S = \theta(A_D)$ [37] and combining (2.51), (2.55), and (2.56) gives

$$A_I = O(NF^2L'^2) \qquad (2.57)$$

For all networks, including those that require connections between processors at opposite ends of the chip, $L' = O(\sqrt{N})$. For networks containing only nearest neighbor connections (e.g., the mesh), $L' = \theta(1)$. Thus, for networks such as the hypercube, crossbar, and perfect shuffle, $A_I = O(N^2F)$ for shared link networks, and $A_I = O(N^2F^2)$ for dedicated links. For a mesh connected network, $A_I = O(NF)$.

2.3.3.2 Fourier Plane Filtering Architectures

The interconnect density capabilities can be further increased by exploiting the space-invariant properties of a particular connection pattern. This can be done by using a coherent imaging system and performing Fourier plane filtering. The transmitter subholograms form the object plane, the detector subholograms form the image plane, and an additional Fourier filter is placed in the Fourier plane (Fig. 2.23). For the worst case interconnection problem for which $L_{MAX} = \theta(\sqrt{A_O})$, the area of each CGH is proportional to the product of the number of facets it contains and the number of facets contained in the succeeding CGH. Thus,

$$A_I = O\{\max(N_I M, MN_O)\} \qquad (2.58)$$

where N_I is the number of subholograms in the image plane (number of detectors), M is the number of facets in the Fourier filter, and N_O is the number of subholograms in the

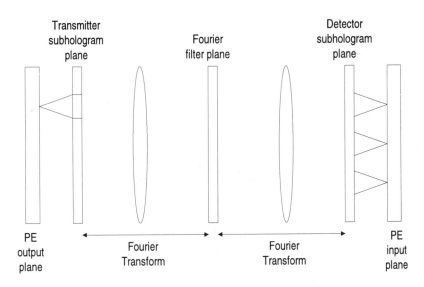

Figure 2.23 Transmissive architecture for implementing interconnections with a basis set or space-variant CGH.

object plane (number of transmitters). Three other interconnect architectures will be discussed in this context.

Space-Invariant Interconnects

Optical interconnect architectures for performing space-invariant interconnects of optical logic gates have been discussed by [40]. Because the Fourier plane filter in this case contains only a single connection pattern, $A_I = O(NF)$, from (2.51). Although none of the networks discussed contain completely space-invariant connections, some of them can be implemented on such an architecture by choosing the invariant connection pattern to be a superset of all the desired connections. This produces extra connections that can be masked off in the detector subhologram plane. For example, a hypercube can be implemented with the invariant connection pattern [41] of Figure 2.24. However, because three-quarters of the interconnects must be blocked in the detector subhologram plane, at least 75% of the light power emanating from each transmitter will be lost. A similar approach to a shared-link crossbar also results in 75% power loss. Space-invariant architectures implement a perfect shuffle with 75% power loss but require a magnification factor of two in the Fourier filter imaging system [42]. This will reduce the interconnect density capabilities of the system by an additional constant factor of four.

Basis-Set Approach

The basis-set approach [40] is similar to the method above, but instead of placing one connection pattern in the Fourier plane, M connection patterns are used, one on each of

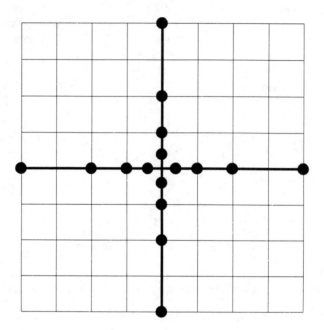

Figure 2.24 A space-invariant superset of all connection patterns for a shared-link hypercube for a 16 × 16 element array ($N = 256$). Hypercube connections only involve PEs separated by 2^j nodes where $j = 0, 1, \ldots, \log(\sqrt{N}/2)$.

M facets. This requires that each connection pattern can be formed by summing together the M recorded connection patterns. The resulting CGH area is $O(MNF)$ (from (2.58)), without loss of light power due to masking out connections as in the space-invariant case. A perfect shuffle can be implemented with only four basis connection patterns, again by magnifying by a factor of two [43]. The basis-set architecture can also be used to reduce the area of hypercube and dedicated link crossbar networks.

Double-Pass Basis-Set Architecture

Further SBWP reductions can be achieved by allowing the beams to pass through multiple Fourier plane filtering systems. For example, the area of the CGH for a dedicated link crossbar can be reduced to $N^{5/2}$ by using two successive basis-set Fourier plane filtering systems. The first performs linear shifts in one direction; the second performs linear shifts in the orthogonal direction. Double-pass basis-set architecture is particularly useful for reflective Fourier plane systems.

2.3.4 Discussion

In the previous sections, equations were derived that describe the interconnect density capabilities of both optical and electrical interconnects. In Section 2.3.2, lower boundaries on the area devoted to computation for electrical interconnects, A_E, and for optical interconnects, A_O, were described. In Section 2.3.3, upper boundaries on the CGH area for optical interconnects, A_I, were derived. In this section, those results are used to compare the interconnect density capabilities of the technologies. Two methods of comparison are used.

In Section 2.3.4.1, the comparison is based on the circuit cost, which, as mentioned earlier, is assumed proportional to the computational area. In Section 2.3.4.2, the comparison is based on cost per performance ($/P$). In both cases, upper boundaries for optical interconnects are compared to lower boundaries for electrical interconnects in order to determine cases where optical interconnects are clearly superior (in terms of these two metrics) for large values of N.

2.3.4.1 Comparison Based on Cost

The cost of a VLSI processor array can be approximated by

$$\$ = NA_{PE}\$P + A_I\$_I \tag{2.59}$$

where NA_{PE} is the silicon area occupied by the PEs, $\$P$ is the cost per unit area of the PEs, $\$_I$ is the cost per area of the interconnects, and A_I is the silicon area devoted to interconnects. Because reduced yield accompanies increased area complexity in VLSI circuits, $\$P$ and $\$_I$ may increase with N. Although, if A_{PE} is kept small enough (fine-grain PEs), and $\$P$ can remain approximately fixed [44], $\$_I$ could have an exponential dependence on A_I due to reduced yield [44]. However, here we will assume that both $\$_I$ and $\$_P$ are approximately constant. Hence, for electrically interconnected processor arrays,

$$\$ = \Omega(A_E) \tag{2.60}$$

where, again, A_E is the area growth rate of an electrically interconnected processor array.

For optical interconnects,

$$\$ = NA_{PE}\$_P + A_I\$_H + A_L\$_L \tag{2.61}$$

where A_I is the hologram area, $\$_H$ is the cost per unit area of the HOE, A_L is the area of all the optoelectronic signal transmitters, and $\$_L$ is the cost of the transmitters per unit area. Because $\$_L$ is independent of N and because A_L grows at a smaller rate than both A_{PE} and A_I, this term will become negligible for large values of N. If a CGH is employed, then $\$_H \approx \$_P$, because a nearly identical fabrication procedure can be used [32,45]. Also, because the number and type of fabrication steps are independent of the complexity, $\$_H$ should remain constant as N increases. In fact, because reduced yield does not accompany increased complexity CGHs (due to the relative insensitivity of holograms to defects), we expect $\$_H \ll \$_I$ for high-complexity, large-area interconnects. Thus, because $\$_P$ and $\$_H$ are independent of N, for VLSIO processor arrays,

$$\$ = O(A_O) \tag{2.62}$$

where A_O is the area growth rate of an optically interconnected processor array.

Because for both VLSI and VLSIO implementations, cost is proportional to the corresponding circuit area, the electrical circuit area lower boundaries in Table 2.2 are lower boundaries on the cost to implement connection networks with VLSI technology. In Table 2.3, the CGH area growth rates required to implement N-node multiprocessor networks with the four optical architectures discussed above are tabulated. The values listed in Table 2.3 are based on (2.54), (2.57), and (2.58), which neglect the PE area growth. The VLSIO circuit cost is given by the maximum of the entries in Table 2.3 (HOE cost) and the entries in column 6 of Table 2.2 (processing area cost). Thus, although a dedicated crossbar can be implemented with $O(N^{5/2})$ area using a double-pass basis-set architecture, the circuit area growth will be $\Omega(N^3)$ if $F{:}F$ PEs are employed.

Note that if one double-pass CGH is used, the computational area growth rate upper boundary is greater than or equal to the circuit area lower boundaries for the corresponding electrical interconnect network (from Tables 2.2 and 2.3 or from (2.41), (2.54), and (2.57)). Thus, it is possible for electrically interconnected circuits to occupy less area than optically interconnected circuits employing one double-pass CGH.

However, the CGH area can be significantly reduced (for many networks) by employing one of the other three architectures mentioned above that contain Fourier filtering operations and exploit the space-invariant properties of a connection pattern. Use

Table 2.3
Upper Bounds on Area Growth Rates for VLSIO Processor Arrays

Architecture	Links	Optics (Upper Bounds)			
		Single Double-Pass CGH	Basis	Double-Pass Basis	Space-Invariant
Crossbar	Shared	N^3	N^3	$N^{5/2}$	N^2
	Dedicated	N^4	N^3	$N^{5/2}$	N/A
Hypercube	Shared	$N^2\log N$	$N/\log^2 N$	N/A	$N/\log N$
	Dedicated	$N^2\log^2 N$	$N/\log^2 N$	N/A	N/A
Shuffle exchange	Either	N^2	N	$N^{3/2}$	N
Mesh	Either	N	N	N/A	N

Note: CGH = computer-generated hologram; N/A = not applicable.

of the basis-set, double-pass basis-set, and space-invariant architectures reduces upper boundaries on optically interconnected circuit area growth rates below lower boundaries on VLSI circuits for high minimum bisection width architectures. Optical interconnect advantages increase with increasing minimum bisection width. For example, an optically interconnected hypercube, with a basis-set architecture, takes area $O(N \log^2 N)$, while an electrically interconnected hypercube requires area $\Omega(N^2)$. The Fourier plane filtering architectures reduce the growth rates of some of the networks (shared link crossbar, shared link hypercube, perfect shuffle, and mesh) to their minimum possible values (Table 2.2).

2.3.4.2 Comparison Based on Cost per Performance

In this section, a performance measure is defined, and the \$/P for optical and electrical interconnects is compared. The \$/P is calculated neglecting power dissipation limitations and speed of light delays. (A modification to account for limited power dissipation per unit area is described at the end of this section.)

The circuit performance, or system throughput, can be expressed as

$$P = S/(\tau_P + \tau_I) \qquad (2.63)$$

where τ_P is the latency of a single PE, defined as the time between the reception of inputs by a PE and the generation of its outputs. τ_I is the rise time of the longest communication link. The system speedup, S, is defined as the ratio of the time required to solve a problem on one PE to the time required on N identical PEs (neglecting interconnect rise time). The speedup can vary depending on the information content of the problem as well as the minimum bisection width of the interconnection network and the parallelism at each node. In general, higher minimum bisection width architectures will have high speedups for a wider variety of problems. Equation (2.63) is a good measure of performance for communication intensive algorithms on fine-grain synchronous SIMD processor arrays that require all PEs to receive their inputs at each step of an algorithm before proceeding to the next step.

For simple PEs, the PE latency, τ_P, is approximately independent of the number of PEs. However, the interconnect rise time, τ_I, may increase with the number of nodes for VLSI implementations. Because speed of light limitations and power dissipation effects are neglected, τ_I is determined by the RC capacitance of the interconnect. Although RC delay has a quadratic dependence on line length, the delay can be reduced to a linear dependence through the use of repeater gates [46]. Thus, we assume

$$\tau_I = \Omega(L_{MAX}) \qquad (2.64)$$

where L_{MAX} is the length of the longest communication link.

Using (2.60), (2.63), and (2.64), the \$/P for a VLSI processor array is given by

$$\$/P = \Omega[A_E(\tau_P + \tau_I)/S] \tag{2.65}$$

Because $\tau_I > \tau_P$ for large N,

$$\$/P = \Omega(A_E\tau_I/S) = \Omega(A_E L/S) \tag{2.66}$$

for VLSI.

For optical interconnects, the communication time is limited by transmitter and detector switching speeds, which are independent of N. (Effects of limited power dissipation are discussed below.) Speed of light delays are not significant for the circuit scales of current interest. (For $A = 10$ by 10 cm, $\phi = 50°$, the speed of light delay is 0.5 ns.)

For a VLSIO processor array, because $(\tau_P + \tau_I)$ is independent of N,

$$\$/P = O(A_O/S) \tag{2.67}$$

The \$/P for VLSI and VLSIO processor arrays for linear speedup is compared in Table 2.4. Lower boundaries for electrically interconnected processor arrays, obtained from (2.66), and upper boundaries for VLSIO processor arrays, obtained from (2.67), are listed for the networks discussed previously. Table 2.4 indicates that VLSIO can provide a lower \$/P than VLSI implementations for large numbers of processors. Again, optical interconnect advantages increase with increasing minimum bisection width.

For fixed power dissipation per unit area, the above expressions for \$/P can be modified by defining the switching energy of an interconnect network, E_{SW}, as the product of the power required for all nodes to communicate, p, and the communication time τ_I.

$$E_{SW} = p\tau_I \tag{2.68}$$

Table 2.4
Growth Rates of Cost-per-Performance for Linear Speedup

| Architecture | Link | Complexity | Optics (Upper Bounds) | | | | Electronics (Lower Bounds) |
			Single Double-Pass CGH	Basis	Double-Pass Basis	Space-Invariant	
Crossbar	Shared	F:1	N^2	N^2	$N^{3/2}$	N	$N^{7/2}$
	Dedicated	F:1	N^3	N^2	$N^{3/2}$	na	N^5
		F:F	N^3	N^2	N^2		
Hypercube	Shared	F:1	$N/\log N$	$\log^2 N$	$\log N$	$\log N$	N^2
	Dedicated	F:1 or F:F	$N/\log N$	$\log^2 N$	N/A	N/A	N^2
Shuffle exchange	Either	F:1 or F:F	N	1	N/A	1	$N^2/\log^3 N$
Mesh	Either	F:1 or F:F	1	1	na	1	1

Note: CGH = computer-generated hologram; N/A = not applicable.

Using (2.68) and the previous definition of $/P$, it can be shown that for fixed power dissipation per unit area, $/P = \theta(E_{SW}/S)$.

For optical interconnects, E_{SW} is proportional to the total number of detectors receiving a signal in one communication cycle (for large N). Thus, for a shared link network $/P = O(NF/S)$. A corresponding dedicated link network (in which each PE receives only one signal per unit time) would have $/P = O(N/S)$.

For electrical interconnects, the switching energy of each interconnect grows linearly with the line length [27]. Thus, for any dedicated link architecture where each PE receives only one signal per unit time and in which the electrical link line length grows with N, optical interconnects will have a smaller $/P$ growth rate.

2.3.5 Conclusions

A model for optically interconnected processor arrays has been developed. This model was used to calculate asymptotic growth rates with the number of processing elements, N. The model restricts the connection elements to a constant number of thin holograms so that the interconnection cost is proportional to the CGH area. Results tabulated in Tables 2.2 and 2.3 show that if a single CGH is employed, (allowing completely space-variant connections) the area of the CGH grows at a rate greater than or equal to a lower boundary on the circuit area of conventional VLSI interconnection networks. Thus, optical interconnections for completely space-variant connection networks might not provide an area advantage over VLSI connections. However, by using the space-invariant properties of an interconnect network in a slightly more complicated architecture (Fig. 2.23), the CGH area can be significantly reduced. Use of the basis-set, double-pass basis-set, and space-invariant architectures reduces upper boundaries on optically interconnected circuit area growth rates below lower boundaries on VLSI circuits for high minimum bisection width architectures. Optical interconnect advantages increase with increasing minimum bisection width, a measure of the global nature of an interconnect topology. Attainment of smaller area growth rates suggests that as the number of PEs increases, the advantages of optical interconnects improve steadily. Furthermore, although only thin holograms are employed, some networks (e.g., mesh, perfect shuffle, shared hypercube, shared crossbar) can achieve area growth rates proportional to fundamental minimum values.

All of the upper boundaries derived in this chapter (Tables 2.3 and 2.4) apply to thin CGHs. However, the lower boundaries on optical interconnect area (Table 2.2) apply to volume holograms as well as thin holograms. Thus, volume holograms could provide no more than a constant factor increase in connection density than a space-invariant architecture employing thin CGHs. However, they would not require the masking of connections in the detector subhologram plane that lowers the diffraction efficiency of thin hologram space-invariant architectures, and they have the potential to provide higher connection densities for space-variant connections.

Optically interconnected high minimum bisection width networks of SIMD synchronous processor arrays achieve even greater advantages in $/P$. Because the $/P$ measure combines the two optical interconnect advantages of compact size and smaller delay time, optical interconnects have a multiplicative advantage. This is illustrated in Table 2.4, where lower boundaries on the growth rate of $/P$ (for linear speedup) for VLSI implementations are compared with optically interconnected processor arrays. Note that optically interconnected arrays can achieve lower $/P$ growth rates than VLSI versions, even if a single double-pass CGH is used, allowing space-variant connections. As an example, consider the problem of sorting N k-bit numbers ($k = \log N$), which have a speedup of $N/\log N$ on a shuffle exchange network [35]. From Table 2.4 and (2.66) and (2.67), a VLSI implementation achieves a $/P = \Omega(N^2/\log^2 N)$, while an optically interconnected space-invariant perfect shuffle network can achieve a $/P = O(\log N)$.

REFERENCES

[1] Goodman, J. W., F. I. Leonberger, S. Y. Kung and R. A. Athale, "Optical Interconnections for VLSI Systems," *IEEE Proc.*, Vol. 72, 1984, pp. 850–866.
[2] Saraswat, K. C. and F. Mohammadi, "Effect of Scaling of Interconnections on the time delay of VLSI Circuits," *IEEE Trans. Electron Devices*, Vol. ED-29, 1982, pp. 645–650.
[3] Gardner, D.S., J.D. Meindl and K.C. Saraswat, "Interconnection and Electromigration Scaling Theory," *IEEE Trans. Electron Devices*, Vol. ED-34, 1987, pp. 633–643.
[4] Bergman, L. A., et. al., "Holographic optical interconnects in VLSI," *Optical Engineering*, Vol. 25, 1986, pp. 1109–1118.
[5] Wu, W. H., L. A. Bergman, A. R. Johnston, C. C. Guest, S. C. Esener, P. K. L. Yu, M. R. Feldman and S. H. Lee, "Implementation of Optical Interconnections for VLSI," *IEEE Trans. Electron Devices*, Vol. ED-34, 1987, pp. 706–713.
[6] Kostuk, R. K., J. W. Goodman, and L. Hesselink, "Optical Imaging Applied to Microelectronic Chip-To-Chip Interconnections," *Appl. Opt.*, Vol. 24, 1985, pp. 2851–2858.
[7] Barakat, R. and J. Reif, "Lower Bounds on the Computational Efficiency of Optical Computing Systems," *Appl. Opt.*, Vol. 26, 1987, pp. 1015–1018.
[8] Sakai, S., H. Shiraishi and M. Umeno, "AlGaAs/GaAs Stripe Laser Diodes Fabricated on Si Substrates by MOCVD," *IEEE J. of Quantum Electronics*, Vol. QE-23, No. 6, 1987, pp. 1080–1084.
[9] Sakai, S., X. W. Hu and M. Umeno, "AlGaAs/GaAs Transverse Junction Stripe Lasers Fabricated on Si Substrates Using Superlattice Intermediate Layers by MOCVD," *IEEE J. of Quantum Electronics*, Vol. QE-23, No. 6, 1987, pp. 1085–1088.
[10] Mead, C. and L. Conway, *Introduction to VLSI Systems*, Addison-Wesley, Menlo Park, Calif., 1980, pp. 11–12.
[11] Quarles, T., A.R. Newton, D.O. Pederson, A. Sangiovanni-Vincentelli, *SPICE Version 3A7 User's Guide*, University of California, Berkeley, Sept. 23, 1986.
[12] Glasser, L. A. and D. W. Dobberpuhl, *The Design and Analysis of VLSI Circuits*, Addison-Wesley, Menlo Park, Calif., pp. 139–141, 1985.
[13] *The MOSIS System (what it is and how to use it)*, Report No. ISI/TM-84-128, Information Sciences Institute, University of Southern California, Marina del Rey, CA. 90292, Mar., 1984.
[14] Haugen, P. R., S. Rychnovsky, A. Husain, L. D. Hutcheson, "Optical Interconnects for High Speed Computing," *Optical Engineering*, Vol. 25, 1986, pp. 1076–1085.
[15] Shibutani, T., et. al., "A Novel High-Power Laser Structure with Current-Blocked Regions Near Cavity Facets," *IEEE J. of Quantum Electronics*, Vol. QE-23, No. 6, 1987, pp. 760–763.
[16] Glasser, L. A. and D. W. Dobberpuhl, pp. 135–136.
[17] Mead, C. and L. Conway, pp. 341–342.

[18] Derry, P. L. and A. Yariv, "Ultralow-threshold graded-index separate-confinement single quantum well buried heterostructure (Al,Ga)As lasers with high reflectivity coatings," *Appl. Phys. Lett.*, Vol. 50, No. 25, 1987, pp. 1773–1775.
[19] Brooks, R. E., "Micromechanical light Modulators on Silicon," *Optical Engineering*, Vol. 24, No. 1, 1985, pp. 101–106.
[20] Bradley, E. and P. K. L. Yu, "Proposed Modulator for Global VLSI Optical Interconnect Network," *Japanese Journal of Applied Physics*, Vol. 26, No. 6, 1987, pp. L971–L973.
[21] Boyd, G. D., D. A. B. Miller, D. S. Chemla, S. L. McCall, A. C. Gossard, J. H. English, "Multiple Quantum Well Reflection Modulator," *Appl. Phys. Lett.*, Vol. 50, No. 17, 1987, pp. 1119–1121.
[22] Lee, S. H., S. C. Esener, M. A. Title and T. J. Drabik, "Two-dimensional silicon/PLZT spatial light modulators: design considerations and technology," *Optical Engineering*, Vol. 25, 1986, pp. 250–260.
[23] Miller, D. A. B., D. S. Chemla, T. C. Damen, T. H. Wood, C. A. Burrus, Jr., A. C. Gossard, and W. Wiegmann, "The Quantum Well Self-Electrooptic Effect Device: Optoelectronic Bistability and Oscillation, and Self-Linearized Modulation," *IEEE Journal of Quantum Electronics*, Vol. QE-21, No. 9, 1985, pp. 1462–1476.
[24] Hillis, Daniel, *The Connection Machine*, MIT Press, Cambridge, 1985.
[25] Seitz, C., "Concurrent VLSI Architectures," *IEEE Trans. Comput.* Vol. C-33, 1984, pp. 1247–1265.
[26] Bowler, K.C., A.D. Bruce, R.D. Kenway, G.S. Pawley and D.J. Wallace, "Exploiting Highly Concurrent Computers for Physics," *Phys. Today*, Vol. 40, 1987, pp. 40–48.
[27] Feldman, M.R., S.C. Esener, C.C. Guest and Sing H. Lee, "Comparison Between Optical and Electrical Interconnects Based on Power and Speed Considerations," *Appl. Opt.*, Vol. 27, 1988, 1742–1751.
[28] Sakai, S., H. Shiraishi and M. Umeno, "AlGaAs/GaAs Stripe Laser Diodes Fabricated on Si Substrates by MOCVD," *IEEE Journ. Quantum Elec.*, Vol. QE-23, 1987, pp. 1080–1084.
[29] Brooks, R.E., "Micromechanical Light Modulators on Silicon," *Opt. Eng.*, Vol. 24, 1985, pp. 101–106.
[30] Boyd, G.D., D.A.B. Miller, D.S. Chemla, S.L. McCall, A.C. Gossard, J.H. English, "Multiple Quantum Well Reflection Modulator," *Appl. Phys. Lett.*, Vol. 50, 1987, pp. 1119–1121.
[31] Bradley, E. and P.K.L. Yu, "Proposed Modulator for Global VLSI Optical Interconnect Network," *Jpn. J. Appl. Phys.*, Vol. 26, 1987, pp. L971–L973.
[32] Feldman, M.R. and C.C. Guest, "Computer Generated Holographic Optical Elements for Optical Interconnection of Very Large Scale Integrated Circuits," *Appl. Opt.*, Vol. 26, 1987, pp. 4377–4384.
[33] Kostuk, R.K., J.W. Goodman and L. Hesselink, "Optical Imaging Applied to Microelectronic Chip-to-Chip Interconnections," *Appl. Opt.*, Vol. 24, 1985, pp. 2851–2858.
[34] Thompson, C.D., "Area-Time Complexity for VLSI," *Proceedings of the Eleventh Annual ACM Symposium on Theory of Computing*, Atlanta, Ga., Vol. 30 Apr. 1979, pp. 81–88.
[35] Ullman, J.D., *Computational Aspects of VLSI*, Computer Science Press, Rockville, Md., 1984, Chapter 2.
[36] Gabor, D., "Light and Information," *Prog. Opt.*, Vol. 1, 1960, pp. 109–110.
[37] Feldman, M.R. and C.C. Guest, "Interconnect Density Capabilities of Computer Generated Holograms for Optical Interconnection of Very Large Scale Integrated Circuits," *Appl. Opt.*, Vol. 28, 1989, pp. 3134–3137.
[38] Moharam, M. G., T.K. Gaylord and R. Magnusson, "Criteria for Raman-Nath Regime Diffraction by Phase Gratings," *Opt. Commun.*, Vol. 32, 1980, pp. 19–23.
[39] Goodman, J. W., "Fan-in and Fan-out with Optical Interconnection," *Acta*, Vol. 32, 1985, pp. 1489–1496.
[40] Jenkins, B. K., P. Chavel, R. Forchheimer, A.A. Sawchuck and T.C. Strand, "Architectural Implications of a Digital Optical Process," *Appl. Opt.*, Vol. 23, 1984, pp. 3465–3474.
[41] Huang, K. S., B.K. Jenkins and A.A. Sawchuk, "Programming a Digital Optical Cellular Image Processor," *J. Opt. Soc. Am.*, Vol. A 4(13), 1987, pp. 87–88.
[42] Lohmann, A. W., "What Classical Optics Can Do for the Digital Optical Computer," *Appl. Opt.*, Vol. 25, 1986, pp. 1543–1549.
[43] Stirk, C. W., R.A. Athale and M.W. Haney, "Folded Perfect Shuffle Optical Processor," *Appl. Opt.*, Vol. 27, 1988, pp. 202–203.

[44] Esener, S. C., "Silicon Device Development for Si/PLZT Spatial Light Modulators," Ph.D. Thesis (Univ. of Calif., San Diego, La Jolla, June 1986), Chapter 1.
[45] Glassesr, L. A. and D.W. Dobberpuhl, *The Design and Analysis of VLSI Circuits*, Addison-Wesley, Reading, MA, 1985, p. 196.
[46] Feldman, M. R. and C.C. Guest, "Automated Design of Holographic Optical Elements for Interconnection of Electronic Circuits," *J. Opt. Soc. Am.* Vol. A 3(13), 1986, p. 80.
[49] Feldman, M. R., C.C. Guest and S.L. Lee, "Design of Computer Generated Holograms for a Shared Memory Network," Proc. Soc. *Photo-Opt. Instrum. Eng.* Vol. 28, 1989, pp. 258–261.

Section II
Current Optical Interconnection Approaches and Limitations

Chapter 3
Bulk Waveguide Routing—Fiber Optics

Ronald A. Nordin

3.1 INTRODUCTION

This decade is seeing an unprecedented need for higher interconnection bandwidth and density technologies to implement the next generation of telecommunication and computer systems. This technological need arises from the demand for enhanced digital telecommunication services (e.g., broadband ISDN, entertainment video distribution, video conferencing, and higher speed data such as modem and fax), and advanced computer services (e.g., graphical animation, image processing, speech processing, and modeling). This increased bandwidth and density for interconnection technologies is exhausting present electrical capabilities [1–4]. As a result, the use of optical technologies [5] to supplant conventional electrical techniques is under investigation.

Figure 3.1 shows the empirical relationship (i.e., Rent's rule) between the number of terminals required versus the high-data-rate integrated circuit (IC) gate count for a large digital system design. It is expected that the 1990s will have ICs with gate counts in the 100K to 1M range with external data rates in the 500 Mbps range and beyond. Rent's rule can be used to estimate the terminal count for this class of ICs as well as the terminal count of higher levels of packaging in the interconnection hierarchy. Hence, this relationship shows a system designer where a potential "weakness" exists (or how much stress is placed) at various points in the interconnection hierarchy in order to implement the overall system. For example, at the 100K gate count IC, the chip requires 311 terminals, the multichip module (MCM) requires 900 terminals (if an MCM with 10 ICs on it was employed), the printed circuit board (PCB) requires 1,464 terminals, and the shelf (or technically the backplane) supports 3,532 terminals. Figure 3.1 is useful in identifying the technologies that require improvement in the interconnection hierarchy.

The level in the interconnection hierarchy at which electronics begin to feel a restriction in performance is at the PCB to another PCB (hereafter referred to as PCB-

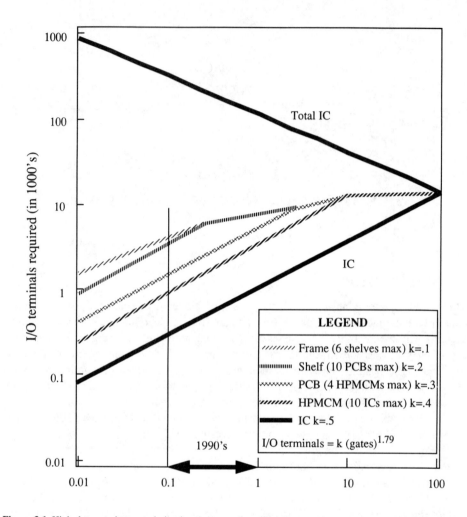

Figure 3.1 High data-rate integrated circuit gate count (in millions).

to-PCB) level within a shelf or enclosure. Here the interconnection varies in length from a few centimeters to approximately a meter. Below this level, electrical interconnection technologies (e.g., flip chip (also referred to as controlled collapse chip connection, or C4), VLSI, MCMs, and special dielectric loaded printed circuit boards) presently meet predicted needs more efficiently than optical technology. Above this level, the performance of the electrical interconnection technology erodes when compared with optical technologies. This performance erosion can be measured with respect to the following parameters: cost, bandwidth, density, skew, power dissipation, cross-sectional area, electromagnetic interference and susceptibility, latency, and system extensibility (space/frequency). Hence, the discussion of this chapter concentrates on what has been identified as optical backplane technologies that provide the PCB-to-PCB interconnection function and the infrastructure to go beyond this level (i.e., intraframe and intrasystem connectivity). These optical backplane interconnection technologies have only recently been realized due to the emergence of two critical items. The first enabling technology is referred to as one-dimensional optical data links (1D-ODLs) or parallel optical interconnection modules [6–9]. An example is shown in Figure 3.2. The driving force behind the development of these types of modules is in obtaining a low-cost, high-performance, parallel interface between peripherals and other computer hosts. Note that the use of integrated optical device arrays and

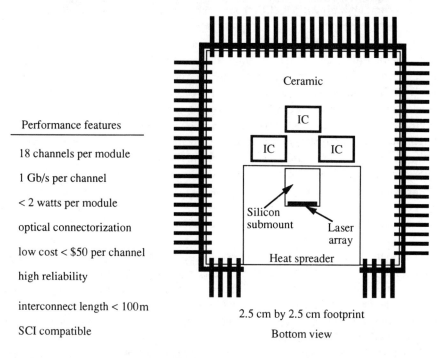

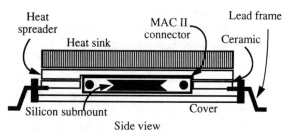

Figure 3.2 One-dimensional optical data links.

electronic amplifier arrays is incorporated into the design. This not only accomplishes low-delay variations in the module (i.e., skew), which is important for parallel interfaces, but does this at a low cost per channel. These modules exemplify what is being considered as a solution to solve the PCB-to-PCB interconnection problem. A high signal density and bit rate performance is evident in these types of modules. 1D-ODL modules are appearing today not only in the research lab but in the market place and should be prolific in high-performance systems beginning in the 1995–96 time frame. These modules are wide (18 channels), have small footprints (on the order of 6 cm2), and have high data transmission rates (on the order of 1 Gbs channel). This technology has emerged due to high laser or light emitting diods (LED) array manufacturing yields and clever packaging technology developments.

Figure 3.3 depicts how printed circuit board designers could use 1D-ODLs in a system. The modules can be surface mounted onto the PCB, and then, at the appropriate point in the manufacturing process, the multifiber array connectors (MAC) (such as the MAC-II family from AT&T) can be mounted and connected to the 1D-ODL. Figure 3.2 shows 18 1D-ODL modules (each 18 channels wide) mounted on a circuit board, allowing an aggregate bandwidth of 324 Gbs off the circuit board. For high-bandwidth connections, this represents a significant improvement over traditional circuit-board-to-backplane options. The low-profile nature and low power dissipation of the 1D-ODLs allow high module densities on PCBs and allow for compact spacing between circuit boards in a shelf of equipment. The remaining challenge is the management of the optical-fiber ribbon that leaves the PCB. This 1D-ODL technological capability elicits the need for higher density optical compatible solutions to implement the backplane function.

The second "enabling" technology is the recent number of developments from material research centers that support the development of fiber-based optical backplane technologies. This is the essence of this chapter's discussion. An optical backplane can be fundamentally implemented in two different ways [10]. They differ in only the method of light transmission. One can develop a free-space system [11] in which the light is collimated and transported to a detector in an unguided medium (e.g., air). The other technique relies on a guided medium (i.e., optical waveguides) to transport the light. This guided medium can be in the form of a fabricated waveguide (e.g., a polymer waveguide on a glass epoxy board or a silicon dioxide waveguide on a silicon substrate) or a traditional optical fiber embedded in a substrate material. This chapter focuses on the latter scheme for providing the infrastructure capability for optical backplane interconnection technologies.

This chapter will begin with a discussion of the needs and requirements of the intrashelf and intershelf interconnection hierarchy. A discussion of the different techniques that are used in implementing fiber-embedded substrates that can efficiently implement intrashelf and intershelf interconnection will follow. Finally, a discussion of each technique's capabilities and limitations, along with a presentation of the connectivity/product grade capability, is included.

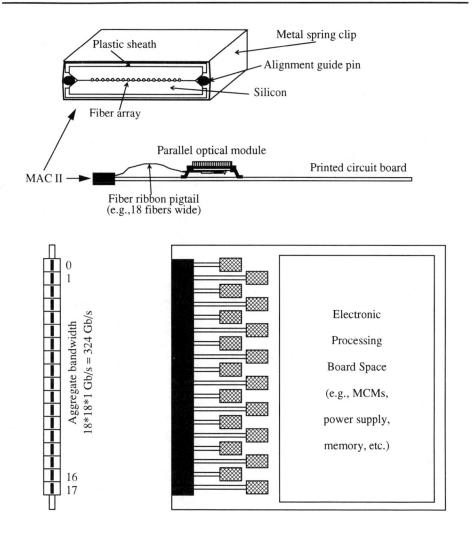

Figure 3.3 One-dimensional optical data links used in a system.

3.1.1 Intrashelf Communication

In the intrashelf level of intercommunication, the problem is to interconnect a PCB to another PCB that resides on the same shelf (i.e., in the electrical domain, the interconnection is implemented in a printed backplane). Typical shelf dimensions for the telecommunication industry are 0.5 to 0.6m across and 0.5m high. This interconnection level is generally referred to as electrical or optical backplane technology. The interconnection between the shelves is deferred until the next section. Ignoring, for the moment, the need to interconnect the PCB to the backplane, the author will first evaluate the backplane needs.

The point at which optics will become competitive with electronics is a function of cost, interconnection distance, signal density, bit rate, and fanout [12]. To provide the comparison, a cost-versus-interconnection length is provided in Figure 3.4 for point-to-point connections. For the electrical case, the connections are implemented on a glass epoxy backplane or in a coaxial form for the longer distances. The optical case assumes the existence of the 1D-ODL module. The interconnection distance is limited by conductor and dielectric loss. The signal density generally dictates the physical dimensions of the microstrip or strip transmission lines used. The operating data rate assumed was 800 Mbps. The cost for both technologies include drivers, receivers, board-to-backplane con-

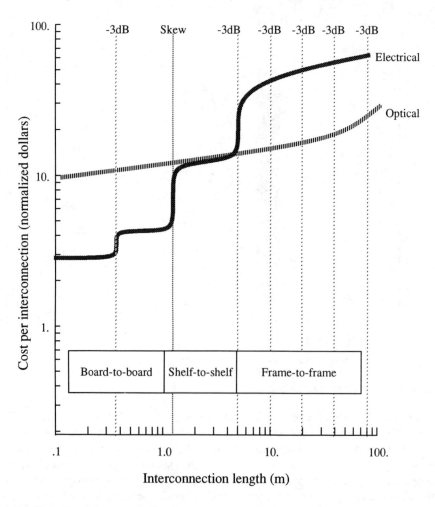

Figure 3.4 Cost versus interconnection length.

nectors, and the interconnecting media (i.e., fiber for the optical case and PCB backplane or ribbon coax for the electrical case).

For the electrical case, when the interconnecting length reaches a distance at which the signal is attenuated by −3 dB, the signal must either be regenerated or demultiplexed at the source (which is the assumed technique) to extend to longer lengths. Additionally, when the interconnecting length reaches a distance at which the skew becomes excessive (the assumed limit value of one third the bit period), then the cost of a clock recovery circuit per channel must be added to extend beyond this length. The results show cost parity for the 800 Mbps rate in a distance range between 2 and 5m (i.e., the shelf-to-shelf interconnection hierarchy level), and as the data rate decreases for example to 200 Mbps rate, the cost parity distance increases to approximately 30m (i.e., to the frame-to-frame level). Hence, as data rates increase in computer or switching systems, the penetration of optics will increase, simply on a cost basis. One would suspect that the optical curve in Figure 3.4 will decrease in cost relative to the electrical curve because of the immaturity level of the optical components and their packaging in high-volume applications as compared with the very mature level of electronics and their packaging technologies.

The present I/O density capability of electrical backplane boards is on the order of 150 to 200 coupled differential transmission lines per centimeter. (Note that this density must be derated according to the type of connector used due to routing track blockage that can occur.) The electrical backplane can technically support the interconnection density needs; however, a question arises as to whether it can support the bit rate demands.

Figure 3.5 shows the simulated performance of an electrical backplane for a point-to-point connection and for a passive/active bus applications (standard PCB glass epoxy FR4 and Teflon-loaded laminate materials were simulated). Figure 3.5(a) describes the model used in the simulation. A capacitive parasitic element was assumed to model the electrical characteristics of the PCB connector that establishes the connectivity between the PCB and the backplane. Its value is typically 2 pF for passive backplanes. For the bus simulation, eight destination boards were assumed. The passive bus is distinguished from the active bus by its destination board's parasitic capacitance and stub length. The active bus contains buffers (e.g., dual gate FETs) mounted on the backplane itself in order to decouple the destination board with the transmission line bus. When an active backplane is employed, the FET gate capacitance must be included in the simulation, and a value of 0.5 pF is assumed. The other parasitic element included in the simulation is the stub length connecting the backplane to the high-input impedance receiver IC on the PCB (or, for the active backplane, the interconnecting length from the buffer gate to the backplane). Typical physical dimensions and material constants for the transmission line structures were assumed. Figure 3.5(b) contains the time domain step response of a data line, and Figure 3.5(c) plots the frequency domain transfer function.

The transfer function for point-to-point applications has a cutoff frequency ($f_{-3\,dB}$) of nearly 1 GHz for the Teflon-loaded line. This suggests that the maximum clock rate that one could use is in the 300- to 400-MHz range (i.e., $f_{-3\,dB}$ divided by three). This

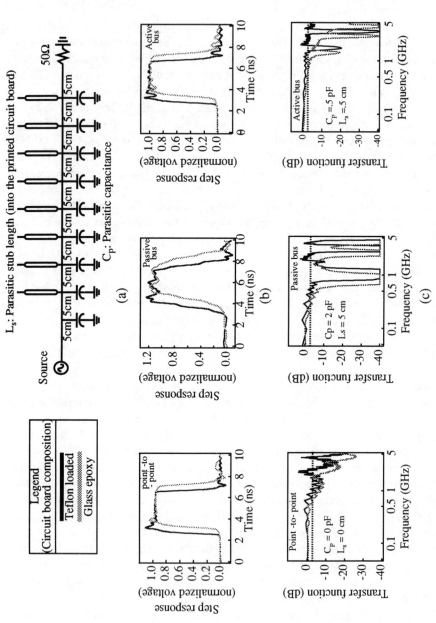

Figure 3.5 Simulated performance of an electrical backplane for point-to-point and passive/active bus applications: (a) simulation model; (b) time domain response of a data line; (c) plot of the frequency domain transfer function.

corresponds to a range of 300- to 400-Mbps data line rate limitation, assuming data encoding is not used.

The transfer function for the passive bus application has a cutoff frequency of only 500 MHz, which is restored to a comparable frequency when it is replaced with an active bus. The results show that for point-to-point and bus interconnection types, a bit rate limitation in the range of 300 to 400 Mbps is the best that designers should expect for electrical PCB backplanes. For data rates that exceed these limits, optics will provide an advantage in the intrashelf interconnection hierarchy as long as the cost is low in comparison to electrical implementation.

For synchronous systems, the interconnection technology must support high connection density and low delay variation (i.e., skew). Low interconnection skew should provide a minimum connection cost because providing per channel clock recovery and complex frame alignment circuitry can be expensive when a clock line is not routed with the data group. Examples of commercial interface standards where the clock is routed with the data channels are the small computer system interface (SCSI), the high-performance parallel interface (HiPPI), and the scalable coherent interface (SCI) standards. Figure 3.6 shows a comparison of the bandwidth and skew performance of several interconnection technologies.

The skew performance for electrical lines is primarily controlled by two parameters: the spatial uniformity of the dielectric material and, to a somewhat lesser extent, the dimension control of the transmission line. For the electrical cases, the skew per unit length, l, can be approximated as:

$$\frac{\text{skew}}{l} \approx \frac{\sqrt{\epsilon_r}}{c} \delta_\epsilon \tag{3.1}$$

where c is the speed of light, ϵ_r is the relative dielectric constant of the material, and δ_ϵ is the manufacturing tolerance of the dielectric material in the transverse direction of the data bus. Hence, the lower the dielectric constant is, the skew tends to be generally lower. For example, when comparing glass epoxy PCB to Teflon-loaded PCB (assuming that the manufacturing process leads to a tolerance control of 1%), the following results are elicited: for the glass epoxy PCB parallel signal lines, the skew/l = 70 ps/m; and for the Teflon-loaded PCB parallel signal lines, the skew/l = 50 ps/m.

Fundamentally, the main advantage of fiber optics over electrical interconnection technology is the material uniformity control of its manufacturing process. There are four parameters that can effect the skew in a parallel fiber-optic media: refractive index variations, physical length and end effect variations, temperature variations, and strain variations from fiber to fiber within the ribbon cable. Assuming that all the fibers in the fiber-optic ribbon cable are exposed to the same environment and have the same lengths, then only the refractive index variations across the array will dominate. The manufacturing control of the refractive index is quite good, on the order of 0.025%. Hence for optical interconnections, the skew per unit length can be approximated as:

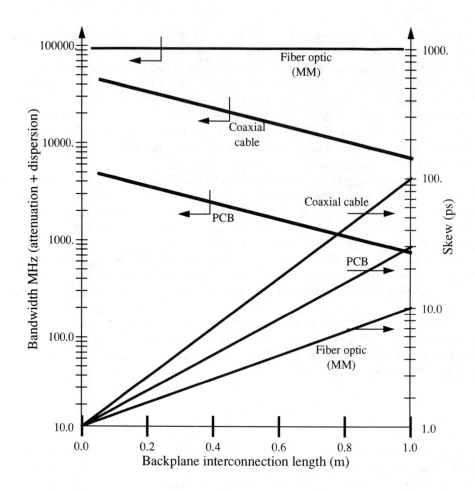

Figure 3.6 A comparison of the bandwidth and show performance of several interconnection technologies.

$$(\text{skew})/l \approx \frac{n}{c}\delta_n \qquad (3.2)$$

where n is the refractive index and δ_n is its manufacturing tolerance. Substituting the tolerance value of 0.025% and a refractive index of 1.47 into Equation 3.2 yields a skew per unit length of 1.225 ps/m. As a result, skew control is at least 40 times better in fiber optics than in electronics.

Another potential limitation of electrical PCB-to-PCB communication that must be addressed is the number of I/O pins on a PCB edge that can connect to the backplane. For a relatively large board (43 by 33 cm), 600 staked pin-type connectors are readily available, but only a relatively low bandwidth (on the order of 500 MHz) can be realized.

If one adds ground blades around the pins in such a way that no reduction in the number of signal I/O pins will occur for grounding purposes, one can increase this bandwidth to approximately 1 GHz. With other connector technologies (i.e., Gold Dot, Cinch's Synapse, Beta Phase's thermal shape memory) the density exceeds 2,000 I/Os per PCB, and bandwidths in the range of 2 GHz can be obtained with an added increase in cost.

Guided wave optics can supply a cross-sectional area advantage. It can provide a factor of 1.3 to 1.5 reduction in the connector area (assuming a MAC type, compared to a staked-pin electrical connector). Hence, for an equivalent electrical connector area of 600 contact pins (which could supply perhaps 120 signal I/Os while the remaining pins are held at ground), the optical ribbon connector could supply 216 I/Os (assuming a MAC-type connector). In addition to an improved connector I/O density, optical packaging could reduce the PCB area required to drive the I/Os by a factor of 4, increase the number of connections as well as the interconnection bandwidth, lower the I/O power requirements, lower the electromagnetic emission, raise susceptibility, and increase fanout possibilities. All of these improvements are key items for digital system designers. (In the future, it is possible that free-space optical interconnection of PCBs might exceed the I/O density of the optical guided wave approach by an appreciable factor and might therefore be an attractive technology.)

In summary, a likely role for optics would be in the substrate on one PCB to a substrate on another PCB (intrashelf communication) when a higher level of performance is required. For example, when the data rate of the interconnecting lines approach 500 Mbps, an optical interconnection may be advantageous. This performance level may be needed for next generation telecommunication [13,14] (implementing multimedia applications) and/or computer systems (in high-performance parallel processors). An optical solution becomes even more competitive in the PCB-to-PCB intershelf communication because the interconnection length is longer.

3.1.2 Intershelf Communication

In this level of interconnection, the backplane must extend into the 2m range, and hence fewer types of electronic backplane technologies are available to interconnect PCBs. The following options exist: coaxial cables (single or ribbon), twinax (single or ribbon), shielded/unshielded twisted pair (single or ribbon), flexible stripline circuits (which form an electronic backplane), and the optical alternatives. In these instances, optics can compete quite well with electrical alternatives because electrical technologies become expensive with respect to cable cost and system cost, whereas the optical cost as a function of distance is relatively flat. As was shown earlier in this section, the crossover point where optics is less costly than electrical is a function of bit rate, distance, density, skew, and fanout requirements. In a point-to-point interconnection, this crossover point is in the 20 to 40m range at 200 Mbps and in the 1 to 5m range at the 800 Mbps rate. Therefore, use of fiber optics presents an attractive alternative in advanced switching (or computer) system designs.

Optics, which uses fiber ribbon, also has improved skew performance in the 2 to 5 ps/m range as compared with electrical coaxial ribbon cable which is in the 100 to 300 ps/m range. This implies that a system can be designed synchronously without the cost burden of providing clock recovery circuits per channel. The data are simply retimed at the destination by a simple digital skew cancellation technique. Additionally, optics provides nearly a factor of 100 times improvement in the signal density (i.e. the signal's cross-sectional area) in the backplane as compared to coaxial ribbon connections. Although a cost advantage is difficult to quantify, optics is clearly advantageous with respect to assembly, maintenance, and repair.

In summary, it has been shown that an optical solution is best utilized as a backplane technology. Optics will have a very difficult time replacing electronic interconnection technologies at the chip-to-chip or at the substrate-to-substrate (within a PCB) packaging levels in the near future. To penetrate this area, optics has to provide revolutionary functionality (i.e., not merely compete in an evolutionary electronic arena). However, in the backplane environment, optics does compete effectively as an evolutionary interconnection technology. The backplane provides interconnection for PCBs on the same shelf, PCBs on different shelves in the same frame, and/or PCBs in completely different frames. The advantages ultimately manifest themselves as a lower cost technology. The additional performance parameters outlined in the preceding section make the optical performance cost ratio far superior than that of the electrical alternatives. The remainder of this chapter will be concerned with optical solutions for backplane interconnection technologies.

3.1.3 System Backplane Requirements

The backplane interconnection system requirements are implementation independent (optical or electronic). They provide the specifications for intrashelf (0- to 1m distances) as well as intershelf (1- to 10m distances) PCB interconnections. Fundamentally, the input/output of the interconnection technology used should behave as a digital interconnect. Because the optical problems that are being solved in switching and in data communication areas differ from traditional telecommunication optical long-haul transmission system implementation solutions, a different paradigm should be sought. Cost is the primary motivator. It mandates simplified transmitter and receiver designs so as to obtain low cost per interconnection.

Based on generic telecommunication central office switching equipment environments, the general system specifications for an advanced synchronous high-performance system include:

- Cost per interconnection < $50 (includes driver, receiver, connectors, and backplane media).
- High reliability: >10^5 hours.
- Zero system interconnect error (appropriate digital coding (e.g., parity or error correcting codes) to ensure system integrity—it does not make sense to discuss a bit error rate figure of merit at this level in the interconnection hierarchy).

- Temperature (ambient): 0°C < T < 70°C.
- Absolute humidity < 0.026 (5% < relative humidity < 95%).
- Interconnect length (L): 1 < L < 10 m.
- Slide detachable connectors.
- Robust connectors (e.g., capable of a large number of insertions).
- Power supplies (if required): +5, ground, −2, and −4.8 (volts).
- Bit rate (BR): 0 < BR < 1 Gb per s (application dependent, useful rates are: SCI = 1 Gbs, SONET OC-12 = 622 Mbps, and HiPPI = 800 Mbps).
- Average group interconnect delay variation ($\Delta \tau_{gd}$): <200 ps (driven by the desire to use a synchronous system architecture).
- Skew (delay variation within a group $\Delta \tau_d$): total interconnect skew <200 ps (driven by the desire not to have clock recovery circuitry nor allow for any jitter accumulation).

The above specifications suggest that when considering an optical backplane implementation, the following additional requirements would be recommended:

- DC coupled system (obviates the need for data encoding).
- Multifiber array connector compatibility.
- Total insertion loss <−10 dB.
- Insertion loss variability <3 dB.
- Total cross talk <−30 dB.
- Support coupled power in the fiber >1 mW.

These requirements provide a path toward a low-cost implementation. To obtain low-cost electronics with the required performance indices, the drivers and receivers must be monolithically integrated into an array form. It is quite difficult to integrate very high gain/bandwidth amplifiers due to crosstalk and power/ground "bounce." However, the integration of low-gain transimpedance amplifiers (e.g., 300Ω) is much more feasible—hence the desire to have low loss in the interconnecting media. From a systems perspective, the even distribution of power dissipation between the transmitter and receiver is also desirable.

3.2 FIBER-EMBEDDED SUBSTRATE TECHNOLOGY

The interconnection of PCBs on a common shelf (or shelves) is the function of the backplane whether the implementation is electrical or optical. Traditional electrical backplanes can support thousands of inter-PCB interconnections. However, as the bit rate of the interconnecting signals increases beyond a few hundred Mbps, parameters such as bandwidth, skew, and cost drive system designers to evaluate the capabilities of optical backplanes.

Interconnections on this type of backplane are accomplished using optical waveguides. These waveguides can be either photolithographically fabricated on the backplane (e.g., polymeric waveguides [15]) or prefabricated waveguides (e.g., optical fiber) that are embedded in the backplane. The discussion on fabricated waveguides will be deferred until the next chapter.

This chapter focuses on optical-fiber backplane technology where the waveguide function is performed by optical fiber. The system design has many implementation options, which include the type of fiber (e.g., single mode (SM), multimode (MM), plastic cladded, plastic core/cladding [16–18], and "D-fiber") and the type of substrate material (e.g., rigid, flexible, and plastic molded). The challenges that must be addressed for a successful implementation include connectivity, material and process compatibility with the fiber, development of a set of design guidelines, and an automated way of routing relatively short lengths of fiber. In the author's opinion, an optical backplane technology capability that is compatible with electrical interconnectivity such that they can both coexist in a hybrid electro-optical backplane would be the most useful. It is doubtful that a backplane that strictly supports optics would be widely accepted. Interconnection of today's vast variety of electrical technologies will also be required in the backplane. This hybrid will provide the most flexible capability, where the higher data rate signals can be optically interconnected, and the lower data rate signals (along with power and ground) can be electrically interconnected. The remaining contents of this subsection will discuss rigid and flexible substrate optical backplane technologies.

3.2.1 Rigid Substrates

Automated optical-fiber routing on rigid substrates had a naturally derived beginning from an existing electrical technology called Multiwire [19]. Multiwire uses discrete polyimide-coated wires that are embedded in a soft matrix of adhesive dielectric that is coated on the backplane. Generally, the backplane substrate has FR-4 or G-10 as the dielectric base. An advantage of this technology over conventionally printed circuit boards is that because the wires are insulated, they can cross over each other. Hence, in one level of the PCB, both X and Y wiring directions can be fabricated. The wires are typically 100- to 150-μm in diameter, and line spacings of 375 μm can be accomplished. Use of a finer wire diameter (i.e., 50 μm) called Microwire technology helps to reduce line spaces to approximately 200 to 250 μm. The line spacing limitations are imposed by the wiring head. The wiring head can be moved in the X-Y plane and also rotated about its own axis under software control. The electronic industry is actively investing in this technology for two reasons: achieving a better interconnection density over PCBs and the fact that the characteristic impedance can be controlled by manufacturing when the wires are embedded over a ground plane. By simply replacing the copper wire with optical fiber [20], one can embed the fiber into the substrate to realize the optical backplane. This technology can provide an infrastructure for the manufacture of embedded optical fibers and/or electrical wire in rigid substrates (i.e., an optical backplane).

An example of an optical-fiber wiring schematic and photograph of an optical test board is shown in Figure 3.7. A photograph of AIT's wiring head is shown in Figure 3.8.

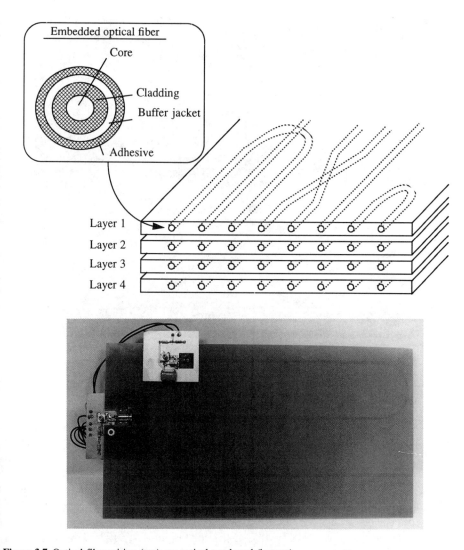

Figure 3.7 Optical fiber wiring (top); an optical test board (bottom).

The coated fiber (generally polyimide with an adhesive overlayer) from a spool is threaded through a feed mechanism and then guided under the stylus of the wiring head. The stylus pushes the adhesive coated fiber into the thermal set (TS) stage epoxy, and it is bonded by the ultrasonic energy supplied from the wiring head. Three parameters control this process: the velocity of the wiring head, the amount of ultrasonic power, and the downward force the wiring head places on the fiber. These parameters are generally computer controlled and are allowed to vary dependent on the type of routing (e.g., straight paths

Figure 3.8 An AIT optical wiring head.

versus curved, crossovers, terminating ends). For straight sections, velocities are on the order of 2 cm/sec; for curved sections, the velocity is reduced (as is the ultrasonic power to maintain a roughly constant energy being sourced to the fiber). The ultrasonic energy applied to the optical fibers during the placement sequence heats the fibers locally to between 240° and 285°C.

One of the more difficult problems challenging backplane designers today is connectivity. There are primarily three concepts that are presently being investigated. The first approach collects and groups the fibers into an equally spaced ribbon section or partition. A multifiber connector (e.g., AT&T MAC-II [21] or Fujikura MT [22]) is then attached to the formed ribbon. Once placed in the connector, glass fibers would be cleaved and polished, while the plastic fibers would be cut clean. The polishing of a glass fiber embossed in a connector should not be understated. Excessive residual stress within the fiber, which occurs in the connector attachment process, poses a challenge for the polishing step. Special techniques are used to prevent fracture of the fiber within the connector.

The second approach couples the signal energy from a connector into the backplane via the fibers evanescent field coupling. This form of connector can be managed by the use of back-to-back D-fiber [23] in both connector halves. Both techniques are being pursued, and are discussed in more detail in subsequent sections.

The third approach is modeled after the multiwire technique called Microdot. In the electrical domain, the wire to be connected is brought to the surface (hence, bending the copper wire) by a reduction of energy being applied to the wire via the insertion head or by routing the wire over another wire (i.e., a crossover). In the optical domain, the latter technique can be implemented by routing the fiber over another fiber or wire forming a crossover. Figure 3.9 describes this last technique of coupling light from one fiber to another. Note that this technique depends on precision polishing, where (depending on the diameter of the crossover fiber) two types of interconnects can be made. If the diameter

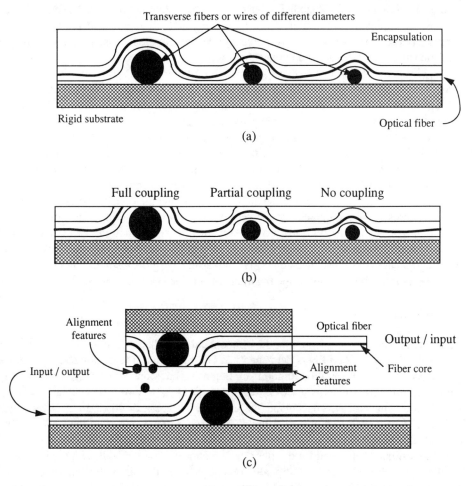

Figure 3.9 Evanescent field coupling: (a) general electromagnetic field distribution; (b) creating various discontinuous fields dependent on optical fiber diameter; and (c) basis features and approach for field coupling between two signal planes.

of the crossover element (the element being either wire or fiber) is large, the entire fiber is brought to the surface, and after polishing, the diameter is reduced by one-half. An elliptical shaped optical tap remains, and all of the light can be tapped. If the diameter of the crossover element is smaller than the previous size, the fiber is only partially polished, and an optical tap is formed, which is accomplished by the coupling of the evanescent field. This can be useful for constructing an optical splitter or a bus.

The connecting side of the connector or tap for either the 100% coupler or the partial coupler is fabricated in an exact fashion. The sides are then aligned together and attached in some fashion (e.g., soldered in the case of termination to a source/detector module, or perhaps clamped together in the case of a backplane connector). The difficulties of these types of connectors come from two categories: alignment and polishing. The alignment issues can be somewhat overcome if large diameter fiber is used (e.g., multimode or plastic fiber) allowing a large contact field to hit.

Additionally, alignment features can be photolithographically placed on the circuit board. These alignment features can be as simple as "fiducials," through-holes, and solder pads, or as complex as using fiber or polymer ridges after the precision polishing is done for passive mechanical alignment features only.

Residual stress is the other difficulty involved in the polishing of the PCB material, and it is a much harder problem to circumvent. Residual stress in the fiber (after bending) can lead to crack formation when polishing. Additionally, there are residual defects in the original fiber, which lead to a probability of cracking during polishing. The overall concern manifests itself as a yield of the circuit board during the polishing process. Unless properly addressed, these defects can present problems when working with hundreds of connections (or taps). Finally, if the crossover element is small enough, the fiber will not be polished at all and, hence, can serve as a normal crossover interconnecting fiber. This gives the optical system designer impressive versatility . Rigid substrates can be used for backplanes, circuit boards, and implementing the connectivity scheme.

3.2.2 Flexible Substrates

The general concept for implementing flexible optical backplane substrates is to automatically route adhesive coated fibers onto sheets of flexible material (e.g., polyester, polyimide, fluorocarbon, and aramid). Random high-density interconnection patterns between PCBs (including crossovers) can be achieved. At the periphery of the backplane, where the PCB interconnects to the backplane, the substrate is "necked" down. In these regions, the fiber can be routed to create a "ribbon" of fiber, wherein an array connector could be attached. The advantage of a flexible backplane arises from its ability to conform to three-dimensional shapes and for the fiber in the connector to be at a different axial angle than the fiber in the substrate. Additionally, optical flex circuits can be used in a backplane environment or in a shelf-to-shelf or longer interconnection scheme. For example, one can visualize an optical harness assembly that interconnects multiple shelves together, not necessarily in the same frame.

The fabrication process is somewhat similar to a rigid optical substrate construction. A somewhat unique processing example is provided by The AT&T Optiflex5 [24] (Figure 3.10). A polyimide film of 0.002 inches provides the flexible substrate. A pressure sensitive adhesive is used to capture the fiber instead of a thermal adhesive. This avoids complexities resulting from supplying thermal energy via an ultrasonic source (i.e., this process reduces stress on fiber). The fibers are then automatically routed with a routing head similar to the multiwire technique (minus the ultrasonic unit). Thermoplastic (i.e., polyurethane) encapsulation is next used to encapsulate the structure. Finally, a .001-inch polyimide cover sheet is applied.

Figure 3.10 shows a fabricated Optiflex backplane. As discussed previously, the advantages of a flexible backplane are its connectivity ability and capability for flexibility. Note that in the figure, ribbons are formed, which allows for the easy attachment of array connectors. Additionally, when applying this structure to a backplane, the connectors do not necessarily have to be linear (Figure 3.10(b)). Connector positions across the shelf do not have to be placed across a straight line, and multiple connectors per PCB can be interconnected. As was shown in the rigid substrate construction, additional electrical or optical layers can be added.

Finally, an example of a flexible electro-optical harness is shown in Figure 3.11. This is the most likely first application of flexible substrates. High-density optical and

Figure 3.10 AT&T flexible optical backplane (OptoFlex).

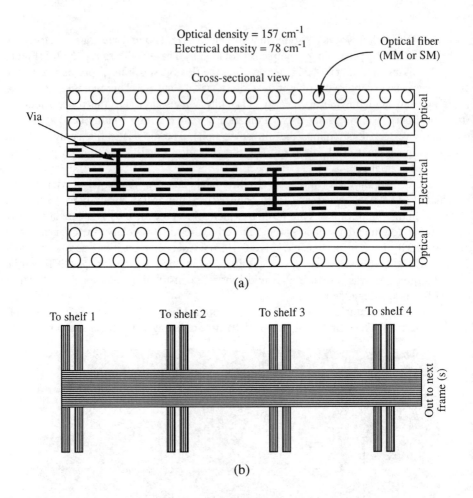

Figure 3.11 Sandwiched optical and electrical interconnection: (a) cross-sectional and (b) planar views.

electrical interconnects can be used for the interconnection of PCBs in one shelf to PCBs in another shelf (not necessarily collocated in the same equipment frame). The long length capability of flexible circuit construction is utilized to fabricate an electro-optic harness. A testimony to the long length capability of flexible circuits is in its "rolled" fabrication process. Flexible electrical circuits [25] (e.g., automotive circuits) are processed in rolls of lengths up to 500 ft. Hence, interconnecting PCBs in this length range should be achievable.

3.3 TECHNOLOGY CAPABILITIES AND LIMITATIONS

In this section, the optical-fiber backplane technology performance capabilities and limitations are explored to quantify the potential for next-generation products. The features a backplane should support include:

1. Connectivity between the circuit boards and backplane.
2. Connectivity between circuit boards.
3. Low cost.
4. Random interconnection patterns.
5. High data rate interconnections.
6. Power and ground distribution.
7. High interconnection density.
8. Variable length interconnection patterns up to 2m.

These desired backplane capabilities create a set of implementation-dependent technology performance indices. Capabilities of the fiber-optic based backplane include:

1. Bandwidth.
    a. Attenuation factor.
    b. Dispersion.
    c. Fiber bend loss and minimum radius of curvature.
    d. Connector coupling loss (butt or evanescent).
2. Fiber delay variation (i.e., skew).
3. Fiber cost and fiber router speed.
4. Fiber dimension and fiber-to-fiber spacing.
5. Multilayer capability with more fiber or copper-based interconnect.

These indices are subsequently addressed in the following order: first, from a connector perspective; second, from a fiber interconnection; and third, from a cost inspection.

3.3.1 Fiber-Array Connector Capabilities

The two relevant performance indices on fiber-array connectors are the fiber density (which is based on the linear distance along the edge of the circuit board) and the loss through the connector. The two types of fiber-array connectors that will be addressed are categorized as butt-coupled and evanescent field-coupled. Examples of butt-coupled fiber-array connectors are the AT&T MAC and Fujikara MT connectors. Figure 3.12 shows an example of the evanescent field connector represented by the technique implemented by British Telecom Research Laboratories (via the D-fiber technique) and Alcatel Bell Telephone teamed with AIT.

The single butt-coupled connector is the most widely used connector type (i.e., biconical, ferrule, capillary, double eccentric, triple ball, and expanded beam). Multifiber connectors can be implemented from single type connectors packaged together, or they can be implemented as in the design approach of the MAC and MT connector. Low-cost plastic molded versions of these array connectors [26] have recently been emerging from research labs. Butt-coupled array connectors rely on the alignment of the two prepared (i.e., polished) fiber-array ends that are in close proximity so that the fiber core axes coincide. Optical power loss [27] through the connector can arise from:

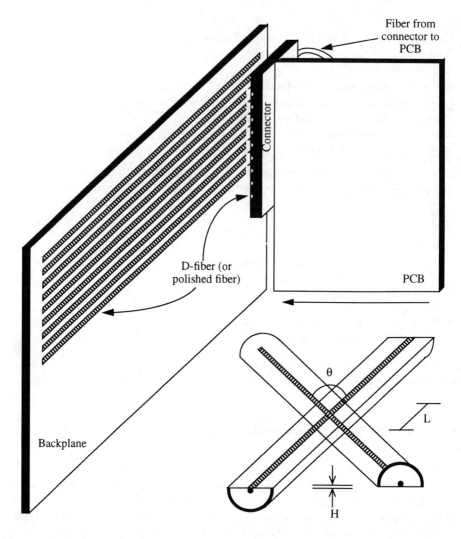

Figure 3.12 Evanescent coupler design developed by British Telecom Research Labs, Alcatel Bell Telephone, and AIT for exchanging energy between planes.

1. Different core and/or cladding diameters.
2. Different numerical apertures and/or relative refractive index differences (Fresnel reflection).
3. Fiber core ellipticity and/or concentricity.
4. Longitudinal misalignment.
5. Transverse misalignment.
6. Angular misalignment.

The last two loss mechanisms tend to dominate array connectors. As described in a paper by Nemoto and Makimoto [28], the coupling efficiency η_c can be approximated by assuming two identical fibers (which make up the connector), both having Gaussian amplitude distributions with no phase aberrations. The result is

$$\eta_c = \frac{16n^2\sigma}{q(n+1)^4}\exp\left(-\frac{pu}{q}\right) \tag{3.3}$$

where

$$u = v + \sigma\left[G^2 + \frac{1}{4}(\alpha+1)\right]\sin^2\theta \qquad K = 2\frac{\pi}{\lambda}$$

$$v = (\sigma+1)F^2 + 2\sigma FG\sin\theta \qquad G = \frac{z}{KW_t^2}$$

$$q = G^2 + \frac{1}{4}(\sigma+1)^2 \qquad F = \frac{x}{KW_t^2}$$

$$p = \frac{1}{2}(KW_t) \qquad \sigma = \left(\frac{W_r}{W_t}\right)^2$$

where n is the refractive index of the fibers, λ is the free-space wavelength, W_t and W_r are the mode field radii of the transmitting and receiving fibers respectively, z is the longitudinal displacement, x is the transverse displacement, and q is the offset angle between the fiber cores. For example, the coupling efficiency for a step index (SI) fiber connector having the following characteristics and misalignments; $n = 1.5$, $l = 1.3$ μm, $W_t = W_r = 6$ μm, $z = x = 5$ μm, and $\theta = 2°$, yields $\eta_c = 34.3\%$. This relationship is an excellent approximation for most applications.

The specifications for the AT&T MAC multimode glass fiber connector are mean loss = .5 dB (.7 dB for single mode) and standard deviation loss = .2 dB (.3 dB for single mode). Hence, a worst case (3σ) analysis results in a 1.1 dB connection loss. The power loss in a connector is related to the coupling efficiency of the connector:

$$\text{loss (dB)} = 10\log(\eta_c) \tag{3.4}$$

Hence, an optical power loss of 1.1 dB results from a coupling efficiency of 77.6%. Plastic fiber has a higher numerical aperture (e.g., $NA = \sqrt{n_{12} - n_{22}} = 0.5 - 0.6$) than glass fiber, which leads one to expect a higher loss (via Fresnel reflection). However, because the core diameters are so much larger, the misalignment loss is substantially lower. This can result in a lower connector loss. For example, the coupling efficiency for a step index plastic fiber connector with the following characteristics and misalignments—$n = 1.5$, $l = 1.3\mu m$, $W_t = W_r = 30\mu m$, $z = x = 5\mu m$, $\theta = 2°$, yields $\eta_c = 88.2\%$.

The evanescent field coupled connector in Figure 3.12, which is still in a research phase of product development, relies on either two D-fibers laid face-to-face or two polished fiber lengths embossed in plastic and laid face-to-face. In both cases, some of the cladding is removed to allow interaction of the evanescent optical fields. When light is launched into one fiber, evanescent coupling will take place between the light propagating in the cores. This results in light power transfer from the first into the second. This coupling, as viewed in the butt-coupled fiber connector, can provide a variable power splitter, a switch, or a demountable connector.

At this point, the connector loss is primarily governed by the interaction length (l), (i.e., the overlap length of the two fiber cores), the core-to-core spacing (h), the wavelength, and fiber refractive index difference. The coupling efficiency for two identical fibers can be approximated as

$$\eta_c = \sin^2\left(\frac{\pi}{2} \times \frac{l}{l_c}\right) \quad (3.5)$$

where l is the interaction length required for complete power transfer. As the core separation, h, increases, l_c must accordingly increase at a rate of approximately twice per micron of core separation increase. Hence, η_c is quite dependent on the parameter h. Additionally, l_c is quite sensitive to wavelength. The interaction length, l, can be expressed as

$$l = \frac{d_c}{\sin \theta} \quad (3.6)$$

where d_c is the fiber core diameter and θ is the angle between the fibers as shown in Figure 3.12. For example, assuming a core diameter of 50 μm and a value of $l_c = 4.7$mm, the angle between the fibers should be 0 and 3° to obtain, respectively, a 90% and a 10% coupling efficiency connector. (Note that the interaction length, l, is calculated to be 3.82 and 0.96 mm, respectively.) By properly controlling the angle, a high-efficiency connector and optical tap (or power splitter for data bus applications) can be synthesized.

The connectorized fiber densities are quite different between the two types of connector families. The butt-coupled glass fiber connector can obtain a density of 40 fibers/cm. For a one-dimensional (linear) connection density, this assumes a 250 μm fiber-to-fiber spacing. Therefore, for a 40-cm board edge, 1,600 connections can theoretically be obtained. This figure must be derated according to the connector package size. For example, the AT&T MAC connector can house 18 fibers that occupy approximately 2 cm of linear distance along the edge of a board. Eighteen of these connectors can realistically be mounted on a board 40 cm high. This results in a fiber density of 324. (Note: The package derating factor turned out to equal 80%.)

If another dimension is used, a higher signal density can be achieved. This could be accomplished by packaging multiple linear connectors in a stacked arrangement or by exploiting alternative technology schemes (e.g., the two-dimensional optical connector). The signal density can be increased by more than a factor of ten for the butt-coupled connector style. The evanescent-coupled connector, on the other hand, already relies on a two-dimensional (or surface area) connector. In one direction, the fiber density is dictated by the fiber spacing. Once again, approximately 40 fibers/cm can be obtained. In the optical interaction coupling direction, the density is dictated by the amount of length required to guarantee flatness for the connection and the minimum fiber bend radius.

Now, assume that 2 cm of length is required for this dimension, and it is in the perpendicular direction of the board edge. This will result in a fiber density exactly comparable to the linear butt-coupled fiber density. Hence, the butt-coupled fiber connector has an advantage in connection density off the circuit board into the backplane. Table 3.1 summarizes the connector performance indices.

3.3.2 Interconnection Density and Fiber Packaging Issues

Random backplane interconnection patterns implemented with fiber are limited by the minimum bend radius. Fiber bends primarily arise from two sources: first, when routing curves on the planar substrate; and second, when a crossover is encountered. The crossover is perhaps the most critical loss element, where the bend radius may vary from approximately 250 to 1,000 μm when crossing over a 125-μm diameter fiber. Fibers suffer from severe radiation loss [29] when the radius of curvature exceeds a critical value. This loss is exponentially related to the radius, R, as

$$\alpha_{bend} = c_0 \exp(-c_1 R) \tag{3.7}$$

Table 3.1
Connector Performance Indices

Butt-Coupled Fiber Connector	Loss (dB)	Connections on 40-cm-High PCB (Linear/Area)	Cost per I/O ($)
Glass SM	1.6	324/3240	10.00
Glass MM	1.1	324/3240	10.00
Plastic MM		162/1620	2.00
Evanescent Coupled			
Glass SM		324/324	N/A
Glass MM		324/324	N/A
Plastic MM		Variable	N/A
Electrical Coaxial	1.0	100/300	3.00

Note: SM = single mode; MM = multimode; PCB = printed circuit board; N/A = not applicable.

where c_0 and c_1 are constants. A critical radius, R_c, can be defined as when the loss becomes excessive (e.g., >1 dB). It can than be estimated as

$$R_c = \frac{3n_0^2 \lambda}{4\pi(NA)^2} \qquad (3.8)$$

Therefore, it is desirable to use a large numerical aperture (NA) fiber and short wavelengths to decrease R_c. For example, at 850 nm, a comparison of R_c with multimode (SI) and POF (SI) results in R_c = 450 μm and 2.7 μm, respectively. More than two orders of magnitude improvement is obtained when the NA is simply increased by use of POF. The lower the critical radius is, the more freedom the layout can have. Also, there will be a wider tolerance range for the fiber router machine when a crossover is encountered. Hence, care in the selection of fiber type and routing guidelines is important for high-performance optical backplane design. Table 3.2 provides a summary for fiber performance indices as well as packaging density and cost.

3.3.3 Passive and Active (Rare-Earth-Doped) Fiber Backplane Networks

Figure 3.13 shows four conventional network topologies for optical backplane applications. Each of these networks will be subsequently addressed.

Table 3.2
Interconnection Technology Comparison

Interconnect Technology	Attenuation (dB/km^{-1})	Dispersion (Mhz-km)	Linear Signal Density/cm	Multilayer Capability	Cabled Cost per Meter ($)
Glass SM	1–5	>1,000	40	Yes	
Glass MM (SI)	5–10	10–30	40	Yes	
Glass MM (GI)	2–10	>500	40	Yes	1.55
Plastic clad,					
MM (SI)	5–50	5–25	20–30	Yes	
MM (GI)	4–15	200–400	30–40	Yes	
All plastic,					
MM (SI)	150–1000	10	10–20	Yes	0.55
MM (GI)	150	200	10–20	Yes	N/A
Electrical					
Coaxial	1,000*	~5	2.5–3.0	No	1.20
Shielded TP				No	1.00
Unshielded	1,000†				
TP (5)	1,000*				0.30
Flex stripline	3,000‡	~1	10°	Yes	0.60

Note: SM = single mode; MM = multimode; SI = step index; GI = graded index; TP = twisted pair; N/A = not applicable.
*1 GHz. †25 MHz. ‡1 GHz. °Differential.

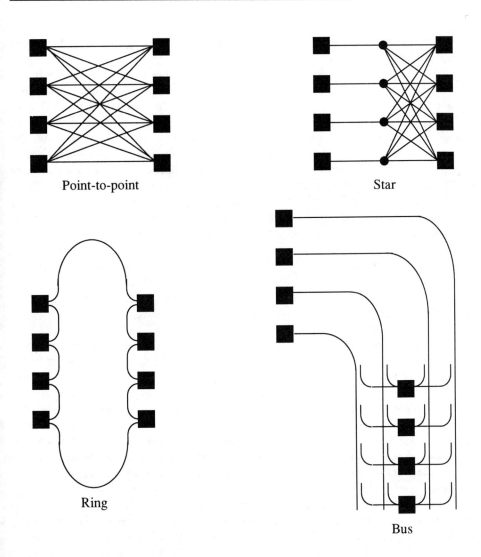

Figure 3.13 (a) Point-to-point; (b) Ring; (c) Star; (d) Bus.

The point-to-point topology is the most straightforward optical approach. A dedicated fiber (or possibly a pair of fibers) is routed to each destination. There are no passive or active components necessary in the backplane. Each fiber has a transmitter and a receiver. This topology requires the most fiber and the highest backplane routing complexity, as well as low-cost transmitters and receivers. There are numerous standard interface protocols that have emerged in the past decade that rely on point-to-point connectivity. Examples like SCSI (ANSI standard X3.131-1986), SCI (IEEE 1596-1992) and HiPPI (ANSI standard

X3.183-1991) represent this type of architecture. It is most likely that growth in these specific standards will occur.

The star topology is the next most straightforward approach. Passive or active splitters are used to fanout to the individual destinations. Time, wavelength, or code division transmitters multiplex the destination data together and pass the data over a fiber to the one-to-n splitter. The splitter may be located in the backplane, in a grouping of destinations, or in the transmitter section itself. Cost issues usually dictate placement. The advantages of this topology include fewer fibers and connectors at the transmitting section, perhaps a cost-efficient use of the transmitter's bandwidth, and fewer transmitters. These advantages rely on low-cost, one-to-n splitters; low-cost, high sensitivity receivers (with the added complexity of channel selection); and higher optical power from the transmitter (to overcome splitter loss). The AT&T Star LAN is an example of an optical star network.

The ring topology shares an optical fiber amongst numerous destinations. Each destination has an optical receiver and an optical transmitter. One can build unidirectional rings or bidirectional rings (e.g., FDDI). Bidirectional rings are used when reliable interconnections are necessary. The advantages of this topology include considerable reductions in the fiber (and connectors) and the number of transmitters/receivers. This topology relies on high-speed transmitters and receivers, where the cost is amortized over the number of destination channels.

Finally, the bus topology uses a shared optical-fiber line for communication with destination points. Passive or active taps (splitters) are used to distribute power to the destinations. The bus topology is positioned somewhere between the ring and star advantage/disadvantage position. It requires less fiber than the star.

The choice of network topology is application dependent and based on a number of parameters. One parameter is the capability to insert and remove circuit packs while the system is operational with no service interruption. Network topologies (e.g., star and ring) must be reevaluated to accommodate this need. Adding a slight complexity, such as a counter rotating ring, to the ring topology would be one way to support insertion and removal of circuit packs. There are other parameters that must be investigated to determine which topology optimally fits a particular application. These parameters include the connector cost, fiber cost, backplane capability, transmitter/receiver cost, fiber bandwidth capability, and passive/active splitter technology capabilities (and respective cost). A discussion of the splitter function is necessary before the determination of a compatible network topology can proceed. For example, in order to implement a star or a bus network topology, signal fiber splitters and combiners (classified as couplers) must be realizable. Because this chapter describes bulk waveguide routing via fiber optics, the use of electro-optic devices (e.g., $LiNbO_2$ or polymer devices) for implementing the splitter function will not be discussed. Only fiber techniques will be addressed.

The fiber splitter function can be implemented in the following ways: fused-fiber coupler; polymeric mixing rod coupler [30]; evanescent field coupler. All of these splitters are essentially passive couplers. An active splitter can be formed by using one of the passive dichroic couplers in conjunction with a specially doped fiber of rare-earth ions (e.g., Erbium or Neodymium). The doped fiber can be used to build an optical amplifier that amplifies the optical signal either before or after the splitter. A discussion of the passive couplers will be addressed first, and then a discussion of the active coupler will follow.

The fused-fiber coupler is schematically shown in Figure 3.14(a). This type of coupler is used in a variety of applications, ranging from the instrumentation to telecommunication industries. In the telecommunication industry, this coupler is planned for use in the passive optical network (PON) [31] implementation of fiber-to-the-home (FTTH). This exciting new capability will allow high-bandwidth services (particularly multimedia) to be transported to the residential customer as well as the business customer. Telephony over PON (TPON) is presently envisioned to be implemented with passive splitters in a star topology [32]. One fiber from the central switching office is shared with up to 128 optical network units (ONU) located on or near the customer premise. A return channel is clearly required for voice service, but this channel is also needed for remote program selection and other customer services such as teleshopping and interactive games. Initially, time division multiple access (TDMA) will most likely be used, but as higher data rates are demanded, techniques such as wavelength division multiple access (WDMA) or even frequency division multiple access (FDMA) must be taken advantage of.

Fused-fiber couplers rely on the evanescent fields from each fiber coupling together. To accomplish this, the cores from each fiber must be brought together (on the order of one wavelength). This coupler is generally formed by taking two glass fibers and applying heat while the fibers under tension are twisted together. The glass melts, elongating the fibers, and the fibers then fuse together. When the fibers are fused (under tension), the cores shrink and form taper regions. In this case, the evanescent field spreads out further into the cladding and couples strongly to the adjoining core. As the signal leaves the tapered region, the fields from each fiber separate and couple back into the guided mode of their respective fibers. The tapering process is usually computer controlled to achieve specific splitting ratios (e.g., 3 dB, 6 dB, 10 dB, and so forth.). An excellent example of fused-fiber couplers in electronic processing systems is described in [14]. To build large fanouts (e.g., 1 to 16), a multistage network can be constructed out of two-by-two couplers, or a multifiber fused coupler can be formed [33]. Note that these couplers tend to be wavelength dependent (i.e., the splitting ratio between one port and another will vary sinusoidal with wavelength). This can be a useful phenomena to exploit when constructing fiber wavelength multiplexers and demultiplexers. Research into plastic optical-fiber variable ratio splitters [34] offers devices at a potentially low cost for future applications.

The polymeric mixing rod coupler (shown schematically in Figure 3.14(b)) relies on a polymer mode mixing region where the input light spreads out uniformly and couples back into the output fibers. The fibers can be glass or plastic, SM or MM. This technique seems to be the most promising for optical backplane technologies because of the possibility of batch processing these mixing regions onto either a flexible or rigid backplane substrate. Additionally, the material cost is quite low. In [30], a polymeric mixing rod was designed from a Teflon tube filled with a transparent resin. The input port fiber used was a large

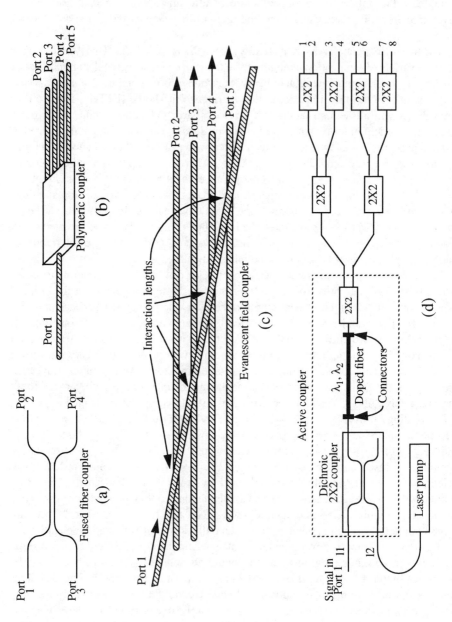

Figure 3.14 (a) A fused fiber coupler. (b) A polymeric mixing rod coupler. (c) An evanescent field coupler. (d) An active splitter.

core plastic optical fiber (POF), while the output port fibers used were plastic clad silica fibers. The output fibers needed to be grouped in a close-pack arrangement to minimize the excess loss of the coupler. The reported loss through the mixing region was 10 to 20 dB/m, which was only slightly wavelength dependent. Because the mode mixing region length is only on the order of a few centimeters, the insertion loss is under 1 dB. When implementing a splitter or a combiner styled coupler, it is important to match the input port fiber diameter to the diameter of the mixing region. The mixing region diameter (or actually its cross-sectional area) is determined by the output port fiber diameter and the number of output fibers needed. The ratio of the cross-sectional area of the mixing region, S_m, to the cross-sectional area of the output fiber core, S_f, dictates the theoretical splitter ratio. For a cylindrical shaped mixing region, this ratio is

$$\text{power (dB)} = 10 \log \frac{S_f}{S_m} \tag{3.9}$$

Clearly, small cladding thickness is as important as the packing density. Reported efficiencies on the order of 60% with ratios of 1:7, 1:19, and 1:37 have been achieved with excellent fiber-to-fiber uniformity. Note that this technique is not necessarily limited to cylindrically shaped mixing regions. It is used because of its optimum shape, which interfaces well with cylindrical fibers. Various shaped mixing regions can be designed to support a mix of output fiber configurations and to conform to backplane restrictions. Generally, the excess loss is the only parameter that will be affected by the choice of fanout size. These couplers can be applied to optical backplanes by either incorporating them as an over-layer structure or integrating right onto the backplane medium. When integrating onto the backplane, low cost should result due to the possibility of batch-processed couplers.

Finally, the evanescent field coupler (shown schematically in Figure 3.14(c)) is simply an extension of the evanescent field coupler. The connector, which can behave as a 3-dB splitter, can be extended into a 1-in-N out coupler as follows. Instead of coupling into one fiber, the interaction length of multiple fibers is exposed to the input fiber. The difficulty of designing this type of coupler is that the split ratio may be held constant across the output fibers, but the intensity in each output fiber will be quite different. This happens due to the subsequent power splitting occurring in cascade. Additionally, the size gets to be quite large (e.g., cm) as the complexity gets large. In this case, complexity refers to the ''connectors'' that must be attached to the backplane to not only construct the coupler but also to merge the fiber or fibers back into the backplane media.

The fundamental problem with passive splitters is the incurred power loss. Given the sensitivity and dynamic range limits of the receiver, the fanout capability tends to be severely limited. (Note that the receivers that would be useful in an advanced system application would have limited sensitivity and dynamic range (e.g., −15 dBm), which enables the integration of many channels together on the same substrate. This limits the fanout capability even more.) This limitation can be overcome by including a fiber amplifier that amplifies the signal either before or after the splitter.

The fiber amplifier is based on the principal operation of the optically pumped solid-state laser [35]. The laser pump operates at a lower wavelength, which corresponds to a higher photon energy, than the signal wavelength. The laser pump establishes an electron density in the doped fiber via photon absorption, which emits photons when stimulated by the photons from the input signal. Hence, laser action occurs. The gain is proportional to the difference in the number of doped electrons in the excited state as compared to the number in the ground state. Hence, the absorption and emission spectra must be carefully tailored to the fiber-amplifier application. For example, an Er-doped fiber has absorption peaks at 800, 980, and near 1,480 nm. The fiber has an emission peak near 1,550 nm. Therefore, a laser pump operating near an absorption peak (e.g., 980 nm) with a signal wavelength near the emission peak will result in a high-gain (e.g., 10 to 40 dB) fiber amplifier. As one might expect, the gain of the fiber amplifier is wavelength dependent due to the absorption/emission spectra dependence of the doped fiber. Typical bandwidths of 40 to 50 nm are obtained for Er-doped fiber amplifiers.

The construction of the fiber amplifier is accomplished by doping rare-earth ions (e.g., erbium, neodymium, holmium, or thulium) into the core of a glass fiber system (e.g., silica- or fluoride-based fiber). The light from the pump laser is coupled with the signal light source by a dichroic coupler as shown in Figure 3.14(d). The length of the doped fiber is measured in the 10s of cm. This inclusion of the fiber amplifier into the passive splitter is then classified as an active splitter. An example is shown in Figure 3.14(d). If a fiber amplifier of gain 11 dB is applied before a 1:8 passive splitter (whose characteristics might be 10 dB loss with a 1 dB excess loss), the net insertion loss would be zero. At this point, there would be no power penalty associated with the splitter, and the receiver characteristics would not have to be overstressed. (Note that the fiber amplifier will saturate when the output power at the signal wavelength approaches that of the pump power.) For the previous example, the pump power must be greater than the signal power by at least 11 dB. This suggests that in some applications, the fiber amplifier must be inserted after the splitter, where the signal intensity is lower. This active splitter offers designers the best potential for building optical backplane network topologies. Although the passive splitters could be integrated (or molded) into the backplane, the doped fiber may have to be externally attached. The fiber backplane could be built out of doped fiber, but the cost would be prohibitive. It may be possible to selectively integrate the doped fiber into the backplane where the ends matchup to polymer mixing regions that form the passive couplers. As optical backplane technology matures, the realization of this and other related concepts will be clearer.

Table 3.3 summarizes the coupler technologies with respect to backplane compatibility, relative size, and cost. In the author's opinion, it is clear that coupler technologies that are compatible with the backplane media and batch-processed (e.g., the polymer coupler) will be prevalent in optical backplanes. The size and cost of these couplers suggest that there will only be a limited number in use in a backplane. Possible uses

Table 3.3
Fiber Coupler Technology Comparison

Coupler Technology	Backplane Compatibility	Size	Cost
Fused fiber	No, must be connectorized	Centimeters	Medium
Evanescent field	No, must be connectorized	Centimeters	High
Polymeric	Yes, possibly batch processed	Millimeters	Low
Active (doped fiber)	No, must be connectorized	Centimeters	Highest

include distributing clock and address/data bus signals. Using a massive amount of couplers to implement a star topology for a large network would simply be prohibitive. The network topologies that are realizable in optical backplane technology are the bus and ring architectures. If the cost of the transmitters/receivers drops significantly (as they should via 1D-ODL concepts), then point-to-point interconnection techniques may dominate. However, the correct determination of which network topology to use is strictly architecture dependent. More than likely, the right choice for a computer parallel-processor application will be different than the choice for telecommunication switch applications. Hence, all this subsection can expect to conclude or accomplish is the presentation of different optical network topologies and their implementation technologies. This should allow the readers to select a particular technology that meets their needs.

3.4 ENVIRONMENTAL RUGGEDNESS ISSUES

This final section will highlight the environmental and reliability issues that must be met for a successful product introduction. The reliability practices already in place for optical connectors and electronic backplanes are expected to be used for the optical backplane. These practices include accelerated-aging tests (via thermal and humidity exposure and cycling), mechanical testing (via insertion, vibration, impact, flexing and twisting tests), and chemical/flammability resistive tests. Each of these three categories will be discussed briefly. Accelerated-aging tests are usually applied to products to ensure an adequate product lifetime. They are classified into the following tests: humidity, thermal, and temperature cycling. For a controlled environment application (such as the one being discussed for an optical backplane), the maximum change in optical loss must be less than 0.3 dB. The humidity test requires an operating temperature of 60°C while the relative humidity is controlled between 90% and 95%. The thermal test requires an operating temperature of 85°C with no humidity control. The temperature cycling test requires 3 cycles per day, where the cycle is defined as temperature excursions between 49° and 2°C. All of the above tests last for 14 days, and the optical measurements are performed daily.

Mechanical testing is perhaps the most difficult test to accomplish for some backplane implementation technologies. The vibration requirement states that the maximum change in the system operational optical power be no more than 0.1 dB. It is also assumed that no physical damage has occurred. The vibration is specified in the frequency range between 10 and 55 Hz with an amplitude of 20g. The resulting positional displacement occurring from this vibration is assumed to be limited to 1.52 mm. This test is meant to simulate vibrations from heavy equipment movement, equipment being dropped nearby, and/or earth tremors and quakes. This specification may have significant implications for free-space optical backplane implementation technologies, especially when vibration waves are specified. The impact requirement is similar to the vibration test. For this specification, the optical change in loss must be less than 0.2 dB (with no damage) after being dropped 8 times onto a concrete floor from a height of 0.6m. The insertion specification is intended to apply to the connector technology being used. The specification states that the equipment is to perform within stated performance specifications after 100 insertions and de-insertions are done.

Finally, if a flexible backplane is used with no supporting structures, or part of the backplane is partitioned into flexible extensions (possibly for array connector attachments), then the flex/twist requirements apply. The maximum change in optical loss must be no more than 0.2 dB, and, once again, no damage can be incurred. Ten cycles of twisting and 300 cycles of flex are applied with a 0.5-kg load attached to the flexible portion. For additional information on this test procedure, see the ASTM recommendation designation D790-86.

The last category to be discussed is the chemical and flammability resistance specification. The highest flammability classification as specified by Underwriters Laboratories is 94 HB. It specifies a burning rate of approximately 5 cm/m over an 8-cm span. (This rate is only slightly dependent on the backplane substrate thickness.) More important, the flame must cease before it reaches a 10-cm span. Care must be used when selecting the material composition of the optical backplane. The chemical resistance specification is applicable when electrical interconnection is integrated with the optical patterns. The electrical (and possibly some of the optical processing technologies) backplane is usually subjected to a variety of chemicals during the manufacturing process. The chemical exposure primarily arises from solder or copper-plating processes or during a number of cleaning operations. Generally, the requirement states that the component (in this case the backplane) must survive an exposure test. The exposure test is one in which the backplane is immersed into a chemical bath for 1 minute and then dried for 10 minute. The sample is then tested (and retested again after 24 hour) to ensure that no compositional breakdown of the material has occurred. The specific list of chemical types used in this test varies from manufacturer to manufacturer and evolves as technological improvements are incorporated into the process. Hence, it is inappropriate to specify a list of chemicals. All that can be recommended in this discussion is to consult the specific manufacturer.

This discussion on environmental ruggedness is included to acquaint the reader with some of the tests and requirements that must be addressed before an optical backplane technology product introduction. The requirements stated in this section should serve only as a guideline when selecting materials and fabrication processes. An awareness of these requirements, while still in the product research phase, is strategic for the successful insertion of the optical backplane technology into industrial applications. For further technical information on this subject, consultation of Bellcore (i.e., specific technical advisories) and CCITT specifications for telecommunication products and Underwriter Laboratory (UL) specifications for consumer products should be thoroughly done.

3.5 SUMMARY

Advanced switching (and computer) system architecture designs require higher performance and lower cost packaging technologies to implement new product capabilities and enhanced services. Presently, electronic processing and interconnection techniques (e.g., MCMs and C^4) offer next-generation systems superior performance at a lower cost for the chip-to-chip packaging level (intra-PCB interconnectivity) over that of alternative methods. These electronic techniques are driving the needs for higher performance backplanes (e.g., high-density, high-bandwidth, and flexible routing). Optical interconnection backplane strategies offer advantages at higher levels of packaging (i.e., board-to-board, shelf-to-shelf, and frame-to-frame levels of interconnection). Furthermore, high-data-rate parallel optical data links can presently offer greater flexibility and lower cost than time-multiplexed serial data links.

Of the numerous techniques for implementing an optical backplane, the fiber guided wave solution has the highest potential for being pervasive in the industrial marketplace. Already, the flexible backplane using a MAC connectivity scheme is being introduced by AT&T into one of its electronic switching system product lines. Other telecommunication companies (e.g., Alcatel, Siemens, and Ericsson) are presently positioning themselves for the introduction of optical backplane technology into their product lines. The flexible optical backplane technology is useful for grooming discrete fibers into fiber ribbons for connectivity; implementing low-cost, high-performance random interconnection patterns; supporting electronic/optical interconnections; and providing a capability for long (multiple meter range) interconnections.

The key concepts that make the fiber-based optical backplane plausible are PCB mounted parallel optical data link modules, multifiber connectors, an existing infrastructure (i.e., automated fiber routing systems similar to the multiwire head technology) for fiber routing, and a long history and experience base established in fiber optics. This chapter discussed the advantages and disadvantages of different fiber-based solutions for the optical backplane. In the near term, glass fiber will probably dominate. However, plastic fiber (and plastic fiber-array connectors) should eventually take over as this technology matures. Similarly, point-to-point network topologies will be the near term architecture of choice, but passive/active splitter technologies (as they mature) will offer the designers more flexibility to match their particular application to an optimized network topology. Fiber-based optical backplane technology is an enabling technology that will allow industries (telecommunication, computer, and CATV) to evolve their hardware platforms and expedite the introduction of high-bandwidth services.

REFERENCES

[1] Nordin, R. A., A. F. J. Levi, R. N. Nottenburg, J. O'Gorman, T. Tanbun-Ek, and R. A. Logan, "A Systems Perspective on Digital Interconnection Technology," *J. Lightwave Tech.*, Vol. 10, No. 6, June 1992, pp. 811–827.
[2] Tummala, R. R. and E. J. Rymaszewski, *Microelectronic Packaging Handbook*, VanNostrand Reinhold, 1989.
[3] Hartman, D. H., "Use of Guided Wave Optics for Board Level and Mainframe Level Interconnects," *Proceedings of the 41st Electronic Components and Technology Conference*, June 1991, pp. 463–74.
[4] Ryckebusch, M., "A High Performance Electrical and Optical Interconnection Technology," *Proceedings of the 40th Electronic Components and Technology Conference*, June 1990, Vol. 2, pp. 974–9.
[5] Midwinter, J. E., *Photonics in Switching, Vols. I and II*, Academic Press, 1993.
[6] Nordin, R. A., D. B. Buchholz, R. F. Huisman, N. R. Basavanhally, and A. F. J. Levi, "High Performance Optical Data Link Array Technology," *Proceedings of the 43rd Electronic Components and Technology Conference*, June 1993, pp. 795–801.
[7] Nagahori, T., M. Itoh, I. Watanabe, J. Hayashi, H. Honmou, and T. Uji, "150 Mb/s/ch 12-Channel Optical Parallel Interface Using an LED and a PD Array," *Optical and Quantum Electronics*, Vol. 24, No.4, April 1992, pp. 5479–5490.
[8] Ewen, J. F., K. P. Jackson, R. J. S. Bates, and E. B. Flint, "GaAs Fiber-Optic Modules for Optical Data Processing Networks," *J. Lightwave Tech.*, Vol. 9, No. 12, Dec. 1991, pp. 1755–1763.
[9] Goodwin, M. J., A. J. Moseley, M. Q. Kearly, R. C. Morris, C. J. G. Kirkby, J. Thompson, and R. C. Goodfellow, "Optoelectronic Component Arrays for Optical Interconnection of Circuits and Systems," *J. Lightwave Tech.*, Vol. 9, No. 12, 1991.
[10] Bristow, J., A. Guha, C. Sullivan, and A. Husain, "Optical Implementations of Interconnection Networks for Massively Parallel Architectures," *Applied Optics*, Vol. 29, No. 8, 1990, pp.1077–93.
[11] Hinton, H. S., *An Introduction to Photonic Switching Fabrics*, Plenum, 1993.
[12] DeRosse, N., M. Ryckebusch, M. Botte and E. Beyne, "Comparison of Different Interconnect Technologies for High Frequency Applications," *IEEE/CHMT '89 Japan IEMT Symposium*, pp. 319–324.
[13] Grimes, G. J. and L. J. Haas, "An Optical Backplane for High Performance Switches," *Proc. Internat. Switching Symposium*, May 1990, Vol. 1, pp. 85–89.
[14] Grimes, G. J., C. J. Sherman, R. W. Gasvest, S. R. Peck, W. K. Honea, J. S. Helton, W. W. Jamison, W. J. Parzygnat, R. Bonanni, R. J. Nadler, K. S. Rausch, J. J. Thomas, and L. L. Blyler, "Packaging of Optoelectronics and Passive Optics in a High Capacity Transmission Terminal," *Proceedings of the 43rd Electronic Components and Technology Conference*, June 1993, pp. 718–724.
[15] Sullivan, C. T., B. L. Booth, and A. Husain, "Polymeric Waveguides," *IEEE Circuits and Device Mag.*, Vol. 8, No. 1, Jan. 1992, pp. 27–31.
[16] Koike, Y., "High-Bandwidth Graded-Index Polymer Optical Fibre," *POLYMER*, Vol 32, No. 10, 1991, pp.1737–45.
[17] Bates, R. J. S. and S. D. Walker, "Evaluation of all Plastic Optical Fibre Computer Data Link Dispersion Limits," *Electronics Letters*, 21 May 1992, Vol. 28, No. 11, pp. 996–998.
[18] Davila, D., "Plastic vs. Glass on Premises," *Lightwave Mag.*, Jan. 1992, pp. 50–53.
[19] Nakahara, H., G. Messner, and T. Buck, "Manufacturing of Complex Systems by a Discrete Wiring Interconnection Technology," *IEEE/CHMT '89 Japan IEMT Symposium*, pp. 40–44.
[20] Delbare, W., L. Vandam, J. Vandewege, J. Verbeke, and M. Fitzgibbon, "Electro-Optical Board Technology Based on Discrete Wiring," *1991 International Electronic Packaging Conference*, Sept. 1991, pp. 604–618.
[21] Weiss, R. E., "A Family of Connectors for Circuit Pack to Backplane Optical Interconnection," *Proc. 9th Internat. Electron Packaging Conf.*, Sept. 1989, Vol. 2, pp.1033–1042.
[22] Yokosuka, H., Y. Tamaki, and K. Inada, "A Low Loss Multifiber Connector and its Applications," *Proceedings of the 40th Electronic Components and Technology Conference*, June 1990, pp. 865–868.

[23] MacKenzie, F., T. G. Hodgkinson, S. A. Cassidy, and P. Healy, "Optical Interconnect Based on a Fibre Bus," *Optical and Quantum Electronics*, Vol. 24, No. 4, April 92, pp. S491–S504.
[24] Holland, W. R., J. J. Burack, and R. P. Stawicki, "Optical Fiber Circuits," *Proceedings of the 43rd Electronic Components and Technology Conference*, June 1993, pp. 711–717.
[25] Gilleo, K., *Handbook of Flexible Circuits*, Van Nostrand Reinhold, 1992.
[26] Robertsson, E., P. Eriksen, B. Lindsrom, H. C. Moll, and J. A. Engstrand, "Plastic Optical Connectors Molded Directly onto Optical Fibers and Optical Fiber Ribbons," *Proceedings of the 43rd Electronic Components and Technology Conference*, June 1993, pp. 498–504.
[27] Lasky, R. C., et al, "Coupling of Light into Single Mode Optical Fiber for Data Communications," *MRS Symp. Proc.*, Vol. 264, pp. 379–393.
[28] Nemoto, S. and T. Makimoto, *Opt. Quant. Elect.*, No. 11, p. 447, 1979.
[29] Senior, J. M., *Optical Fiber Communications—Principles and Practice*, Prentice Hall, 1985.
[30] Blyler, L. L. and G. J. Grimes, "A Molded Polymeric Optical Coupler," *Proceedings of the 41st Electronic Components and Technology Conference*, May 1991, pp. 38–41.
[31] Oakley, K., "Passive Optical Networks: The Low Cost Road to the Future," *Fiber Optics Reprint Series*, Vol. 12, pp. 209–14.
[32] Lin, M.Y., K. M. Lin, D. R. Spears, and M. Lin, "Fiber-Based Local Access Network Architectures," *IEEE Communications Magazine*, Oct. 1989, pp. 64–72.
[33] Arkwright, J. W., D. B. Mortimore and R. M. Adnams, "Monolithic 1X19 Single Mode Fused Fiber Couplers," *Electrons Lett.* Vol. 27, No. 9, 1991, pp. 737–8.
[34] Kagami, M., Y. Sakai, and H. Okada, "Variable-Ratio Tap for Plastic Optical Fiber," *Applied Optics*, 20 Feb. 1991, Vol. 30, No. 6, pp. 645–649.
[35] Midwinter, J. E., and Y. L. Guo, *Optoelectronics and Lightwave Technology*, Wiley, 1992.

Chapter 4
Free-Space Routing

Rick Morrison

4.1 INTRODUCTION

Free-space optical routing presents a revolutionary opportunity to enhance the communication throughput of information processing and switching systems. Free-space optical routing systems constitute an interconnection network based on light beams that are guided autonomously of conventional optical fiber. In this new technology, both conventional optical components (such as lenses, mirrors, and polarization-sensitive elements) and evolving technologies (such as holography and binary optics) serve to collect, split, direct, and focus the light beams between arrays of electronic processing devices. Free-space optical routing has the potential to supplement electronic interconnections by exploiting the volume immediately surrounding planar electronic circuits to generate large numbers of parallel, high-density, high-bandwidth interconnections. These architectural advantages become valuable as electronic systems become more susceptible to the escalating communication bottlenecks enhanced by the limitations of their planar format.

The motivation for free-space optical routing derives from both the potential advantages of this technology and the success of guided wave optical interconnections demonstrated in the long distance telecommunications network. The primary advantages that fiber optics provides over electronic communications include its abundant bandwidth and highly energy-efficient distribution of signals at an economical price.

Because the fundamental advantages of optical interconnectivity scale over a large range of distances, it is advantageous to explore smaller scale integration, even to the level of logic gate interconnections that occur within a complex circuit. Free-space optical connections become more practical than fiber connections when densities grow beyond a few tens of connectors per chip package and more practical than electronic connections beyond densities of a few hundred. Prototype photonic device arrays have been fabricated that are capable of making thousands of optical connections per chip [1]. For comparison, current electronic microprocessor packages contain around a hundred pin connectors.

Dense, free-space optical interconnectivity permits a flexibility that has no counterpart in electronic communications at this scale. For example, at distances greater than a few meters, electronic communication is established by either conducting media, such as wire and cable, or via unguided means, such as microwave transmission. Unguided electromagnetic transmission is generally not practical or cost effective at very short ranges. With light, however, due to its inherently smaller wavelength, unguided communication is viable to much smaller distances.

Because the field of free-space optical routing is still in its infancy, we shall principally highlight system issues in this section. Prototype systems will be discussed as illustrations of how proposed techniques could be implemented. The primary purpose will be to examine optical routing architectures that connect groups or arrays of photonic devices. Our chief assumption is that information processing is best provided by electronics and that communication should ideally be elevated to the optical domain. We will disregard point to point connections between isolated processing systems, such as satellite links, and also analog implementations associated with pattern recognition and classification systems. We will concentrate on optical technologies that would supplement electronic networks associated with a variety of commercially applicable systems such as digital computing systems, analog neural networks, and telecommunication switches.

Free-space optical routing benefits digital computing applications as new architectures are created based on arrays of parallel processors, as processing cycles become faster, and as connectivity between processors and system components grows. Neural-network-based systems are envisioned as solutions to speech recognition, visual perception, knowledge acquisition, representation, and processing problems. These networks, still in their infancy, would benefit from massive and dense optical interconnectivity. Finally, telecommunications switching systems will benefit as services venture from conventional audio channels into image, video, and high-speed data traffic.

Our fundamental interest is the interconnection system itself; however, the novelty of this technology introduces dependencies between the various system constituents. For example, although fabrication techniques for complex silicon processors are mature and might thus depict silicon as a prime candidate for an optical system integration, the indirect bandgap of silicon makes it unsuitable to directly construct an efficient optical source or modulator. On the other hand, gallium arsenide (GaAs) and indium phosphide (InP) yield excellent photonic devices, yet the level of processor circuit complexity currently falls short of that of silicon. Such limitations require tradeoffs between optimal and practical designs.

A hypothetical view of the future predicts optical interconnectivity will be incorporated into processor arrays or communication systems principally at those points where it serves to enhance the capabilities of electronics. The advantages that free-space optical interconnections offer for the near-term are:

1. Efficient power utilization—Signals on high-speed electronic transmission links suffer severe power attenuation due to parasitic impedances and termination resistors. When the connection is over a few hundred microns in length, it is beneficial to convert a signal to an optical format [2].
2. Interconnection density and parallelism—Wires between electronic circuits can require bonding areas that are thousands of square microns in size, whereas optical detectors and receivers can be fashioned with areas of only tens of square microns due to the high resolution of optical lens systems. Interconnection of over 8,000 optical beams into a chip having an area of about $1.7mm^2$ has been demonstrated in a simple photonic system [3]. When read using a light-beam array generated by a single light source, thousands of information channels can be extracted in parallel.
3. Processing density—Removing a large fraction of the electrical transmission lines and the associated electronic receivers and drivers will result in a higher processor density. Also, a larger percentage of the power could then be committed to processor associated activities.
4. Nonlocal interconnections—Nonlocal electronic connectivity between processing cells is expensive from an energy and area utilization standpoint. However, nonlocal optical connections are practical because photonic beams travel through each other without interference.
5. Volume utilization—Electronics is primarily a planar technology. Free-space optical interconnectivity exploits the volume surrounding the substrate. Unfortunately, some conventional optical frameworks may require considerable space to achieve this connectivity.
6. Signal crosstalk—Neighboring electronic paths are prone to inductive coupling unless they are well separated or appropriately designed as transmission lines. In an ideal free-space optical routing system, light from separate channels is isolated at the input and output connections.
7. Signal skew—In high-speed systems, signal groups must often propagate between cells within a fixed interval and are highly sensitive to path length. A set of parallel optical paths has, for practical purposes, equivalent lengths due to the routing format.

Because there are many physical advantages to incorporating free-space optical interconnections, it becomes the task of the researcher and engineer to demonstrate that these advantages can be blended into a commercially successful system. In the near term, it may be far more realistic to design very simple electronic processors that operate within an optical interconnection framework. These islands of logic, also referred to as "smart pixels," and optical neural networks represent a favorable path for introducing optical interconnection into the highly developed electronic market.

Let us begin our discussion by identifying the elements of a generic system based on free-space optical routing. Our processing system can be categorized into general units as illustrated in Figure 4.1. The principal modules, for the purpose of this study, are the processor array and the communication network. The processor array is a matrix of

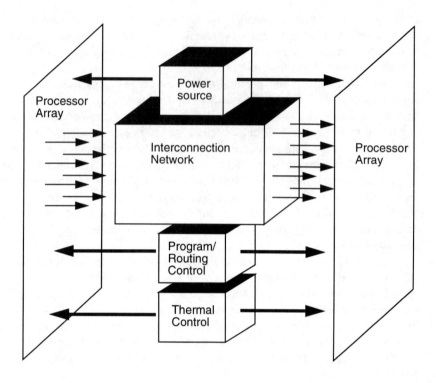

Figure 4.1 Some of the fundamental units of a free-space optical routing system. The lighter arrows represent high-bandwidth optical information channels; the darker arrows are low-bandwidth channels or power connections.

devices whose duties may range from simple functions, such as regenerating data to more specialized tasks where one or several data streams are surveyed. Within the communication network, light beams are collected and routed to other processors common to the array or routed to the following processor module, as shown in Figure 4.1. An optical route may have an identical form for each processor (i.e., be space-invariant) or, differ in a space-variant manner. As we proceed, we must continue to realize that the functional separation of processing and routing components does not imply that system design can also be compartmentalized.

Several other modules are also needed to complete the system. An input preparation stage (not shown in figure 4.1) is required for providing communication paths into the system. An output communication module is needed as well. These two interfaces present serious issues for free-space routing whenever large numbers of input and output channels are connected because the information must either be converted from electrical to optical format using an array of spatial light modulators or inserted optically via a fiber bundle having a two-dimensional coupler.

Also shown in Figure 4.1 are three other modules. The program control unit can install the stored program in a digital computing system or inject the network routing configurations for arrays of switching nodes. In a neural network application, it may provide the predetermined connection strengths between analog neural processors. The control unit can transmit intermittently at a low bandwidth because it is expected to provide information that is seldom modified. A second unit is the power source for producing the optical and electrical power. For some implementations, the optical power source could feasibly be integrated into each node of the processor array. Lastly, a thermal management system may be necessary for densely packaged and/or high-bandwidth systems.

Figure 4.2 shows the processor module and optical interconnection components in finer detail. The processor node is typically conceived as having a two-dimensional, planar configuration that can subdivided into smaller units. These units are the receiver, the processor, and the transmitter or modulator.

The receiver acquires information directed to it by a routing module. It may function as the optical-to-electronic signal converter and if so may also provide signal amplification. This amplification can be beneficial because it can overcome optical power loss in the routing stage and potentially enhance the throughput of the optical relay.

The second unit is the processor itself. The processor may be responsible for controlling the deflection of an incident light beam, as in the case of mechanically deformable mirror arrays, or it may be capable of signal format conversion and data manipulation.

Finally, the transmitter prepares the processed data and encodes it for optical transmission through the next routing stage. To generate optically encoded data, the transmitter must either act as a light modulator when illuminated via an external light source or else a local light source must be integrated. In some simple systems, the receiver, processor and transmitter could be functionally integrated within one device. Other processor array components that have been ignored include the thermal management system, the electronic power network, and, if needed, a local interprocessor electronic communication system.

The optical routing stage is a three-dimensional volume incorporating refractive, reflective, and/or diffractive optical elements. It can be static (i.e., the network connectivity remains constant throughout time), or it can be dynamic, such as when photorefractive materials are used to redefine connections or modify connection strengths in neural network applications. Some primary functions of the optical routing system are to (1) collect light from the transmitter and direct the beams, (2) achieve fanout by beam splitting, (3) combine various signal beams, and (4) focus the light onto a receiver.

Before free-space optical routing systems can become a highly regarded interconnection technology, a number of issues must be studied.

1. Energy utilization—Although the transmission of information using light is more energy efficient than using an electronic transmission line at distance greater than a fraction of a millimeter, a high percentage of the light can be lost in transit from

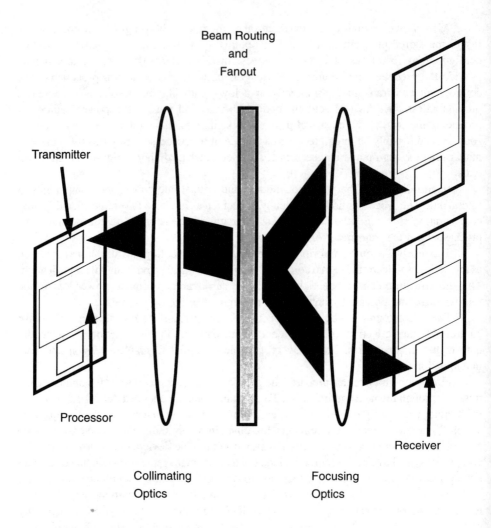

Figure 4.2 Components associated with processor array and optical interconnection network.

the transmitter to the subsequent receiver. Reflection losses from sequences of lenses, lower efficiency diffractive components, and lossy beam combination techniques all combine to extinguish the optical power. A poor utilization of optical energy places greater demands on lasers or light sources and accentuates the inherent problem of limited optical power availability.

2. Thermal management—The process of computing and communicating consumes energy and generates waste heat. Practical considerations governed by the optical framework keep this system compact due to a limited optical field of view resulting

in a relatively small processing area. This configuration generates a considerable amount of local heat. Initially, the energy advantages of optical communication may overcome this issue. However, because a significant amount of high-speed functionality will eventually be concentrated in a small area, an effective solution must be developed.
3. Commercial viability—The market for electronic products has steadily grown due to a vigilant effort to reduce the price per component of each element. To succeed, system requirements for high-quality optical operation must also be tempered by the need to construct economical systems. Key elements, such as high-power semiconductor laser diodes and custom optical lenses, must be reduced in cost, or alternatives must be developed.
4. Packaging—The alignment and supporting infrastructure of the free-space photonic system should be capable of maintaining micron-scale positioning accuracy and stability in a volume that could measure hundreds of millimeters in each dimension. Although submicron positioning accuracy is standard in the planar lithography of electronics, this issue must still be addressed when volumetric systems are considered. Thermal expansion must be well understood and corrected.
5. Optoelectronic integration—The majority of complex electronic processing is fabricated in silicon-based technology, while optoelectronics integration is centered in GaAs and InP materials. It will be necessary either to develop GaAs and InP integration to a point where it is competitive with silicon or develop a technique to fabricate hybrid systems.
6. Light-beam manipulation—The multitude of processor input and output beams and the optical power distribution beams must be successfully combined, divided, and directed using economical, power-efficient components.
7. Extensibility—Although the demonstration of an isolated device or component provides promise, it must be followed by a successful integration into a moderate size system. Even the demonstration of array integration is insufficient if the surrounding infrastructure required to exploit the technology is incomplete. An array of optically addressable processors depends on the optomechanic alignment system, the external optical power supply, and even the particular implementation of the network interconnection framework. Thus, it is important to judge a new technology not just on its own merits but by how compatible or available the necessary support is.

The principles necessary to assemble an optical interconnection network based on free-space concepts are investigated in the following pages. In the next section, many of the issues connected with components of a free-space optical routing system are examined. The following section explores network architectures and demonstrates how two practical implementations address some of the issues. The third section examines some of the physical devices being developed for the processor node. The last section discusses issues related to external laser optical power supplies and some requirements for diagnostic equipment.

4.2 FREE-SPACE OPTICAL ROUTING COMPONENTS AND ARCHITECTURES

Various principles and interconnection architectures that are advantageous for assorted applications for free-space optical routing will be examined in this section. Practical system issues include the difference between regular and highly irregular interconnection networks, the ability to communicate from both surfaces of a processor substrate versus one side, local versus remote optical power supplies, beam array generation, and the ability to combine and separate beam arrays.

Analog and digital based processing systems can benefit from free-space optical routing even though each type of system may require significantly dissimilar routing architectures. Analog systems are typically modeled after the neural systems of living organisms that often achieve a high connectivity per processor. The network connectivity can appear highly irregular as shown in Figure 4.3. There may be only one or a few

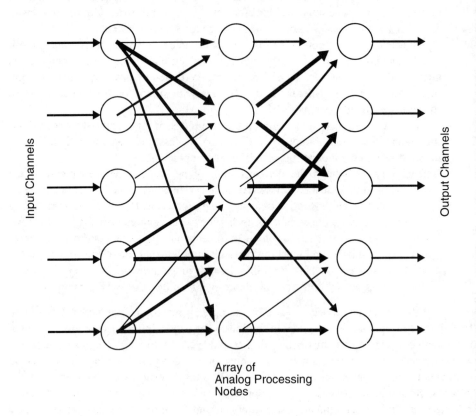

Figure 4.3 The irregularity and analog nature of a multilayer neural network configuration.

stages between input data and processed output. The primary motivation behind using optical routing for this application is the abundant connectivity.

Digital systems, however, more often resemble the design of telecommunication switches or computers that relay quantized information. Their networks are usually much more regular and may consist of many stages. The processing power is assembled in a hierarchy growing from transistors, to logic gates, to chips, to boards, to frames, to networks. The primary motivation for using free-space optical routing is the introduction of high-interconnection density and high-bandwidth potential into the lower ranks of this structure. Some examples of the regular structure of digital interconnection networks are shown in Figure 4.7.

The method of integrating the input and output communication channels and, potentially, the optical power beam array has a significant influence on packaging and the optomechanical framework. An idealized system might appear schematically as in the diagram of Figure 4.4. Input is received on one side of a processor substrate, and the output signals are relayed from the opposite side. Light for the output signal is generated by integrated lasers [4] that efficiently convert electrical power into optical power. Between stages, the optical framework forms the interconnection network. If micro-optical elements are used and connections are restricted to neighboring elements, the system can form a fairly compact package.

Unfortunately, the concept of dual-sided communication cannot be casually implemented. Microelectronic circuit fabrication has been primarily limited to a single side of a substrate because precise registration between both surfaces is difficult. Dual-sided communication with single-sided processing is still feasible if the substrate is transparent so that a light beam can propagate through the substrate. However, this restricts the substrate material that can be used and the range of possible wavelengths. Therefore, many initial system prototypes have been developed with reflective modulators or top-surface light emitters that limit substrate access to one surface. Figure 4.4(b,c) illustrates two methods of assembling this type of optical architecture.

In Figure 4.4(b), the bulk optomechanical framework is composed of a variety of isolated optical components that may be adjusted to attain optimal positioning. This flexibility is advantageous for system experiments and development, but it can prove costly in production if extensive labor-intensive adjustments are required and long-term stability is compromised.

Figure 4.4(c) shows a planar architecture [5] where optoelectronic elements are solder bonded onto an optically transparent substrate. This framework provides micron-scale alignment due to the lithographic definition of both the optoelectronic components and the alignment features on the substrate. Many optical functions are served by diffractive optic elements and reflective surfaces. This approach, although restrictive in the types of optical components that can be used, could eventually address many of the packaging issues.

Another important issue is the question of how to best incorporate the optical power supply into the system. Although light emitting diodes could be used, it is very likely

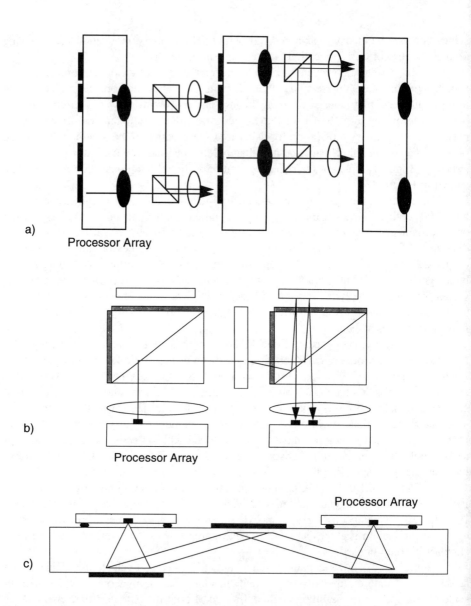

Figure 4.4 (a) Dual-sided communication between transparent substrate device arrays; (b) single-sided communication; (c) planar integration using a diffractive and reflective optics.

that the energy-efficient laser diode will be the favored optical source. However, the ability to integrate several surface-emitting lasers within an electronic processing array is currently limited. Surface emitting lasers have only recently been developed. Many researchers are working on improving the efficiency (to eliminate thermal problems) and the fabrication process (to integrate with other electronic circuits).

During the near-term, systems have been developed using a remote optical power source, such as a high-power semiconductor laser diode and a Dammann [6] grating or similar technique [7,8] to generate an array of beams to illuminate a matrix of transmitters. Each beam is then amplitude modulated or the polarization state toggled to encode information. This approach leads to modular design, favoring easier maintenance because a defective laser or processor array can be replaced.

If the technique of localized light generation is abandoned and a single-sided communication approach is embraced, it becomes necessary to design an optical framework to manipulate beams of light from several sources. For example, it may be necessary to combine and separate a set of input channels, output channels, and an array of illumination beams used for processor readout. Several methods are illustrated in Figure 5.5. The principal consideration is to accomplish the task with a minimum of light loss, a minimum of optical components that increase expense and can introduce excessive beam aberrations, and a minimal addition of position-sensitive hardware.

One approach shown in Figure 5.5(a) is to maintain spatial separation of all optical channels, such as by using microlenses to relay individual beams between processor arrays, so that beam combination is not a critical issue. However, because the light beams will likely be Gaussian beams, which quickly diverge when relayed by micro-optical elements, the distance between processing nodes must be practically limited to a few millimeters.

Figure 4.5(b) shows how two beams can be combined using partially reflecting components, albeit with a high penalty for power loss. Such a scheme should be used only where the power margin of the system is sufficient.

Beams with differing wavelengths may also be combined using a dichroic mirror. The advantage of this scheme is that it is highly efficient. The disadvantage is that the detectors must be sensitive to two wavelengths that may differ by a few tens of nanometers. This is difficult for some semiconductor devices incorporating quantum wells tuned to a narrow wavelength range. Also, the optical transport framework would need to be fairly wavelength insensitive.

One ideal beam combination technique is the use of polarization-sensitive optics. Using a polarizing beam splitter, one beam polarization state is predominantly transmitted while the orthogonal polarization state is reflected. Polarization states can be efficiently interchanged using a properly oriented half-wave retarder or two quarter-wave retarders. This scheme is highly suitable for systems employing lasers as optical sources because the light is typically highly polarized. The disadvantage is that the polarization properties of the elements must be fairly insensitive to the angular spread of the beam arrays. Because

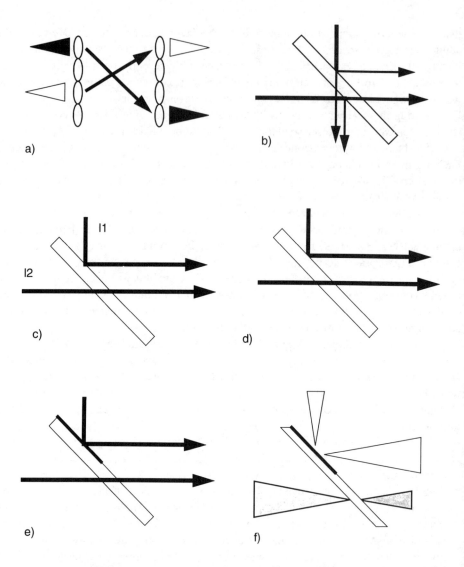

Figure 4.5 Beam combination (and separation) schemes. Arrows represent collimated beams; cones represent converging or diverging beams (a) microchannel; (b) lossy; (c) diachroic; (d) polarization sensitive; (e) pupil division; (f) image plane.

thin films are sometimes used to produce polarization-sensitive coatings, many polarization beam splitters have a limited range of optimal angular response.

Another scheme based on pupil division is to divide an optical aperture into two regions to combine two collimated beams. The disadvantage of this scheme is that the apertures of the objective lens are effectively reduced for each beam leading to larger focused spots.

Finally, Figure 4.5(f) shows how beams can be combined in an image plane by focusing the beam arrays and using an array of mirrors to interleave the images. The disadvantage of this approach is the need for accurate alignment, and the additional lens systems that will increase the system complexity and expense and may lead to additional wavefront aberrations.

There are several styles of interconnection techniques that are useful for optical routing. A few are illustrated in Figure 4.6. The simplest method is based on the idea of a bus as shown in Figure 4.6(a). The routing is a collection of single channels that are broadcast to a series of processors or circuits. Only one device on a channel may transmit during a time interval, while the remaining connections receive or wait. The system can either be configured with channels that are sequentially tapped by each circuit, or the signal can be regenerated at each stage. One disadvantage of this technique is the bottleneck that may develop as the bus is extended and contention for bus ownership increases.

The second interconnection method shown in Figure 4.6(b) is reminiscent of the crossbar routing of telephone switching applications and the vector-matrix product multiplication schemes of analog optical computing. In this scheme, the routing system is capable of generating a path from each processor in one stage to every processor in the subsequent stage. An intermediate spatial light modulator serves to block all but the desired path. The advantage of this system is that communication between all processors is established in one stage. Unfortunately, this system is highly inefficient when extended to large arrays. If n processors must be able to connect to any of m processors, then an array of $n*m$ modulators is required, and at best, only $1/m$ of the light is distributed to the target processor.

The third interconnection system is a multistage network that has a lower fanout per device than the crossbar and requires several cascaded stages to ensure data propagation between any two locations. This regular system is highly desirable for telecommunication switching networks, where the objective is primarily to establish a route between input and output channels. In such a case, each processor is considered to be a simple switching element. A multistage network can use simple interconnections and routing nodes and, because energy is utilized more efficiently through lower fanout, can theoretically support a higher bandwidth. This advantage must be balanced against the additional hardware required for several stages.

There are several network architectures that can serve as multistage networks [9]. The important issues for such schemes are nonblocking connectivity of the network and the size. The size is determined by the complexity of each switch node. If a simple switch node is used that handles at most two channels, then for N channels, there will be on the order of $\log_2 N$ stages and about $N/2 \log_2 N$ switching nodes. The functionality of each node may vary as shown by the various sets in Figure 4.6(d).

For certain network configurations and node functionality, situations may occur where only one of the two input channels into a node can be processed. In these cases a

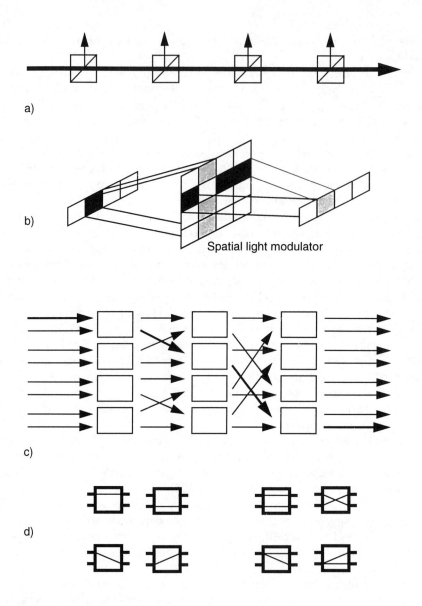

Figure 4.6 (a) Optical bus; (b) optical crossbar; (c) multistage interconnection network; (d) two sets of two input, two output switching nodes.

blocked route can be formed. Blocked routes could also be created when nodes malfunction. To reduce the probability for blocking, it is desirable to expand the number of nodes in each stage via signal fanout to provide alternative routes.

Figure 4.7 shows three fully interconnected, multistage networks that have been demonstrated optically. Each has a regular layout, and as shown, each is space variant.

The crossover and banyan networks have an interconnection scheme that is slightly different for each link stage. In the crossover network, each processor is connected to a processor in a similar location in the next stage and to one other processor that is symmetrically located about a reflecting line. The banyan includes the straight connection and one that is a power of two locations away. Unfortunately, neither connection topology is space invariant, i.e. the nonstraight connection is either up or down depending on the location. This lack of space invariance manifests itself in more complex interconnection modules or the need to disregard certain channels at the processing array.

The perfect shuffle interconnection network shows one highly desirable attribute. As can be seen from Figure 4.7(c), the interconnection scheme is identical for each link stage. This is valuable because it leads to a modularity of interchangeable optical stages. For the system where each processor connects to two new processors, a total of $\log_2(N)$ link stages are needed to connect N input channels. For all three cases it is possible to generate a three-dimensional topology sharing the general aspects of the two-dimensional architectures shown.

As we have shown from the discussion, there is richness in the variety of implementations that free-space optical routing can incorporate. The optimal format will be highly application specific, and the state of the technology will determine many aspects. In summary, beam combination is a necessary consideration, and the layout of the network will be influenced by complexity tradeoffs between the optical framework and power demands of the processor nodes.

4.3 NETWORK IMPLEMENTATIONS

In this section, we will examine more closely the categories of optical components that can be used in a free-space optical routing network. In addition, we will highlight some prototype systems developed for a photonic switching application to illustrate some early systems designs.

The free-space optical routing stage has three primary technologies that it may employ for handling the optical beams: (1) conventional geometric optical elements, (2) binary optics, and (3) holographic elements. The general operation of these elements is shown in Figure 4.8. Geometrical optics refer to the category that contains standard lenses that have been applied in imaging systems for many decades. These refractive and reflective optics can be modeled by geometric means of tracing light rays through a system, and they rely on a particle-like description of light.

Alternatively, holographic elements and binary optics operate based on the wave-like nature of light and its diffraction and interference caused by features close to the size

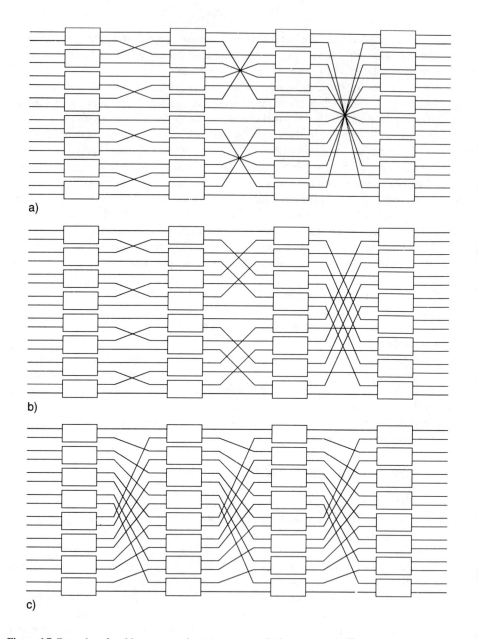

Figure 4.7 Examples of multistage networks: (a) crossover; (b) banyan; (c) shuffle.

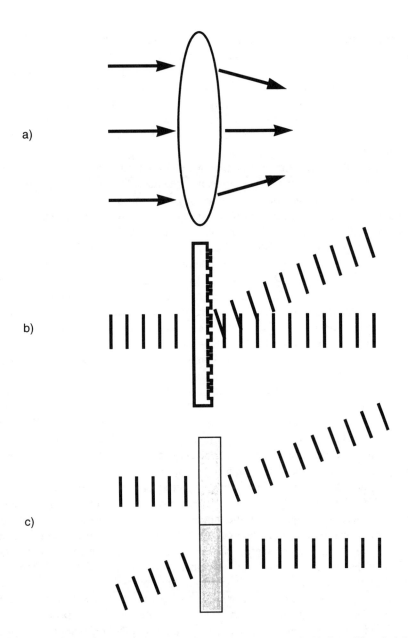

Figure 4.8 Examples of (a) refractive (geometrical); (b) binary (diffractive); and (c) holographic optical elements.

of the wavelength. Binary optics here refers to computer-generated holograms, kinoforms, and diffractive lenses that are fabricated by microscopically modifying just the surface of an element. The generic term ''hologram'' will be applied to those elements where the microscopic diffractive features are contained throughout a volume.

These components function in an optical interconnection system to collect the light, possibly split and direct the beams, and then focus the beam onto the next receiver. A network implementation could be achieved using either one category or a hybrid collection of elements from any of these classifications.

Geometrical optics are conventional optical components based primarily on refractive and reflective principles. The components are normally fashioned by grinding and polishing optical glasses. Although elements based on polishing are usually on the scale of several millimeters, microlithographic processes are capable of defining refractive microlenses with apertures of a few tens of microns. Thin film layers can be applied to these geometrical elements to reduce reflection loss, to create polarization-sensitive elements, and to produce dichroic reflective elements.

Geometrical optics are well understood theoretically because they have been manufactured for several decades. One of their advantages is that a large number of industries support this technology. One basic disadvantage is that the conventional construction results in larger scale pieces. These bulk elements are not entirely consistent with the scale reduction required for small-scale integration.

When using geometrical optics, it should be remembered that the resolution of a lens is ultimately limited by diffraction effects. The best achievable radius of a spot focused by a collimated beam of wavelength, λ, and diameter, D, that uniformly illuminates a lens with focal length, f, is given by $r = 1.22 \lambda * f/D$. If the beam diameter matches that of the lens, the number f/D is then referred to as the $f/\#$ of a lens. Therefore, to improve the resolving power of the system may require small $f/\#$ lens that are typically more difficult and expensive to manufacture. Thus, tradeoffs between spot size and lens design will be required based on system price objectives.

Lenses and lens systems based on geometric optics are well suited to collect, collimate, and focus light. A lens system may comprise a single glass element or a collection of several elements. The multielement lens is typically larger and more expensive to manufacture than its single-element counterpart; however, the multielement design is frequently better corrected for aberrations at lower f numbers and has a much greater field of view or a greater number of resolvable pixels. The spatial resolution is often referred to as the spatial bandwidth. It important to reduce the amount of aberration in lenses used in some optical routing architectures because each beam may pass through twice when reflective modulators are incorporated in the processor arrays.

Diffractive based elements, such as binary optics and volume holograms, can duplicate the performance of most conventional optical elements and provide additional capabilities for beam array generation and multibeam routing. Inherent in their fabrication is the ability to create faceted (i.e., space-variant) elements. Diffraction occurs because the phase profile of the beam is modified or because portions of an illuminating wavefront are selectively blocked by either microscopic areas or by interference formed by reflection from microscopic fringes recorded in a volume. The amplitude hologram (which absorbs light) is the most easily fabricated; however, it has the undesirable property of absorbing a substantial fraction of the available power. Thus, phase holograms that shape the wavefront are preferred in optical routing systems.

The phase-altering process can either occur throughout a volume of variable index material, or it can be localized to the surface of a substrate. The volume hologram has the potential for 100% efficient coupling 100% into the desired wavefront. It is usually produced using photosensitive materials that record the interference between a reference light beam and light that has been manipulated by a set of optical components. When the hologram is replayed by illuminating it with a single light beam, the diffracted beam matches the operation of the beam originally generated by the optical setup. On the other hand, the lower efficiency of the surface-etched binary optical element is more easily fabricated and replicated. Also, surface diffractive elements can be designed using computers and diffraction theory to create optical elements that may have no analog geometrical configuration.

The operation of the typical binary optical element is governed by the grating equation, $P(\sin \theta_m - \sin \theta_i) = m\lambda$, where θ_i is the angle of incidence relative to the surface normal, θ_m is the mth diffraction order, and P is the periodicity of the repeated pattern. The appearance of λ accounts for the strong wavelength dependence of these elements.

The term, "binary optics," was coined to classify diffractive optical elements based on fabrication techniques originally developed for the microelectronics industry [10]. Figure 4.9 illustrates how a computer-generated hologram or fresnel zone lens is replicated on a substrate. The design is first parametrized as a series of masks with opaque and clear regions that are often generated via electron beam writing technology. Each mask represents a binary decomposition of the surface into sets of phase levels so that, at most, L masks are required to produce 2^L levels. Each mask is brought into contact with a substrate covered with a thin layer of photosensitive material known as photoresist. When the mask is illuminated by ultraviolet light, the photoresist in the clear regions changes its chemical composition and is removed in the subsequent development step. This profile is made permanent in the substrate by eroding, in a corrosive environment, material not covered by the protective mask. Typically, reactive ion etching is used because the ions within the plasma etch the material anisotropically and thus preserve the stairstep topology. If a finer representation of the surface is needed, the process is repeated with several etch steps of varying thickness to produce a higher efficiency, multilevel surface profile that better approximates a continuous surface.

Diffractive optics can also emulate polarization-sensitive elements when one-dimensional subwavelength size gratings are created. In addition, subwavelength features can be used to emulate an antireflection coating. Practical components have, to date, been limited to the far infrared.

Holograms are similar to binary optics in that their operation is based on mutually interfering light waves. Whereas binary optics diffracts light due to a surface topology,

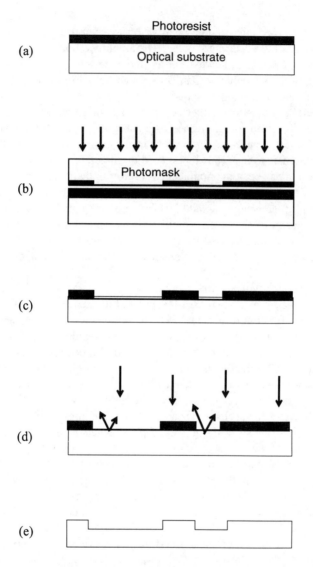

Figure 4.9 Steps in fabricating a binary optic element: (a) photo-resist application; (b) exposure; (c) resist development; (d) reactive ion etch; (e) resist removal.

holographic materials typically contain volumetric features acting as layers of Bragg reflectors. One unique feature of the volume hologram is that a number of holograms can be stored within a common volume, where each can be addressed separately. For example, the angle of incidence is a parameter that may be used to select a particular hologram.

Holograms can be further distinguished as being either static or dynamic. In static holograms, the features are normally recorded in a photosensitive layer a few microns thick covering a substrate. Once the material is processed, its operation is permanently defined. Alternatively, certain electrorefractive crystals have the capability of recording interference fringes for several holographic patterns. This recording capability is attractive in neural network applications because the routing network can be modified during the process of programming the system. The disadvantages of dynamic holograms include (1) the crosstalk that occurs between the multiple holograms contained in the volume, (2) the poorer diffraction efficiencies of these materials compared with static holograms, and (3) the special methodology required to record sequential holograms without erasing earlier recordings.

There are various issues involved with working with volume holograms. Because the hologram has a narrow chromatic range, the material responsible for recording the interference pattern must either be photosensitive at the proposed system wavelength or else a procedure for shifting to that wavelength must be available. Although a variety of holographic materials exist, the wavelength range, diffraction efficiency, and high resolution do not sufficiently cover the full spectrum. Because the features within the hologram are comparable to the wavelength of the light and because of the low photosensitivity of many holographic materials, an ultrastable environment is critical during the time exposure.

Usually, one will find that any of the three categories can be used in common with another to construct optical interconnection networks. Figure 4.10 illustrates how a geometrical optical system [11] and a micro-optical system could generate the shuffle interconnection network. In the geometrical approach, the interconnections are constructed in a space invariant manner by splitting the beams into two parts, reflecting each path with a slight tilt, and then recombining them by interleaving the beams. Polarizing elements could be used to achieve the beam splitting and recombination. The disadvantage of this scheme is the power lost to light directed outside the useful aperture and the subsequent magnification step required to restore the image to the original spacing.

The holographic approach provides a space-variant interconnection by establishing a micro-channel for each light beam. This microchannel can be defined by: micro-refractive lenses or binary lenses coupled with microprisms, off axis lenses, or faceted holograms. One challenge of the microchannel approach is the need to precisely align the array components.

Figure 4.11 is a schematic of one stage of demonstration photonic switching system based on the crossover interconnection network [12,13]. It is shown to illustrate in greater detail some of the techniques used to implement an interconnection network based primarily on geometrical optics. The design philosophy used in this system is to use polarization components and spatial separation at image planes to combine beam arrays. The crossover interconnection is implemented using an array of prismatic mirrors.

In this system, beam arrays are amplitude modulated by quantum-well based S-SEED arrays to communicate information. As seen from Figure 4.7(a) of the crossover network, each light beam from a processor cell must be directed to two new processors. One connection, regardless of the stage, is a one-to-one mapping to an equivalent position in the next stage. The second connection is made to a position reflected symmetrically

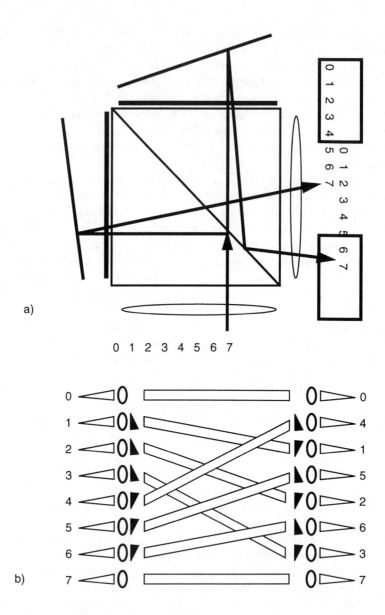

Figure 4.10 Shuffle interconnection as implemented via (a) bulk optics and (b) micro-optics or faceted holograms. The lens and prism sets can also be replaced by off-axis microlenses.

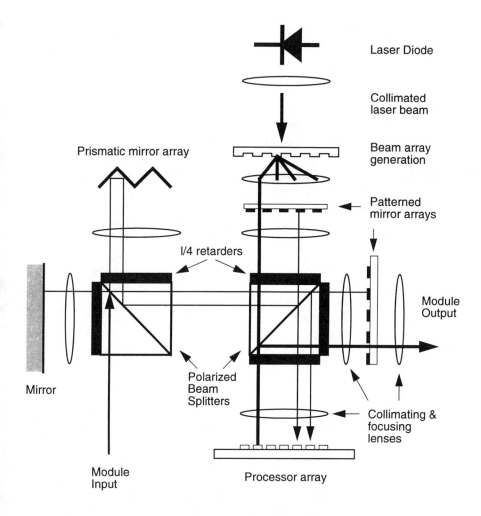

Figure 4.11 Optical components used to implement a crossover network in a prototype photonic switching system demonstration.

about a line through either the middle, one of the quarter, or one of the eighth dividing points, depending on the stage. The mirror in Figure 4.11 has two reflection points.

As shown for one system module, input data from the previous system module enters at the lower left in the diagram as an array of circularly polarized light beams. When the circular polarized light reaches the first polarized beam splitter, one linear polarization state is reflected to the left in the figure while the orthogonal polarization is transmitted. The reflected beams pass through a quarter-wave retarder plate, are focussed by an objective lens onto a mirror, and are recollimated during the second pass through the lens. Returning through the quarter-wave retarder, the beams possess the orthogonal polarization and are relayed toward the second beam splitter. These beams form the one-to-one position mapping of the crossover interconnection. The upwardly moving beams in the figure are similarly relayed, except that the special prismatic mirror array effectively reflects each set of beams about a symmetry point, thus implementing the crossover connection. A unique prismatic mirror array is used in each stage. Upon reflection, the beams enter the beam splitter with the orthogonal polarization, are reflected, and are recombined with the other beam set.

The second portion of this module must spatially combine the interconnection beams with the readout beams that will be generated during the second phase of the cycle. This is accomplished using polarization and image-plane beam combination techniques. Each set of input beams, whether reflected or transmitted by the second beam splitter, is focused onto small mirrors that form an array on a transparent substrate. Upon reflection and relay through the quarter-wave retarder for the second time; the polarization is changed to the orthogonal state; the beams are recombined by the splitter; and, finally, they are imaged onto the device array.

Device readout is accomplished by directing the collimated beam from a semiconductor laser diode through a binary phase grating to diffract the beam into a spot array. These beams are imaged into the transparent areas in the patterned mirror substrate, relayed through the polarized beam splitter, modulated by the state of each S-SEED, reflected on the second pass through the polarized beam splitter, and are finally relayed through transparent regions in the patterned mirror array and onto the next module.

The experiment demonstrates the feasibility of aligning a large number of optical components. The large number of image planes require a sophisticated optical design of the image relay system to suppress the accumulation of damaging wavefront aberration to maintain the ability to sharply focus the spots. In addition, the large beam arrays passing through the systems demand that polarizing beam splitters be manufactured to operate with large angular tolerances.

A second example of a free-space optical system that employs diffractive optics to form the interconnection network is shown in Figure 4.12 [14]. The system uses a computer-generated hologram etched onto a fused silica substrate to split each readout beam into three equally spaced interconnection beams. This task is performed in the Fourier plane using a space-invariant hologram. Then, by selectively blocking one of the three beams at the following image plane, the space-variant banyan interconnection is produced.

This prototype system was designed to demonstrate the relatively high density of optical interconnection that can be achieved in a small area. Each device array is composed of a 32×32 matrix of slightly modified S-SEED. When the four input beams (each signal is represented in a differential format), the two optical power beams, and the two output beams linked to each cell are totaled, over 8,000 input/output connections are generated.

The system requires two clock phases to operate: one to receive the inputs and set the logic state of the processor and a second state to read that logic state and transfer the information to the next system module. Data beams enter the system module from the left side in Figure 4.12. The beams are polarized and thus reflected by the polarized beam

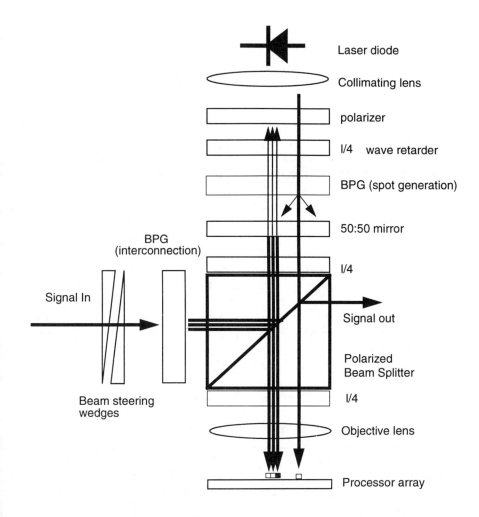

Figure 4.12 Optical components used to implement a banyan interconnection network based on binary optics in a free-space photonic switching demonstration system. BPG = binary phase.

initially away from the SEED device array. The beams pass through a quarter-wave plate and are incident on a partially reflecting mirror. The mirror reflects about half of the power back through the quarter-wave plate so that it is now of the orthogonal polarity. The input beams are then transmitted through the beam splitter and are focused by the objective lens onto the device array.

In the second phase of the cycle, the semiconductor laser that acts as the optical power supply is turned on. The emitted beam is collimated, passes through polarization components, and reaches the first binary phase grating. This grating generates an array of 64 × 32 regularly spaced beams. The beams pass through the partially reflecting mirror (undergoing loss), through a quarter-wave retarder, a polarizing beam splitter, another quarter-wave retarder to form circularly polarized beams, and are imaged onto the device array. The beams are amplitude modulated and reflected by a mirror that forms part of the S-SEED structure. The second pass through the quarter-wave retarder converts the beams to the polarization opposite that of the first pass, and the beams are then reflected toward the input port of the next module. Before reaching the next module, however, each beam is diffracted by another binary phase gratings into multiple beams. This special grating establishes the interconnection scheme. Selected metalization of windows blocks one of the three beams to establish the space-variant banyan interconnection.

Several issues determine which of the three types of optical components is optimal for a optical routing system. In general, bulk geometrical optical elements provide a large spatial bandwidth but are difficult to fabricate at a size where single pieces can be devoted to a single channel. On the other hand, diffractive optics and holograms are very well suited to space-variant applications.

The wavelength sensitivity is a critical concern for many applications. If the illumination spectrum is broader than a few nanometers, then geometrical optics should usually be chosen. It is possible to optimize their chromatic performance over a much broader range than diffractive elements, due to the selection of glasses from a range of refraction indices. Alternatively, if lasers are used to provide monochromatic illumination, diffractive optics are well suited.

The optimal choice among geometrical, binary, and holographic optical elements is highly dependent on the application. The high number of irregular interconnections needed in a neural network system favors the use of binary optics and holograms or, possibly, arrays of refractive microlenses. In addition, training a neural network using dynamic holograms can be advantageous. Alternatively, photonic switching systems and computer systems typically use static, regular interconnection schemes. Because they are competing directly with data capacity of electronics and must efficiently use volume and energy, these optical systems will most likely rely on binary optic and static holograms for larger systems and arrays of refractive microlenslets for microchannel approaches.

4.4 SPATIAL LIGHT MODULATORS AND PROCESSOR-RELATED ISSUES

There are many issues associated with the planar elements composing the processing array that significantly affect the implementation of the free-space optical routing network. The processors are envisioned as a two-dimensional matrix of electronic elements (covering functionality from simple logic gates or memories to smart pixels) whose primary high-bandwidth communication channel is via optical data receivers and transmitters dedicated to each individual processing cell. The primary functions of the processor array envisioned for early commercial applications are to either establish the routing path between digital input and output channels within a switching network or to process the information (digital or analog) received from previous processor nodes and communicate the results to a subsequent node. When necessary, the receivers and transmitters serve to convert information between optical and electrical format for either storage or analysis. Depending on the node complexity, the three functions could be integrated into a single device or split among several components.

Many issues related to the processor node components are still the subject of considerable research effort. Thus, it is by no means clear what possible form the ultimate architecture will resemble. It was noted earlier that the optimal materials for sophisticated information processing and efficient optoelectronic conversion are different. However, work is in progress to either integrate the materials or develop both functions within a single material. The fundamental issue of integrated optical sources versus remote optical sources has been discussed and additional issues associated with laser sources will be examined later. Other issues associated with these node components are (1) the physical phenomenon exploited for beam influence or optoelectronic conversion and, when necessary, the ability to integrate this with some form of electronic processing; (2) whether the processor node passively relays the information or actively processes it; (3) the system throughput and network reconfiguration time; and (4) the format in which data are represented.

There are two issues associated with operational speed of the processor cell: (1) the bandwidth of the data stream itself and (2) the reconfiguration time the processing elements need to process the data or establish a new optical routing path. Depending on the processing mechanism, the two speeds can differ significantly.

The processor node can be categorized as either passive or active depending on how the information in the channel is treated. In a passive system, the incoming light is either blocked, transmitted, or deflected by each cell so that the signal is either relayed to the next node or stopped. In an active processor node, the information is typically converted from an optical to electronic format and then processed with other data that accumulate in the node. The results are then returned to an optical format. An active device array may divide time into various read, process, and write intervals to synchronize the operation of each stage.

Circuit-based switching networks provide an example of an application that could use passive nodes. Such switches route large bandwidths of information, yet the network rearranges its connection scheme on a much longer time scale. These long duration connections are typical of the circuit-type connection of typical telecommunications networks. For such applications, it may be desirable to maintain the optical format for the information throughout the system and use low-speed electronic links to provide control data to relatively simple transceiver elements. However, whenever data interconnections must be frequently and rapidly rearranged, approaching the time scale of the fundamental data bit, the passive node approach may not be ideal. Such situations occur in packet-type connections characteristic of data communications and evolving broadband telecommunications switches.

One example of a passive device array is an array of deformable mirrors with a binary set of deflection states. Typically, information is divided into several channels, and the passive array is configured so that only one possible route is fully connected. The light is not regenerated, and therefore, significant loss can be sustained traveling through a multistage network unless optical amplifiers are introduced. One significant advantage of passive processors is that because typically no conversion is made between the optical and electrical domains, the system bandwidth is determined by the initial transmitter and can thus be quite large. The disadvantage is that the reconfiguration time to redefine the routes can be relatively long when mechanical elements are employed.

The second classification is the active processor cell that senses the incoming information, stores and processes the data, and then transmits or modulates a fresh light beam. The chief advantages are the ability to regenerate and thereby amplify the signal and the ability to process a number of input signals within a cell. Such a scheme promotes the processing advantage of electronics with the communication capability of photonics.

In computation-driven or signal processing applications, the information channel must ideally be converted to an electronic format where the processor arrays will actively modify the contents of the channel. The receiver/transmitter components may represent a relatively small area of the smart pixel. The receiver bandwidth and the operation rate of the processor determine the ultimate throughput of the channel. In such a system, high bandwidth is desirable for the optical channel because it is most probably competing against the electronic alternative, and data are typically regenerated.

Spatial light modulators use spatially separated device arrays to convert electrical or optical information into optical data streams. In these units, an impinging light beam has one of its parameters modulated based on control information. Because data are being encoded ultimately through an optoelectronic and electro-optic conversion, the data rate is determined by the operational characteristics of the device.

As noted in the introduction, optical interconnections have great potential for transferring information in an energy-efficient manner. To fully exploit this advantage, the optical-to-electronic and electronic-to-optical conversions must be optimized. It is thus desirable to position the receiver as close to the processor as possible to avoid the need for lengthy electronic connections. It is critically important to avoid the need for matched impedance transmission lines characteristic of large-scale, high-bandwidth data because the terminating resistor consumes a significant fraction of the signal power.

A typical receiver will probably bear several similarities to a normal high-bandwidth photosensor. The light beam will be directed into a small photosensitive window or region, and the photons will be absorbed near the surface. The photons will liberate electrons, creating a small photocurrent that can be tapped using electrical connections on the surrounding window surface. The device may be connected in series with a resistor to a voltage supply so that the photocurrent generates a voltage change. The subsequent current or voltage change could directly drive a modulator, or electronic amplifiers may increase the signal strength for further electronic processing. A modulator is operated in a similar manner, except the voltage or current will change the reflection, transmission, or deflection properties of a light beam falling within the modulator window. A transmitter based on a localized light source will operate similar to a LED or laser and will require integrated electronic driver circuitry.

The operating bandwidth of a free-space channel is directly proportional to the available optical power, the energy required to switch the receiver (or modulator) to an alternative state, the quantum efficiency of converting photons to electrons, and the parasitic electronic impedances of the receiver circuit itself. Given the limitations of optical power generation and thermal management, it is critical to optimize power utilization.

The switching energy of a particular active photonic device is an especially important consideration. If several thousand optical pinouts are expected to perform at greater than 100 MHz operating speed (to compete against electronics), then each device should require only a few picowatts of combined optical and electrical power per processor, or else special thermal management modules will also be needed. This need for efficient power utilization eliminates some devices from consideration for more universal applications.

Many of the active processing nodes generate photocurrent in a window assembly that resembles a capacitor. The energy needed to charge the capacitance of the window is proportional to CV^2, where V is the induced voltage, and C, the capacitance, is proportional to the area of the window and the length of the electronic connection to the next circuit element. This leads to a design favoring a device needing only a small voltage to switch states or a device having smaller windows. Unfortunately, smaller windows demand highly focused beams and, therefore, lens systems well corrected for aberrations, which are thus more expensive. The modulators share the same energy-related concerns.

It is also possible for the highly focused light beams to saturate the physical effect exploited by the modulators. For example, a semiconductor-based intensity modulator of fixed size has a limited number of photoelectrons available for absorbing and thereby modulating a readout beam. This and the transmission efficiency of the optical framework can limit the fanout and/or operating speed potential for the system.

The active processor cell can either be separated into separate receiver, processor, and transmitter/modulator sections or serve all functions in one area. Bistable devices are often very useful for the second function. The light from either a single beam or several beams is collected during one clock cycle, and the photons are converted to electric charge or thermal energy, which ultimately change parameters such as the absorption or index of refraction. Typically, a bistable device has two distinct historically determined absorption states that are retained when the device is not illuminated or being read out in an appropriate manner. Bistability effectively provides a memory function. During the next clock phase, the retained information is extracted.

The optimal representation of digital data is a critical aspect of the digital processor system. In electronic systems, binary data are often represented by a voltage level or a current that can be compared against a reference standard. In a free-space photonic system, optical channels can be amplitude modulated or their polarization states changed. If a single amplitude modulated beam is used to represent a data channel, then either some power reference is required or else the power must be integrated over a standard data period to determine its value. The first alternative adds complexity by adding an additional electronic or optical energy standard and the second decreases the flexibility of operating the system at a wide range of speeds.

An alternative approach is to use a scheme implemented in many electronic systems; that is, to represent a single bit by the power differential between two signals. For example, if channel A is optically brighter than channel B, the data represent a value of one. If, however, channel A is dimmer than channel B, it represents a value of zero. This scheme removes many of the restrictions of a single-channel approach at the expense of doubling the number of optical ports. It also decreases the sensitivity to optical crosstalk that can occur from light scattering in the optical framework.

The degree of complexity associated with a processor cell can vary dramatically among applications. If the cell function is to serve only as a bistable memory, then closely packed cells can be formed. Such systems will optimally use the space bandwidth of the optical system, providing thousands of optical channels at the expense of low processing capability. Alternatively, smart pixels having complex processing islands with fewer optical interconnections can be used.

A variety of physical effects can be used to detect and modulate a beam of light. Some examples are electro-optic, electroabsorptive, magneto-optic, acousto-optic, mechanical, etalons, photorefractive, and thermal effects. A few samples of devices that can be manufactured as arrays using these effects are shown in Figure 4.13. Criteria that determine the optical choice for photonic applications are (1) the integration of processor and receiver/transmitter, (2) the switching energy required to register an input or output data state, and (3) the bandwidth of the information stream.

If the node must simply deflect the light so that it either reaches or misses the subsequent node, an optomechanical device such as a mirror array with binary state positions is suitable. In this case, the throughput is independent of the node. However, the reconfiguration time of the node can be lengthy. Also, because the signal is not regenerated, the number of cascaded mirror arrays is severely limited.

An acousto-optic cellular array can also be used to deflect light beams. Light is deflected based on the frequency and intensity of sound generated by the transducer, much as light is deflected by a diffraction grating. Similar to the optomechanical devices, the information is not typically regenerated. Both the acousto-optic and optomechanical effects could be used to modulate a remote light source to encode data. However, because both rely on mechanical properties of the materials, they have a limited response time. Their use as a spatial light modulator limits data throughputs to levels that are probably not competitive with electronic processing systems.

Liquid crystal arrays represent a type of modulator array that has already seen commercial success in display applications. Liquid crystal arrays are composed of elongated molecules in a fluid state. Similar to normal crystals, the anisotropic property of liquid crystals leads to birefringence, a different index of refraction along orthogonal axes. The birefringence influences the polarization state of an optical wave transmitted through a liquid crystal cell. Due to its fluid-like nature, the orientation of molecules can be altered (e.g., by using an electric field created by electrodes surrounding two sides of the cell).

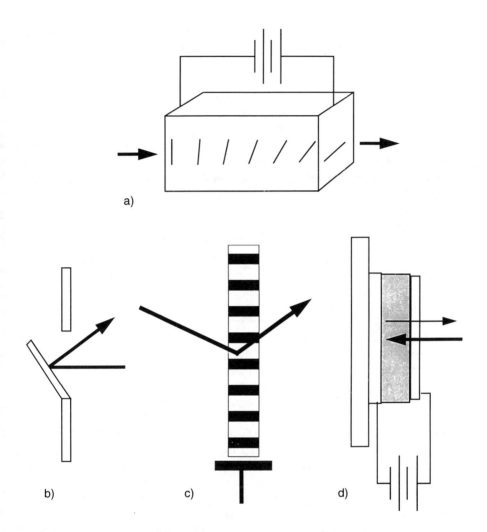

Figure 4.13 Examples of spatial light modulators: (a) liquid-crystal cell (polarization); (b) deformable mirror array (deflection); (c) acousto-optic cells (diffraction); (d) SEED multiquantum well (absorption).

A liquid crystal designed with the appropriate thickness and positioned with the correct orientation can act as a controllable wave retarder, thereby switching a light beam between two polarization states. This polarization switch is highly desirable for optical systems that represent binary data as polarization states. When placed between two crossed polarizers, the system becomes a controllable intensity modulator, thus making it also suitable for systems that communicate data via amplitude modulation.

The crystal orientation is typically controlled by electrodes forming an electric field across the gap containing the liquid. Because the system effectively forms a capacitor and because the molecules must mechanically realign in the viscous fluid, it takes time to change between states. Typical twisted nematic liquid crystal devices operate at a few volts and at frequencies from about 100 Hz to a few kilohertz. Ferroelectric liquid crystals operate at multikilohertz rates. Although their switching time is slow, liquid crystals would work well as a passive element in a switching network needing only intermittent reconfiguration. Because liquid crystals have been shown to be easily integrated with such silicon electronic circuits [15], they have very strong advantages for use in optical interconnection networks. They also can be packaged with a photoconductor to produce optically addressed spatial light modulators.

The use of quantum effects has been successfully realized in semiconductor laser diodes due to the sharply defined energy states created by alternating thin layers of slightly differing semiconductors in quantum wells. The absorption properties are highly wavelength dependent and can be quickly shifted by applying an electric field.

The quantum-well self-electro-optic effect device (SEED) [16] is effectively a *p-i-n* diode with several thin layers of alternating GaAs and AlGaAs material in the intrinsic region. The SEED operates on the basis of the quantum confined Stark effect, whereby an electric field shifts the energy states of the quantum system. The layers of quantum wells result in a well defined energy transition separating absorbing and transmissive regions. The SEED functions as an efficient photodetector for converting photons into electronic current. In addition, when a voltage is applied across the SEED, the absorption peak can be shifted, thereby leading to a modulator. To function as a modulator, reflective materials are fabricated directly under the quantum wells.

The symmetric SEED (S-SEED) is a set of two SEED devices connected in series and attached to a voltage source. It can function as a bistable device. When the S-SEED windows are illuminated by beams of equal intensity, the state of the device remains unchanged. Thus, because one modulator is an absorptive state and the other is in a transmissive state, the reflected light forms a differential signal. The ratio of intensities between absorptive and reflective windows is typically 2 to 1 and greater. In addition, SEED-type devices have been demonstrated to switch states at speeds ranging from a few hertz up into the gigahertz region. This large operational range is valuable for creating experimental systems for exploring photonic system issues.

Finally, there are also systems that integrate surface-emitting LEDs or laser diodes with photodiode, phototransistors, or other photodetectors. These devices share many of the same issues as the previous quantum-well devices, with the additional complication of requiring additional device-processing steps.

In summary, several physical phenomena have been practically introduced in creating several photonic device arrays. However, none has been sufficiently developed or is optimal for all foreseeable free-space applications. Quantum-well SEEDs are fast and have low optical and electronic energy requirements, but they are currently limited to GaAs circuitry. Surface emitting laser-based systems require further research but show great promise. Liquid crystal arrays have already found their niche in the commercial display market and can be integrated with silicon VLSI technology, but they may not possess the speed to compete against electronic interconnections. Many other examples fail to regenerate the light, limiting their applications. Fortunately, the wealth of device arrays allows system engineers to explore the issues associated with the optical interconnection framework, semi-independently of the processor.

4.5 LASER OPTICAL POWER SUPPLIES AND SYSTEM DIAGNOSTIC TOOLS

The previous sections have discussed the theory and the implementation of free-space optical routing networks and the devices responsible for modulating the light-beam information carrier. Let us quickly examine the issues associated with remote optical power sources. Then, we will highlight diagnostic tools that can aid in characterizing the operation of free-space systems.

There are two means to configuring the optical power supply into the system: (1) integrate a surface-emitting light source within each processing cell, or (2) relay the requisite optical energy into the system from an external source. From a practical standpoint, the external optical source is desirable. In this configuration, the waste heat generated from electro-optic conversion is isolated from the device array. Also the lower manufacturing yield of integrating novel laser structures into a processor array is avoided. However, as discussed previously, the difficulties of beam array combination must then be addressed.

The optical power supply is a required element in interconnection architectures that do not incorporate a light emitter in the processor cell. The illumination provided by the optical power source is typically amplitude modulated to encode information destined for the next module. Currently, the choice of the optical power supply, the information modulator, and the interconnection scheme is correlated because the range of wavelengths these elements operate over is restricted and does not always overlap one another.

If the system is small and energy efficiency is not a major concern, then arrays of light emitting diodes (LED) could serve as the optical power supply. LEDs are attractive because they are simple to fabricate. However, LEDs are not as energy efficient. They emit light over a fairly large solid angle, and their spectra are broad, leading to difficulties for diffractive optical elements.

Optical systems need to be energy efficient to compete successfully with high-speed electronics. The optical power supply should also share this efficiency. The optical power, P, required for each processor array depends on the number of transmitters, N; the transmitter's encoding efficiency, (h_1); the interconnection stage transmission efficiency, (h_2); the detector's switching energy threshold, (E); and the time allotted for each information bit, (t). This is expressed by the formula $P = N E/h_1 h_2 t$.

As an example of a module of a representative optical system that would compete against a high-speed electronic system, let us consider an array of 1,000 devices. If the optical switching energy for each device is 1pJ, $h_1 h_2 = 0.01$, and for a rate of 100 MHz, a time interval of $t = 5$ns during a two phase cycle is allowed, then the required power is about 20W. Because of these fairly large power requirements and the desire for efficient light production, the laser will be the preferred remote optical source for larger systems.

The requirements of energy efficiency and economy suggest the semiconductor laser diode as a candidate for the optical power supply. The wavelengths accessible to laser diodes range from visible through the near infrared. Unfortunately, the beam quality of the high-power semiconductor lasers and the power output is limited. Single longitudinal and transverse mode lasers are dictated by the optical demands of the system. Currently, powers of only about 100 mW for narrow stripe lasers are commercially available. In the future, laser arrays and laser amplifiers could extend the power range to a few watts.

Another alternative for the optical source is the solid-state laser, in particular a laser diode pumped system. There are several advantages that using a solid-state laser, such as one based on a Nd:YAG crystal, can provide. High powers have been demonstrated and commercial products are available. The external cavity configuration results in a high-quality beam. Also, the external cavity provides the ability to control the modal structure and wavelength. In addition, some crystals operate naturally at a single wavelength, making the use of diffractive optics less problematic.

The optical pump for the crystal would ideally be an array of laser diodes, due to their energy conversion efficiency. In this scheme, the crystal behaves as a filter that redistributes light from one wavelength to another. In spite of the loss in efficiency in this two-step process, such power sources have the ability to produce beams in the watt or multiwatt power range. One drawback of a solid-state laser is the lack of a laser diode pumped system that can generate high power near 850 nm, the operating wavelength region of many semiconductor devices. Also, the question of amplitude modulating lasers that illuminate a system having a multiphase cycle requires consideration of modulating either several amperes of current or synchronizing a set of mode-locked lasers.

One final note on the optical power supply: The source must typically supply a regularly spaced array of beams, usually of uniform intensity. Thus, a one- or two-dimensional semiconductor laser array would naturally match the format if suitable devices become available. However, a single-plane wave can be split into a number of beams by a variety of methods [7]. One of the most desirable methods is the use of Dammann gratings, periodic reproduced patterns etched into transparent substrates, that diffract the light into an array of beams. The Dammann grating works on the principle of Fourier optics and uses an objective lens to form the spot array. Large spot arrays with over tens of thousands of spots have been demonstrated.

Any communication system is incomplete without a set of design and diagnostic tools to evaluate its operation. Free-space photonics systems require the development and adaptation of several such diagnostic tools to provide insight and thereby promote advancement. Three categories of tools are required: design and simulation tools to evaluate the proposed optical and electronic characteristics of the system, static diagnostic tools that evaluate isolated system components and system alignment, and dynamic diagnostic equipment that examines the flow of information through the system.

A set of tools that would prove invaluable for system designers includes the following:

1. Interferometers to measure the transmission quality of optical components.
2. Lens design programs to evaluate the system optical transfer characteristics.
3. Scalar diffraction models to design high-performance binary optic elements.
4. Polarimeters to examine polarization characteristics in systems highly dependent on polarizing components.
5. CCD video cameras integrated to system viewports to magnify and examine the registration of the spot arrays with the device array because systems could likely operate in the near-infrared.
6. Spectrometers to measure the wavelength of the laser diode optical source, although the optimal adjustment is to align the spots with the arrays.
7. Electronic analyzers, such as oscilloscopes and bit-error rate characterization systems, to evaluate the signals transported by the system.

One important tool that awaits development is the analog of the digital electronic logic analyzer. This electronic diagnostic system allows probes to monitor various sites between modules in a circuit so that logic values of several channels at various time intervals can be examined. It is difficult to electronically probe the logic state of an optically interconnected processing system because a probe could easily block the light from other processors, thereby interrupting communication. Also, because of their nature, the channels of many of the free-space systems are not spatially distinct at locations other than at the device array. Thus, it is necessary to sample a small portion of the light without appreciably introducing aberrations or disturbing alignment and then remotely sample the focused image of the device array. In even a moderately sized system, this could mean an array of several hundred spots. Sampling the image could be done by coupling light into fibers that are then routed to photosensors. Also, the use of a specialized CCD camera with an integrated computer control system to designate regions of interest is one possibility, although such cameras must operate and process images at the high speeds that will be typical of photonic systems.

4.6 SUMMARY

It is evident that free-space optical routing holds the potential to dramatically enhance information processing systems. The relaxation of the constraints of planar electronic architectures allows new small-scale interconnection networks to be envisioned. Developments in the following areas should be monitored because they will influence the fate of free-space optical routing:

1. Integration of electronic processing and optoelectronic functionality.
2. Binary optics and holography.
3. High-power semiconductor laser diodes, solid-state lasers, and surface-emitting microlasers.

4. Network interconnection architectures favoring optics.
5. New photonic devices and spatial light modulators.
6. Optical interconnection system packaging.
7. Photonic system diagnostic tools.

The explosion of consumer products oriented toward information providing services and the increasing trend toward interconnectivity are driving the need to explore alternative connection techniques at all scales. The need to overcome the communication bottlenecks in digital and analog electronic multiprocessor networks has set the stage for the emergence of free-space optical routing systems. Considerable effort has been made in establishing the feasibility of many of the principles described in this section. It is also evident that many issues must be resolved before this technology can establish itself in a broad range of cost-effective applications.

REFERENCES

[1] Lentine, A. L., F. B. McCormick, R. A. Novotny, L. M. F. Chirovsky, L. A. D'Asaro, R. F. Kopf, J. M. Kuo, G. D. Boyd, "A 2 kbit array of symmetric self electro-optic effect devices," *Photonic Technology Letters,* Vol. 2, No. 1, 1990, pp. 51–53.

[2] Miller, D. A. B., "Optics for low-energy communications inside digital processors: quantum detectors, sources, and modulators as efficient impedance convertors," *Optics Letters,* Vol. 14, No. 2, 1989, pp. 146–148.

[3] McCormick, F. B., F. A. P. Tooley, J. M. Sasian, J. L. Brubaker, A. L. Lentine, T. J. Cloonan, R. L. Morrison, S. L. Walker, and R. J. Crisci, "Parallel interconnection of two 64x32 symmetric self electro-optic effect device arrays," *Electronic Letters,* Vol. 27, No. 20, 1991, pp. 1869–1871.

[4] Yamanak, Y., K. Yoshihara, I. Ogura, T. Numai, K. Kasahara, Y. Ono, "Free-space optical bus using cascaded vertical-to-surface transmission electrophotonic devices," *Applied Optics,* Vol. 31, No. 23, 1992. pp. 4676–4681.

[5] Jahns, J. and A. Huang, "Planar integration of free-space optical components," *Applied Optics,* Vol. 28, No. 9, 1989, pp. 1602–1605.

[6] Dammann, H. and K. Gortler, "High efficiency in-line multiple imaging by means of multiple phase holograms," *Optical Communications,* Vol. 3, 1971, pp. 312–315.

[7] Streibl, N., "Beam shaping with optical array generators," *Journal of Modern Optics,* Vol. 36, No. 12, 1989, pp. 1559–1573.

[8] Vasara, A., M. Taghizadeh, J, Turunen, H. Westerholm, E. Noponen, H. Ichikawa, J. M. Miller, T. Jaakkola, S. Kuisma, "Binary surface-relief gratings or array illumination in digital optics," *Applied Optics,* Vol. 31, No. 17, 1992. pp. 3320–3336.

[9] Wu, C., T. Feng, *Tutorial: Interconnection Networks for Parallel and Distributed Processing,* IEEE Computer Society Press, Los Angeles, CA , 1984.

[10] Veldkamp, W. B. and T. J. McHugh, "Binary Optics," *Scientific American,* May, 1992, pp. 92–97.

[11] Brenner, K.-H. and A. Huang, "Optical Implementation of the perfect shuffle interconnection," *Applied Optics,* Vol. 27, 1988, pp. 135–137.

[12] Cloonan, T. J., M. J. Herron, F. A. P. Tooley, G. W. Richards, F. B. McCormick, E. Kerbis, J. L. Brubaker, A. L. Lentine, "An all optical implementation of a 3D crossover switching network," *Photonic Technology Letters,* Vol. 2., No. 6, 1990, pp. 438–440.

[13] McCormick, F. B., F. A. P. Tooley, T. J. Cloonan, J. L. Brubaker, A. L. Lentine, R. L. Morrison, S. J. Hinterlong, M. J. Herron, S. L. Walker, J. M. Sasian, "Experimental investigation of a free-space optical

switching network by using symmetric self-electro-optic-effect devices," *Applied Optics,* Vol. 31, No. 26, 1992, pp. 5431–5446.
[14] McCormick, F. B., F. A. P. Tooley, J. L. Brubaker, J. M. Sasian, T. J. Cloonan, A. L. Lentine, R. L. Morrison, R. J. Crisci, S. L. Walker, S. J. Hinterlong, and M. J. Herron, "Design and tolerancing comparisons for S-SEED-based free-space switching fabrics," *Optical Engineering,* Vol. 31, No. 12, 1992, pp. 2697–2711.
[15] Jared, D. A., R. Turner, K. M. Johnson, "Design and fabrication of VLSI ferroelectric liquid-crystal spatial light modulators," *Optical Computing,* 1991, *Technical Digest Series,* Vol. 6, Optical Society of America, 1991, pp. 55–58.
[16] Lentine, A. L., H. S. Hinton, D. A. B. Miller, J. E. Henry, J. E. Cunningham, and L. M. F. Chirovsky, "Symmetric self electro-optic effect device: optical set-reset latch, differential logic gate and differential modulator/detector," *Journal of Quantum Electronics,* Vol. 25, 1989, p. 1928.

FURTHER READING

Arsenault, H. H., T. Szoplik, B. Macukow, *Optical Processing and Computing,* Academic Press, San Diego, CA 1989.
Athale, R. A., *Digital Optical Computing,* conference proceedings designated as a critical review of optical science and technology, SPIE Optical Engineering Press, Bellingham, Washington, 1990.
Born, M., and E. Wolf, *Principles of Optics,* Pergamon Press, Oxford, England, 1987.
Goodman, J. W., *Introduction to Fourier Optics,* McGraw-Hill, Inc, New York, 1968.
Hecht, E., *Optics,* Addison-Wesley Publishing Company, Inc., Reading, Massachusetts, 1987.
McAulay, A. D., *Optical Computing Architectures,* 1991, Wiley, New York.
Optical Computing, 1987 Technical Digest Series, Vol. 11, Optical Society of America, Washington, D.C. 1987.
Optical Computing, 1989 Technical Digest Series, Vol. 9, Optical Society of America, Washington, D.C. 1989.
Optical Computing, 1991 Technical Digest Series, Vol. 6, Optical Society of America, Washington, D.C. 1991.
Optical Computing, 1993 Technical Digest Series, Optical Society of America, Washington, D.C. 1993.
O'Shea, D. C., *Elements of Modern Optical Design,* Wiley-Interscience, New York, 1985.
Photonic Switching, 1987 Technical Digest, Optical Society of America, Washington, D.C. 1989.
Photonic Switching, 1989 Technical Digest Series, Vol. 3, Optical Society of America, Washington, D.C. 1989.
Photonic Switching, 1991 Technical Digest Series, Vol. 8 Optical Society of America, Washington, D.C. 1988.
Photonics in Switching, 1993 Technical Digest Series Optical Society of America, Washington, D.C. 1993.
Saleh B. E. A., and M. C. Teich, Fundamentals of Photonics, John Wiley & Sons, Inc 1991.
Smith, W. J., Modern Optical Engineering: The Design of Optical Systems, McGraw-Hill Book Co., New York, 1966.
Spatial Light Modulators and Applications, 1993 Technical Digest Series Optical Society of America, Washington, D.C. 1993.
Streibl, N., K.-H. Brenner, A. Huang, J. Jahns, J. Jewell, A. W. Lohmann, D. A. B. Miller, M. Murdocca, M. E. Prise, and T. Sizer, "Digital Optics," *Proc. of the IEEE,* Vol. 77, No. 12, Dec. 1989.

Chapter 5
Integrated Optical Waveguide Routing— Micro-optics

Robert Shi and Tomasz Jannson

5.1 INTRODUCTION

Wave propagation in extremely thin dielectric layers is a major problem when studying integrated optical phenomena. This closely resembles quantum-mechanical phenomena, and a number of nonintuitive effects need to be explained. These concepts include the following:

- Wave propagation in thin films is based on wave vector projection rather than on the velocity projections that are typical in classical mechanics.
- Optical waveguides (thin dielectric films) involve total internal reflection (TIR) effects. This means that the wave reflection coefficient, which goes inside the waveguide, is precisely 100%. In addition, coupling into the waveguide can go through side walls (so called longitudinal coupling). These effects are not generally known in microwave electronics because this discipline usually deals with metallic walls.
- Although guided mode structures are well known in microwave electronics, they are usually simpler than those found in integrated optics and have simpler boundary conditions.
- The propagation properties of single-mode waveguides are fundamentally different from those of multimode waveguides and are of different spatial coherence. This additional degree of freedom is virtually unknown in microwave electronics.

A comprehensive discussion of these effects would require voluminous textbooks. Moreover, many books and comprehensive references have already been written that emphasize either the physical or technological aspects of integrated optics [1–7]. The purpose of this book is to introduce the major design and modeling aspects of integrated optics to an audience ranging from optical and optoelectronic engineers to electronic engineers and physicists.

Integrated optics made its appearance in the late 1960s [8]. It was an exciting technology because it offered the promise of compact, environmentally stable micro-optical systems that would replace bulk optical systems. The planar geometry of integrated optical devices and interconnects made them attractive candidates for hybrid or monolithic integration with electronic systems (e.g., with CMOS electronics). This promise still holds, and it is possible to envision future computers, backplanes, boards, multichip modules (MCM), or even chips, which will be composed of a combination of electronic processing units and optical interconnects, and which will provide an integrated optical waveguide routing. This architecture represents the first generation of optoelectronic processors. In the second generation, micro-optics will be even more visible in integrated optic circuits that will not only provide communication between electronic processing units but will also provide some kind of optical processing. It can also be foreseen that third-generation optoelectronic processors will be either all optical or almost all optical.

There must be a friendly coexistence between optics and electronics in order to make the this scenario possible. Therefore, optics, as a newcomer in the processor world, must be particularly careful in developing friendly relations with electronics (which can live without optics quite well).

To fully utilize this advantage, integrated optics should possess two "friendly-to-electronics" features: transparency and compatibility. These features will be fully explained on the basis of point-to-point waveguide optical interconnects as an elementary example of waveguide routings. Such an integrated waveguide consists of two electro-optic interfaces and a waveguide as a data communication link. The electro-optic interface, or transceiver, consists of a transmitter (e.g., laser diode) and a receiver (e.g., photodetector). Transparency to electronics means that the introduction of an optical interconnect will not require modifications to the electronic processing unit hardware. This means all signal processing functions specific to modulation, transmission, and detection of optical signals will be performed by a transceiver (T_x/R_x) module, as the input/output (I/O) device transforms electrical signals into optical signals and vice versa. Transparency (and intermodularity) features are a major factor in successfully replacing electronic interconnects with optical interconnects. This can already be done in some fiber optic related, interprocessor interconnect applications. In the case of waveguide interconnects that are intraprocessor interconnects (i.e., board-to-board, module-to-module, and chip-to-chip), however, it is more difficult to satisfy this feature, particularly for cases in which there are a large number of transceivers, among other performance related factors. This difficulty arises because of the excessive electrical power consumption of the T_x/R_x modules.

The feature of compatibility is even more difficult to satisfy. Compatibility means that two technological processes are similar (for optical integrated waveguides and electronic integrated circuits) so that existing electronic integrated circuit technologies and architectures do not need to be modified when integrated optics is introduced. This requirement does not mean that relations between electronics and optics may not change in the future, particularly for the envisioned second- and third-generation optoelectronic processors. By the year 2000, however, the optical interconnect should serve electronics as a perfect servant, providing consistently good service in a virtually invisible way.

5.2 OPTICAL WAVEGUIDES

5.2.1 Introduction

Optical waveguides are transmission lines with axial symmetry and rectangular cross sections. By confining wave propagation to a higher refractive index dielectric medium, the waveguides provide communication between the electro-optical (E-O) transmitter, T_x, and receiver, R_x, creating optical interconnects. This is due to the TIR phenomenon. If the aspect ratio of the rectangular cross section is close to 1, the waveguides are channel waveguides. If the aspect ratio is equal to infinity, the waveguides are slab (planar) waveguides. These two types of waveguides are discussed in Sections 5.2.4 and 5.2.3, respectively.

In channel waveguides, the optical guided wave is bound in two directions (x and y) and propagates freely in the axial z direction, similar to the functioning in optical fibers. In the case of slab waveguides, the guided wave is bound only in one direction and can propagate freely in two directions. This case is then a planar medium, which is a formal two-dimensional analog of a three-dimensional medium. This analog provides propagation conditions for two-dimensional planar waves and constitutes a physical ground for integrated optical circuits.

The slab waveguide geometry discussed in Section 5.2.3 is a particular case of planar medium geometry in which propagation of guided waves is limited to one direction (z). Thus, the guided wave is propagated in a two-dimensional waveguide. Traditionally, slab waveguides are defined as two-dimensional waveguides in which wave propagation is limited to one plane (x,z). This elementary geometry is important when explaining the basic properties of guided mode propagation, including the eigenmode structure and single-mode propagation conditions.

To understand integrated optics phenomena, one must understand such physical effects as TIR, evanescent waves, quantization of guided wave propagation (with eigenvalues represented by a discrete set of modal effective indices), eigenfunctions, orthogonality, and spatial coherence of single-mode and multimode structures. These effects will be explained in a general context, independent of specific waveguide geometry, in Section 5.2.2.

The most promising channel waveguide fabrication technologies, based on $LiNbO_3$, GaAs, glass, and polymers are discussed in Section 5.2.6. For guided wave propagation, these technologies vary in the context of waveguide refractive indices, index profiles (step-index versus graded-refractive-index (GRIN)), waveguide size (especially important for the single-mode case) and waveguide propagation loss. It is shown in Sections 5.2.3 and 5.2.4 that differences in geometrical and material waveguide parameters, lead to differences in waveguide eigenmode structures. Knowing the specific waveguide parameters, the reader can approximately evaluate the mode eigenvalues using standard model dispersion curves presented in these sections.

5.2.2 Properties of Guided Waves

5.2.2.1 Plane Waves, Total Internal Reflection, and Effective Refractive Index

Optical waveguides have rectangular symmetry, meaning that their cross sections are rectangular. This is in contrast to optical fibers, which have a circular symmetry. The characteristic cross-section size of an optical waveguide is approximately a few microns (i.e., a few wavelengths) for the single-mode case or 50 to 200 mm for the multimode case. Because of their rectangular symmetry, all propagation properties of waves, guided in optical waveguides, can be described in terms of a superposition of monochromatic plane waves.

To further explain optical waveguides, it is useful to start with the properties of monochromatic plane waves propagating in a uniform medium with a refractive index of n. Such a wave has a plane wavefront and a spatial period equal to λ/n, where λ is the optical wavelength in a vacuum. Its optical frequency, v, is very high (i.e, in the air) for $\lambda = 1$ μm, $v = c/\lambda = 3 \times 10^{14}$ Hz, where c is the velocity of light in a vacuum. Its phase velocity, v_{ph}, is smaller than c,

$$v_{ph} = \frac{\omega}{k} = \frac{\omega}{k_o} \frac{1}{n} = \frac{c}{n} \tag{5.1}$$

where k is its wavenumber, k_o is the wavenumber in vacuum, and $\omega = 2\pi v$ is the angular frequency.

These elementary properties of the monochromatic plane wave are introduced to demonstrate one interesting (and rather nonintuitive) property of wave motion. When a plane wave is projected onto a plane perpendicular to this page, the velocity of the projected wave motion will not be smaller but will be larger than the phase velocity, v_{ph}, as is illustrated in Figure 5.1.

To pass the distance AB, which is larger than the spatial period, λ/n, the projection of the plane wave needs the same time, $1/n$, as does the plane wave itself, while propagating in the direction of wavevector, **k**. Its phase velocity, $v_{ph//}$, is

$$v_{ph//} = \frac{AB}{1/v} = \frac{\frac{\lambda}{n \cos \gamma}}{1/v} = \frac{\lambda v}{n \cos \gamma} = \frac{c}{n \cos \gamma} = \frac{v_{ph}}{\cos \gamma} > v_{ph} \tag{5.2}$$

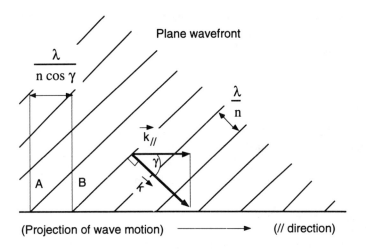

Figure 5.1 Projection of wave motion onto // direction.

This rather nonintuitive result is a consequence of the fact that in the case of wave motion, the wavevector rather than the wave velocity vector is projected. This is in contrast to classical mechanics, in which one sees just the opposite result and which has been the general basis for our understanding of velocity.

To introduce the effective refractive index, a basic parameter in integrated optics, we rewrite (5.2) in a form analogous to that of (5.1).

$$v_{ph//} = \frac{\omega}{k//} = \frac{\omega}{k \cos \gamma} = \frac{\omega}{k_o n \cos \gamma} = \frac{c}{N} \tag{5.3}$$

where

$$N = n \cos \gamma \tag{5.4}$$

is the effective refractive index, representing the projection of wave motion in Figure 5.1. We can consider the effective refractive index as a formal analog of the regular refractive index characterizing the wave phase velocity. The effective index represents wave projections into the waveguide axis direction and plays a major role in integrated optics. To describe the phenomena of guided waves in optical waveguides, it is also important to understand TIR phenomenon. This effect occurs when the plane wave, refracted on the interface between two dielectric media, with indices n_1 and n_2, has an angle of incidence, α_1, which is larger than the critical angle, α_c. In such a case, the incident wave is fully reflected from the interface (100%), which provides transmittance of waveguide waves with virtually no loss.

The TIR effect constitutes the phenomenological base for wave guiding in optical fibers and waveguides, while the effective index is the basic parameter characterizing guided wave propagation. The effect can be readily explained using Snell's law:

$$\frac{\sin \alpha_1}{\sin \alpha_2} = \frac{n_2}{n_1} \qquad (5.5)$$

See Figure 5.2. It is important to note that Snell's law (5.5) is a consequence of a continuity of wavevector tangential components:

$$k_{1x} = k_1 \sin \alpha_1 = k_o n_1 \sin \alpha_1 = k_o n_2 \sin \alpha_2 = k_2 \sin \alpha_2 = k_{2x} \qquad (5.6)$$

The property exhibited in (5.6) distinguishes dielectric waveguide propagation from metallic waveguide propagation. Hence, (5.6) is equivalent to (5.5).

Assuming that the first medium is optically denser ($n_1 > n_2$), we have, according to Snell's law (5.5), $\alpha_1 < \alpha_2$. Therefore, the α_2 angle can approach 90%, while α_1 is still smaller than 90%. In such a case, for $\alpha_2 = \pi/2$, the TIR occurs, where the critical angle, α_c, is defined by (5.5). For example,

$$\sin \alpha_c = \frac{n_2}{n_1} \qquad (5.7)$$

Although, for $\alpha_1 < \alpha_c$, a significant part of the energy is refracted into the second medium (see Fig. 5.2(a)), for $\alpha_1 \geq \alpha_c$, all of the wave energy is reflected (Fig. 5.2(b)). This drastic difference in the behavior of a refracted wave for $\alpha_1 < \alpha_c$ and $\alpha_1 \geq \alpha_c$, can be explained only by using complex number algebra, discussed below and illustrated in Figure 5.2(a,b).

Because the boundary condition (5.6) holds for any α_1, we obtain from $\alpha_1 = \alpha_c$, that and $k_{2x} = k_1 \sin \alpha_1 = k_2$. While for $\alpha_1 > \alpha_c$, we have $k_{2x} = k_1 \sin \alpha_1 > k_2$; and according to (5.6), the transversal wavevector component of the refracted wave is:

$$k_{2y} = \sqrt{k_2^2 - k_{2x}^2} = \sqrt{k_2^2 - k_{1x}^2} = i\sqrt{|k_{1x}|^2 - |k_2|^2} = i\sqrt{k_1^2 \sin^2 \alpha_1 - k_2^2} = ik_o\sqrt{n_1 \sin^2 \alpha_1 - n_2^2} \qquad (5.8)$$

where $i = \sqrt{-1}$ is an imaginary number. Therefore, for $\alpha_1 < \alpha_c$, the amplitude of the refracted wave in medium (5.2) can be described in the form of a regular homogeneous plane wave:

$$U_{(x,y)}^{(h)} = \exp(i\mathbf{k}_2 \cdot \mathbf{r}) = \exp[i(k_{x2} \cdot x + k_{y2} \cdot y)] \qquad (5.9a)$$

where $\mathbf{k} = (k_x, k_y)$. However, the amplitude of the refracted wave in medium (2), for $\alpha_1 > \alpha_c$, is described in the form of an inhomogeneous plane wave.

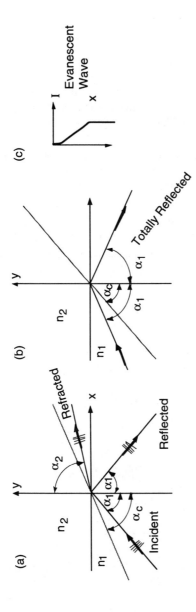

Figure 5.2 (a) Snell's Law and total internal reflection (TIR) ($n_1 > n_2$); (b) for $\alpha_1 \geq \alpha_c$, however, all of the wave energy is reflected; (c) only the weak tail propagates in the lighter medium (n_2), along the x axis, as an evanescent wave; evanescent wave intensity, I, decays on submicrometer distances from the interface.

$$U_{(x,y)}^{(i)} = \exp[(ik_{x2} \cdot x)]\exp\{-k_o|y|\sqrt{n_1^2 \sin^2\alpha_1 - n_2^2}\} \qquad (5.9b)$$

which decays in submicrometer distances from the interface (see Fig. 5.2(c)).

5.2.2.2 Guided Wave Structure for Single-Mode and Multimode Waveguides

If a third medium is added to the two-layer geometry illustrated in Figure 5.2, and if the film layer is optically denser than the two others, then a guiding wave structure is developed in the film [9]. This assumes that the TIR effect occurs on both interfaces. In such a case, the wave cannot escape from the film, which produces the zigzag wave illustrated in Figure 5.3, and which is guided along the axis of the waveguide. From a practical perspective, the first question to be answered is what range of guided wave velocities can be allowed in order to guarantee guided wave propagation?

For the index of substrate, n_s, which is larger than the refractive index of cover, n_c (i.e., for $n_c < n_s < n_f$, where n_f is the refractive index of the film (waveguide)), the critical angle of the substrate-film interface, $\alpha_c^{(s)}$, is larger than that of the cover-film interface, $\alpha_c^{(c)}$. Therefore, to satisfy the TIR effect for both interfaces, the only satisfactory condition (see Fig. 5.3) is:

$$\sin \alpha \geq \sin \alpha_c^{(s)} = \frac{n_s}{n_f} \qquad (5.10)$$

(5.10) is, in fact, the answer to our first question. Indeed, applying the definition of an effective refractive index to the geometry of Figure 5.3, we obtain:

$$N = n_f \sin \alpha \qquad (5.11)$$

Using (5.10), we have:

$$N > n_f \sin \alpha_c^{(s)} = n_s \qquad (5.12)$$

Because $\alpha < \pi/2$ in (5.11), we finally obtain:

$$n_s \leq N \leq n_f \qquad (5.13)$$

which indeed defines the range of phase velocities for guided waves propagated along the waveguide z-axis.

A second, more difficult question to be resolved is: Whether the set of guided wave effective indices are continuous or discrete, and if discrete, what are the specific N numbers? The answer is difficult because it leads to an analog of bound states in quantum

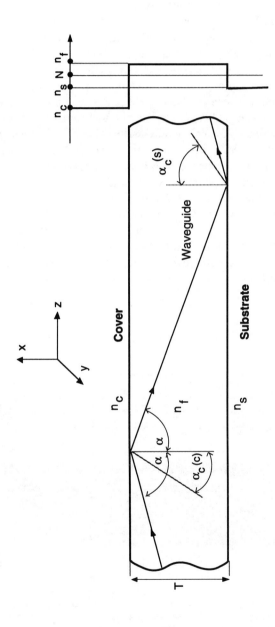

Figure 5.3 Zigzag wave propagating in waveguide, $n_c < n_s < n_f$; thus $\alpha_c^{(s)} > \alpha_c^{(c)}$.

mechanics [10]. Fortunately, a preliminary answer to the second question can be given using the very general, yet simple physical arguments presented below.

Referring again to Figure 5.3, the zigzag wave model presented in this schema is an illustrative model of the superposition of plane waves, which are reflected from interfaces. To expand this model to represent guided waves, elementary reflection waves that have the same a value are made to interfere constructively for specific angles α only. It can be shown that the condition of constructive interference is equivalent to the phase change of a zigzag wave between two equivalent points by 2π or by a multiple of 2π. Therefore, the set of angles, α, and in turn, the set of effective indices, N, must be discrete, with each N value representing a single guided waveguide mode. An analogous situation may be seen in wave mechanics, where the potential energy of a particle is treated as an analog of $(-N)$ providing the existence of a bound quantum mechanical state. This is illustrated in Figure 5.4. For $N < n_s$, however, there are no guided waves; thus, the continuous (radiation) states exist and all N values are acceptable (see Fig. 5.4).

As a result of constructive interference, discussed previously, each guided mode is spatially coherent (i.e., all areas of guide wave structure spatial distribution are permanently phase synchronized). By contrast, different guided modes interfere with each other with random phases. Sometimes, a constructive interference between a number of modes may be obtained, and sometimes, a destructive interference occurs. In a statistical average

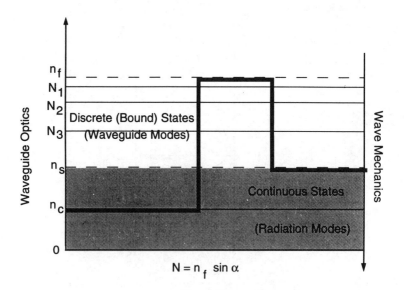

Figure 5.4 Analogy between waveguide optics and quantum (wave) mechanics. The negative potential energy of the particle, $-V$, can be treated as an analog of the effective index, N. Discrete (bound) states can only exist in the potential deep, that is, for $n_s \leq N < n_f$. For $N < n_s$, continuous states, representing radiation modes, exist. These modes are not confined to the waveguide.

sense, however, the mode structures are statistically independent, or orthogonal, and the total power of a multimode waveguide beam is the sum of its modal powers. Each mode power represents a separate single-mode structure.

5.2.3 Slab Waveguides

Further analysis of waveguide mode structure cannot be continued without a specification of waveguide geometry. To begin this specification, it is useful to focus on the simplest waveguide geometry, which is slab waveguide geometry. To find specific modal structures for slab waveguides, it is necessary to solve Maxwell equations with certain boundary conditions. These boundary conditions represent the continuity of electrical and magnetic tangential field components on both interfaces (a reminiscence of these boundary conditions is Snell's law). They also represent the continuity condition for wave vector tangential components. As a result, six electromagnetic field components (E_x, E_y, E_z, H_x, H_y, H_z) are separated into two independent groups representing two polarizations, TE and TM, which are discussed in the next section. Each polarization has a separate, parallel procedure for solution. Continuity conditions represent the behavior of electromagnetic field components on the boundaries between two dielectrics. Thus, these conditions are fundamentally different from analogous conditions in metallic waveguides because those conditions are related to dielectric-metal boundary. Guided wave amplitude continues beyond the waveguide boundary, in the form of an evanescent wave that can be coupled to a neighboring parallel waveguide if the distance between the two waveguides is approximately 0.1 to 0.2 mm. This phenomenon cannot occur in metallic waveguides.

For each polarization, the mathematical problem reduces to a scalar wave equation, solvable only for a discrete set of effective indices, constituting a set of modal effective indices (N', N'', N''', and so on). Such an equation is called an eigenequation, and its modal effective indices constitute the eigenvalues of this equation. This is a very familiar mathematical structure in quantum mechanics. Each modal field spatial distribution, called an eigenfunction, is represented by a single eigenvalue. As an introduction, this chapter will present eigenvalue solutions in graphic form and will provide interpretations when they are useful for practical purposes.

5.2.3.1 Polarization of Slab Waveguide Modes

The slab waveguide geometry, illustrated in Figure 5.6, is symmetrical with respect to any axis located in the y,z-plane. Maxwell equations with cylindrical symmetry (of which slab waveguide geometry is a particular case) can be separated into two independent groups representing two different polarizations. The two groups are TM, with an electrical field axial component, and TE, with a magnetic field axial component.

Assuming a z-axis as an axis of symmetry (see Fig. 5.5), TM polarization is represented by (H_y, E_x, E_z) with an electrical field axial component (E_z), while TE polarization

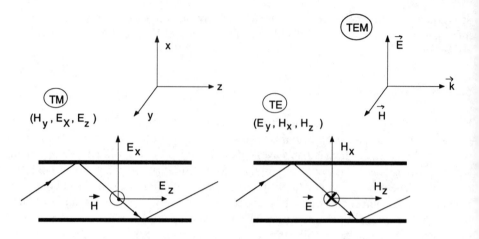

Figure 5.5 Mode polarization for slab waveguides: TM (magnetic field transversal to direction of propagation) and TE (electric field transversal to direction of propagation), compared with TEM-polarization of the wave propagating in free-space.

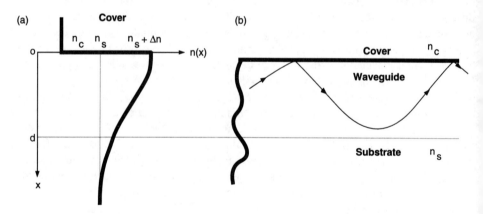

Figure 5.6 Typical graded-refractive-index (GRIN) waveguide, with: (a) GRIN profile; (b) curved zigzag wave trajectories.

is represented by (E_y, H_x, H_z) with magnetic field axial component (H_z). Both polarizations are illustrated in Figure 5.5. The knowledge of mode polarizations is important in asymmetric waveguides ($n_s \neq n_c$), in which mode structures for both polarizations can be very different. This knowledge is also important for electro-optic modulators and switches that are highly sensitive to polarization.

In the case when $n_s < N < n_f$ (see (5.13)), the discrete modal structure, illustrated in Figure 5.4, can be presented in the form of modal effective indices, N_m, where $m = 0$,

1, 2, 3, and so on, for each polarization separately. Those eigenvalues N_0, N_1, N_2, N_3, and so on, are functions of all waveguide geometrical parameters (n_c, n_f, n_s, T) and optical wavelength λ. This characterizes the step-index slab waveguide case as illustrated in Table 5.1) which also presents, the GRIN slab waveguide parameters. In this case, illustrated in Figure 5.6, waveguide thickness, T, is replaced by diffusion length, d, and the film index, n_f, has nonuniform distribution, $n(x)$, typically in the Gaussian form (see Fig. 5.6).

$$n(x) = n_o + \Delta n \, \exp\left(-\frac{x}{d}\right)^2 \tag{5.14}$$

where, $n(x) = n_o$, for $|x| \gg d$, and $n(0) = n_f = n_o + \Delta n$. In the case of diffusion GRIN waveguides, $n_o = n_s$, and $\Delta n \ll 1$. For example, for LiNbO$_3$ waveguides, n_s is equal to 2.3, and $\Delta n \sim 10^{-3} - 10^{-4}$. On the other hand, in the case of polymer GRIN waveguides [11] where the GRIN effect is obtained by polymer processing, the polymer film thickness is much larger than d, and the guided wave cannot penetrate the substrate. Then, $n_o \neq n_s$ and the substrate index can be arbitrary (see Fig. 5.7).

In general, for GRIN waveguides, the zigzag wave trajectories are not straight lines, particularly those close to the waveguide-substrate interface area (see Fig. 5.6). Thus, the critical angle of waveguide-substrate interface, $\alpha_c^{(s)}$ is not well defined.

Table 5.1
Normalized Parameters of Dispersion Curves for Slab Waveguides

	Parameter	Step-Index Slab Waveguide	GRIN Slab Waveguide
TE polarization	Normalized thickness	$V = k_o T \sqrt{n_f^2 - n_s^2}$; $k_o = \dfrac{2\pi}{\lambda}$	$V = k_o d \sqrt{n_f^2 - n_s^2}$
	Normalized modal index	$b_E = \dfrac{N_m^2 - n_s^2}{n_f^2 - n_s^2}$	$b_E = \dfrac{N_m^2 - n_s^2}{n_f^2 - n_s^2}$
	Normalized asymmetry measure	$a_E = \dfrac{n_s^2 - n_c^2}{n_f^2 - n_s^2}$	$a_E = \dfrac{n_s^2 - n_c^2}{n_f^2 - n_s^2}$
TM polarization*	Normalized thickness	$V = k_o T \sqrt{n_f^2 - n_s^2}$ †	$V = k_o d \sqrt{n_f^2 - n_s^2}$
	Normalized modal index	$b_M = b_E \left(\dfrac{n_f}{n_s q_s}\right)^{2\ddagger}$	$b_M = b_E \left(\dfrac{n_f}{n_s q_s}\right)^2$
	Normalized asymmetry measure	$a_M = a_E \left(\dfrac{n_f}{n_c}\right)^4$	$a_M = a_E \left(\dfrac{n_f}{n_c}\right)^4$

*These formulas can be used only for $n_f \cong n_s$.
†For GRIN polymer waveguides, n_s is replaced by n_o.
‡$q_s = (N_m/n_f)^2 + (N_m/n_s)^2 - 1$.

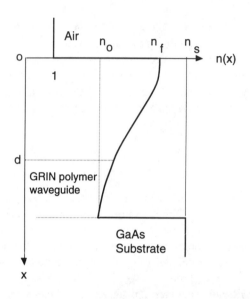
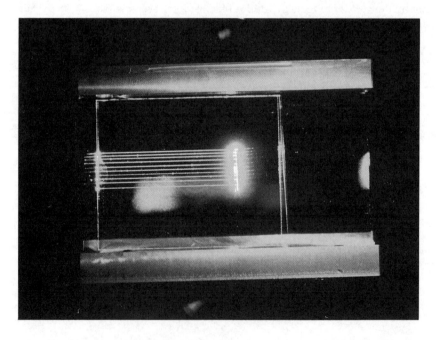

Figure 5.7 GRIN polymer waveguide coated on exemplary high-index GaAs substrate. Top: GRIN profile; bottom: GRIN polymer slab waveguide. $n_f = 1.55$, $n_o = 1.5$, and $n_s = 3.5$.

Table 5.1 summarizes all normalized parameters of slab waveguides to effectively interpret the mode dispersion curves for the step-index waveguide (Fig. 5.8) and the GRIN waveguide (Fig. 5.9). The first parameter, V, is called the normalized thickness (or normalized frequency), and it is proportional to T/λ. Thus, we see that the V parameter is also proportional to the approximate number of modes.

$$V \sim \pi M \qquad (5.15)$$

The second parameter, b, is called the normalized modal index, as related to modal index, N_m. The third parameter, a, is called waveguide asymmetry measure and also contains the cover index, n_c. For symmetric waveguide ($n_s = n_c$), $\alpha = 0$.

Those parameters are normalized in such a way that the modal dispersion curves illustrated in Figures 5.8 and 5.9 represent all possible slab waveguide geometries. The parameters are given for both polarizations, TE and TM, separately. For TM polarization,

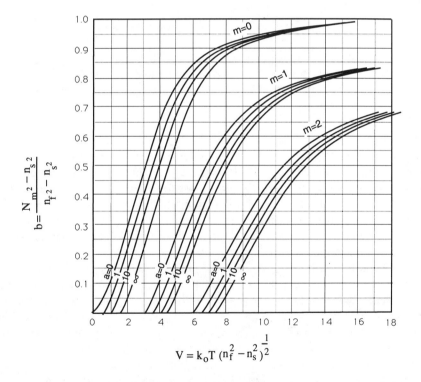

Figure 5.8 Normalized TE modal dispersion curves $b(V, a_E)$ for step-index slab waveguides. The parameters are defined in Table 5.1. They can be also used for TM case for n_f at n_s by replacing the a_E parameter by the a_M parameter. The case b_E at 1 is equivalent to N at n_f (well-confined case) and the case $b_E = 0$ is equivalent to $N = n_s$ (cut-off case).

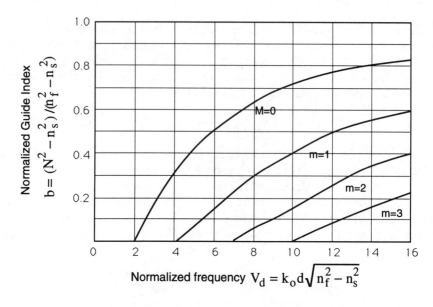

Figure 5.9 Normalized GRIN slab waveguide dispersion curves for TE and TM modes, where $n_f = n_s + \Delta n$; $\Delta n \ll n_s$; d is diffusion depth or effective GRIN waveguide thickness. The average slope of these curves characterizing waveguide dispersion is $db/dv \sim 0.1$; that is, twice lower than that for step index waveguides.

however, they can be used only for $n_f \cong n_s$. The name "dispersion curves" results from fixed waveguide geometry and provides the relation $N_m(\lambda)$. This relation describes waveguide dispersion, which is similar to material dispersion (e.g., $n_f(\lambda)$). Material dispersion is typically ignored because it is a much smaller effect than waveguide dispersion.

The dispersion curves illustrated in Figure 5.8 provide full information about slab waveguide propagation properties in terms of their modal effective refractive indices. Based on these curves, it is possible to obtain the required value of the modal index, using only a simple scientific calculator.

To obtain the value of the modal index, the first step is to choose the waveguide geometry, mode number and polarization (i.e., n_c, n_s, n_f, T), m-number, and TE/TM polarization, respectively. Let us assume that we choose TM polarization for a step-index slab waveguide, with $n_f \cong n_s$. Then, by calculating a_M parameter, we select one dispersion curve in Figure 5.8. The second step is to select an optical wavelength and calculate the V parameter. The third step is to find the value of the b_M parameter, and using its definition (see Table 5.1), calculate the value of the modal effective index, N_m.

The normalized dispersion curves have a monotonic profile with a normalized modal index, b, changing between 1 and 0. As indicated in Table 5.1, this is equivalent to the modal index change between n_f and n_s. Because $N = n_f \sin \alpha$, then increasing the modal index, N, is equivalent to increasing angle α, as illustrated in Figure 5.10.

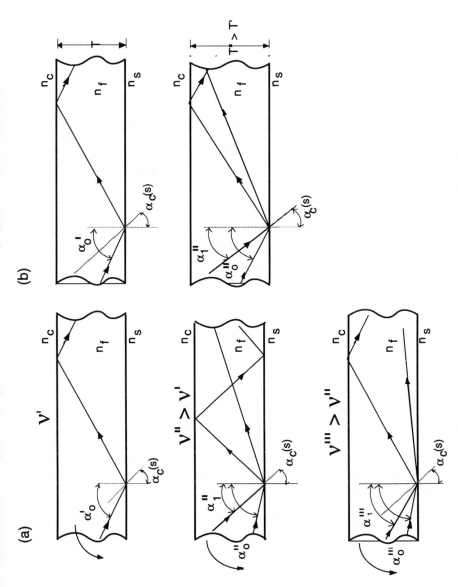

Figure 5.10 TE guided mode zigzag waves for step index slab waveguide: (a) optical frequencies increases from "red to blue" ($v''' > v'' > v'$) while remaining parameters do not change; (b) only waveguide thickness increases ($T'' > T'$) while the other parameters remain unchanged.

In many applications of integrated optics, it is important to preserve single-mode propagation. In such a case, assuming TE polarization, the normalized frequency for fundamental mode TE_0 should be located between the cutoff normalized frequencies for TE_0 and TE_1 modes (i.e., for $b = 0$). We have:

$$V_o \leq V \leq V_o + \pi$$
$$V_o = \tan^{-1}\sqrt{a_E} \tag{5.16}$$

In such a case, only the TE_0-mode will be propagated. Moreover, for the symmetrical waveguide case, we have $a_E = a_M = 0$. Both polarizations propagate identically, with $V_o = 0$ (i.e., there is no cutoff for the fundamental TE_0 mode).

For anisotropic substrates and waveguides, the model eigenequation is much more complicated. (For a comprehensive review of this problem see [2].) However, for $n_f \cong n_s$, and anisotropic substrate with low anisotropy (which is usually the case), we can still use the isotropic case analysis and obtain a good approximation.

In the case of GRIN waveguide, we obtain the cutoff normalized in the form:

$$V_{dm} = \sqrt{2\pi}\left(m + 3/4\right) \tag{5.17}$$

This means that to preserve at least the TE_0 mode propagation, we have: $V_d > 0.75\sqrt{2\pi}$.

Using Table 5.1 and (5.17) for $m = 0$ and $m = 1$, we obtain the following single-mode condition.

$$1.88 < k_o d\sqrt{2n_s \Delta n} < 4.38 \tag{5.18}$$

where the condition $\Delta n \ll n_s$ was assumed. This relation has been illustrated in Table 5.2 for $n_s = 1.55$, and $\lambda = 1.3 \ \mu m$.

The acceptable range of diffusion lengths is 3 μm for $\Delta n = 0.01$ and shrinks to 0.5 μm for $\Delta n = 0.2$. Therefore, rather small index modulations should be used, in order to preserve reasonable GRIN waveguide fabrication tolerances.

Table 5.2
Single-Mode Condition for GRIN Slab Waveguide

Δ	0.01	0.05	0.1	0.15	0.2
$d_{min}(\mu m)$	2.2	0.98	0.69	0.57	0.49
$d_{max}(\mu m)$	5.1	2.28	1.61	1.32	1.14

Note: $n_s = 1.55$, and $\lambda = 1.3 \mu m$.

5.2.3.2 Mode Eigenfunctions for Slab Waveguides

The Maxwell equations are solved separately for both polarizations TE and TM, including the boundary conditions at two waveguide interfaces representing the continuity of tangential components E_t and H_t. (For comparison, in the case of perfectly conducting metallic microwave waveguides, $E_t = 0$.) As a consequence, we obtain mode eigenfunctions $E_{ym}(x)$ and $H_{ym}(x)$ for TE and TM polarizations, respectively, where the mode number is $m = 0$, 1, 2, 3. For each polarization (e.g., TE), the specific eigenfunction ($E_{ymo}(x)$) corresponds to one and only one eigenvalue (e.g., $N_{m_o}^{(E)}$).

The spatial distributions of modal structures (eigenfunctions) are characteristic for each mode. They are illustrated in Figure 5.11(a) for TE cases in which they are rapidly attenuating outside the waveguide at 0.1- to 0.2-mm distance. However, in the case of

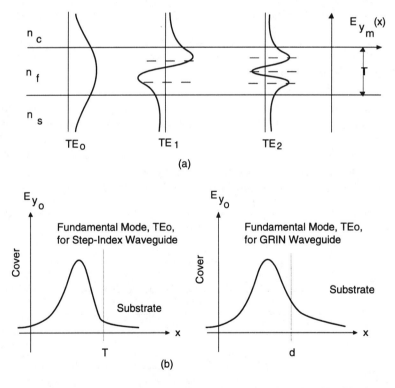

Figure 5.11 (a) TE mode structures for fundamental mode (TE$_o$) and higher modes (TE$_1$, TE$_2$) for a step index slab waveguide. The number of zeroes is equal to the modal number, m. (b) Comparison of fundamental mode structures for step index and GRIN waveguides. The GRIN waveguide mode structure has a longer evanescent tail, up to 3D.

GRIN waveguides, the guided wave is not reflected sharply from the waveguide-substrate interface. Thus, the evanescent tail is longer than in the case of a step-index waveguide.

Both fundamental mode ($m = 0$) structures TE_o and TM_o have similar spatial profiles, with a single extreme at the central part of the waveguide. They can be approximated by Gaussian distribution. In the higher mode structures ($m = 1, 2, 3$, and so on), the number of zeroes is equal to the mode number, m.

5.2.3.3 Slab Waveguides With Metallic Covers

Slab waveguides with metallic cover TE modes propagate reasonably well (with losses of ~1 dB/cm), through a waveguide polarizer effect [12]. This effect is important because many applications require metallic electrodes that are relatively close to the waveguide in order to modulate typical TM modes. To avoid this attenuation, we need to make the dielectric buffer sufficiently thin (~1 μm) to preserve the access of the metallic electrodes into the waveguide and sufficiently thick (>0.2 μm) to prevent significant attenuation (see Fig. 5.12).

5.2.4 Channel Waveguides

Channel (three-dimensional) waveguides are of immense importance in integrated optics, particularly in the context of integrated optical interconnects. Unfortunately, the modal structure for these waveguides is much more sophisticated than that for dielectric slab waveguides. For example, we are not able to further separate modes on TE and TM as in the case of the slab waveguide. Rather, we observe some kind of hybrid mode structures in two forms (p and q are mode numbers in x and y directions, respectively):

- E_{pq}^x-hybrid mode, with an E_x main component, and $H_x = 0$, remaining TM-mode (which has also E_x main transversal component; see Fig. 5.16(c)).

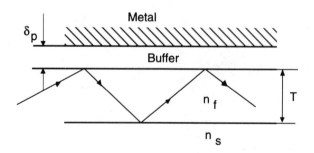

Figure 5.12 Geometry of metallic slab waveguide with buffer cladding. The buffer thickness should be larger than 0.1–0.2 μm.

- E_{pq}^y-hybrid mode, with E_y main component, and $H_y = 0$, remaining TE-mode (which has also E_y main transversal component; see Fig. 5.14(c)).

There are several types of channel waveguides, as illustrated in Figure 5.13. Five are the step-index type and one (buried) is the GRIN type.

Both eigenvalues (i.e., modal indices) and eigenfunctions (i.e., modal field distributions) of the channel waveguide equation can only be solved numerically [13]. The modal field distributions are important for any coupling problems because the input fields should match the modal field accurately to obtain high coupling efficiency. Yet, in the single-mode case, which is most important for practical applications, the fundamental mode field E_{00}^x or E_{00}^y distributions are approximately of Gaussian profile, with axial symmetry, rapidly attenuating within a 0.1 to 0.2 mm distance outside the waveguide. This could simplify the approximate analysis. On the other hand, for the highly multimode cases, a four-dimensional phase-space volume occupied by a channel waveguide beam is defined by a waveguide cross section (x and y) and a two-dimensional numerical aperture.

This section concentrates on determining the single-mode propagation conditions for channel waveguides, both GRIN and step-index. We emphasize the effective index method based on modeling channel waveguides in terms of much simpler slab waveguides. This is useful for eigenvalues evaluation. For a more detailed analysis of this method and an evaluation of eigenfunctions, please refer to literature regarding Marcatili's method [8].

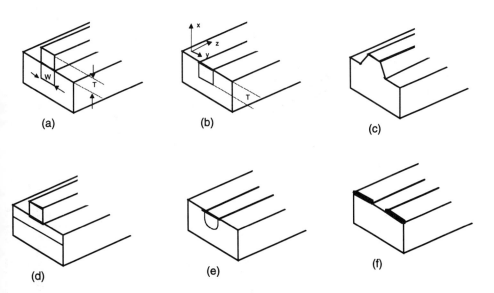

Figure 5.13 Channel waveguides: W = width and T = depth of the waveguide (a) raised step; (b) embedded strip; (c) ridge; (d) strip loaded; (e) buried type; (f) metal loaded.

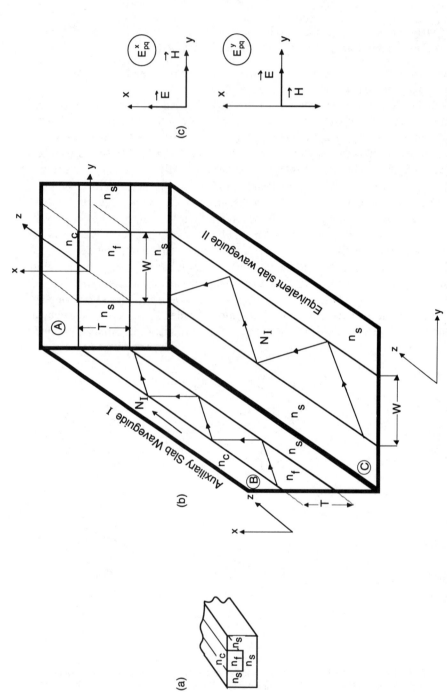

Figure 5.14 Effective index method and hybrid modes. The embedded strip channel 3-D waveguide with (a) a cross section in plane A (x,y) is modeled as superposition of slab I—auxiliary asymmetric slab waveguide in plane B (x,z) and (b) equivalent symmetric slab waveguide in plane C; (c) the transversal structure of basic $\mathbf{E}$, $\mathbf{H}$, components is presented, representing hybrid modes E^x_{pq} and E^y_{pq}, respectively.

The effective index method is illustrated in Figure 5.14. According to this schema, the exemplary embedded strip channel (three-dimensional) waveguide has a cross section in plane (x and y) with width, w, and depth, T. The waveguide is modeled as a superposition of an auxiliary slab waveguide (I) with thickness, T, and indices, n_c, n_f, and n_s, and an equivalent symmetrical slab waveguide (II) with thickness, W, and indices, n_s and N_I.

In the case of an E^x_{pq}-hybrid mode, as illustrated in Figure 5.14, the polarization of guided mode propagated in the auxiliary slab waveguide (I) is a TM type because the transversal electric component, E_x, is parallel to the waveguide. However, assuming $n_f \sim n_s$, we have (see Table 5.1):

$$b_I = b_M \cong b_E = \frac{N_I^2 - n_s^2}{n_f^2 - n_s^2} \tag{5.19}$$

Thus,

$$N_I = \sqrt{n_s^2 + b_I(n_f^2 - n_s^2)} \tag{5.20}$$

According to Table 5.1,

$$V_I = k_o T \sqrt{n_f^2 - n_s^2} \tag{5.21}$$

Using Table 5.1,

$$a_I = a_M = \left(\frac{n_f}{n_s}\right)^4 \left(\frac{n_s^2 - n_c^2}{n_f^2 - n_s^2}\right) \tag{5.22}$$

The auxiliary normalized effective index can be found using modal dispersion curves from Figure 5.8 for the TM case. The formulas for the TM case coincide with the TE case. However, if $n_f \approx n_s$, and $n_c = 1$ (air), then, $b_M \approx b_E$, and $a_M = a_E = \infty$.

The next step is to solve the equivalent symmetric waveguide (II) with film index, n_f, replaced by N_I. We assume TE-polarization because, this time, the electrical component is perpendicular to the waveguide. Thus, we have:

$$b_{II} = b_E = \frac{N_{pq}^2 - n_s^2}{N_I^2 - n_s^2} \tag{5.23}$$

$$V_{II} = k_o W \sqrt{N_I^2 - n_s^2} \tag{5.24}$$

with $a_{II} = a_E = 0$, for symmetric waveguide. Using (5.16) for $a_E = 0$, we obtain the cutoff V_{II} value for fundamental mode ($q = 0$) in the form:

$$V_{II} = \pi \tag{5.25}$$

According to (5.16), with a_E replaced by a_M, the single-mode propagation condition is in the form:

$$\tan^{-1}\sqrt{a_M} \leq V_I \leq \pi + \tan^{-1}\sqrt{a_M} \tag{5.26}$$

This relation, for $a_M = 0$ ($n_c = 1$), reduces to:

$$\frac{\pi}{2} \leq V_I \leq \frac{3}{2}\pi \tag{5.27}$$

Using modal dispersion curves (Fig. 5.8) twice, first in the form,

$$b_I = b_I(V_I, a_M) \tag{5.28}$$

Second, in the form,

$$b_{II} = b_{II}(V_{II}, 0) \tag{5.29}$$

We can find the effective index of the channel waveguide, N_{pq}, as a function of waveguide depth, T, waveguide width, W, optical wavelength, λ_o, and n_f, n_c, n_s indices:

$$N_{pq} = N_{pq}(k_o, T, W, n_f, n_s, n_c) \tag{5.30}$$

This graphical method, for fundamental mode E_{00}^x, is illustrated in Figure 5.15, with the following sequence:

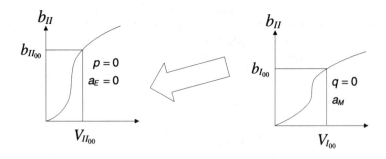

Figure 5.15 Graphical method of finding effective index, N_{00}, for E_{00}^x-hybrid mode of step-index channel waveguide.

$$V_{I00} \to b_{I00} \to V_{II00} \to b_{II00} \to N_{00} \tag{5.31}$$

To obtain the single-mode propagation conditions, we substitute (5.21) and (5.19) into (5.24) and (5.25) to obtain:

$$V_{II} = \frac{W}{T}\sqrt{b_I}V_I \leq \pi \tag{5.32}$$

or

$$0 \leq \frac{W}{T} \leq \frac{\pi}{\sqrt{b_I}V_I} \tag{5.33}$$

This relation, together with (5.38), for $n_c = 1$ ($a_M = \infty$), is illustrated in Figure 5.16. This analysis holds for $W/T > 1$, and aspect ratio (W/T) should be close to the upper limit of (32) to obtain good mode confinement.

For E^y_{pq}-hybrid mode, the I-waveguide propagation is the TE type, while II-waveguide propagation is the TM type. Thus, $b_I = b_E$, $a_I = a_E$, $b_{II} = b_M$, and $a_{II} = a_M = 0$, and we can

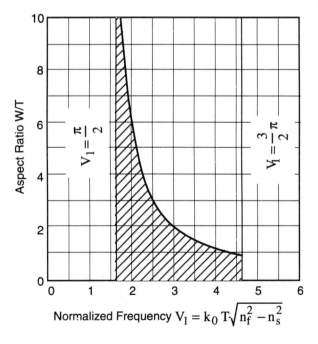

Figure 5.16 Aspect ratio, W/T, versus V_I for step-index channel waveguide, in order to preserve single E^x_{00}-mode propagation conditions (hatched area).

use (5.19) through (5.30) by replacing the values. Using a similar analysis for the GRIN channel waveguide, asymmetric in respect to the x-axis, we obtain in Figure 5.18 the following analog of Figure 5.18, where $n_{fx} = n_s + \Delta n_x$ and dx and dy are diffusion lengths illustrated in Figure 5.17.

5.2.5 Planar Medium

We have previously discussed the guided waves propagating in a single direction. In the case of a channel waveguide, there is obviously only one direction in which a guided wave can propagate. However, slab (planar) waveguides can propagate in an infinite number of directions. Considering a single-mode TE_m guided wave characterized by the modal index, N_m, we introduce a new coordinate system with the vertical z-axis emphasizing the symmetry of the slab waveguide with respect to vertical axis. Of course, for any direction of propagation in the (x- and y-) plane, the represented guided wave is characterized by the same eigenvalue N_m, and by the same longitudinal wave number.

$$\beta_m = k_o N_m \tag{5.34}$$

Omitting index, m, we have introduced the planar wave vector, where

$$\beta_x^2 + \beta_y^2 = \beta^2 = k_o^2 N^2 \tag{5.35}$$

Now we can introduce the generalized planar wave with a curve planar wave front, as illustrated in Figure 5.19. The planar wave fronts are (x and y) projections of guided waves propagating in a two-dimensional planar medium with planar medium index, N_m.

It should be noted that for each mode, we have different planar media in the same slab waveguide. Of course, the planar medium represented by the fundamental mode,

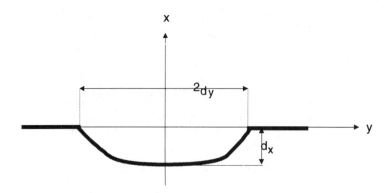

Figure 5.17 Cross section of buried-type GRIN channel waveguide with diffusion lengths dx and dy.

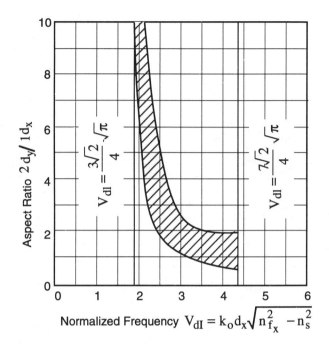

Figure 5.18 Aspect ratio, $2dy/dx$, versus d_{dl} for buried type GRIN channel waveguide, in order to preserve single E^x_{00}-mode propagation conditions (hatched area).

$m = 0$, is the most important. In Figure 5.20, we have two different examples of guided waves propagated in a single-mode planar medium. Such single-mode planar waves can be refracted or diffracted by planar holograms, as in Figure 5.20 [14].

5.2.6 Waveguide Fabrication

There are various types of channel optical waveguides. These types include buried, ridge, dielectric-loaded, and metal strip. The basic idea of channel waveguide fabrication is to fabricate a higher index guiding region that is surrounded by lower index materials. The most popular substrate materials include $LiNbO_3$, glass, GaAs and InP, and polymers [15–19]. The reason for choosing a particular material system depends on the ease of fabrication; loss characteristics; functional device flexibility; and the possibility of monolithic integration with sources, detectors, high-speed electronics, and various passive as well as active devices.

Fabrication of waveguides starts with planar geometry, which confines light only in the vertical direction. The conventional methods include metal diffusion, ion exchange, sputtering, epitaxial growth, and ion implantation. Lateral confinement can be provided by a mask to allow local implementation of these techniques. However, semiconductors

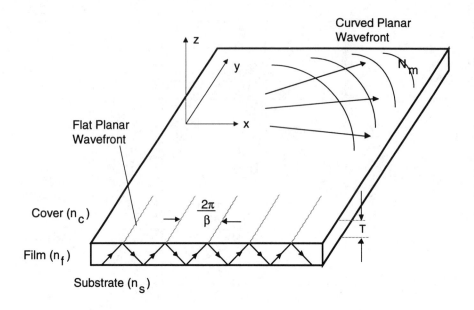

Figure 5.19 Slab waveguide as 2-D planar medium, with medium index, N_m, specific for M^{th}-mode. Flat end curved planar wave fronts represent (x,y)-projection of guided waves.

such as GaAs- and InP-based waveguides are usually fabricated by wet and dry etching methods after epitaxial growth.

In the following sections, waveguide fabrication techniques for a selected integrated optic material system are described. These techniques include an ion exchange technique for glass, Ti-indiffusion for LiNbO$_3$, epitaxial growth and etching for semiconductors, and various types of waveguide fabrication for a polymer-based system [20–27].

5.2.6.1 Ion-Exchanged Glass Waveguides

The ion-exchange technique is a common waveguide fabrication method for glass and LiNbO$_3$ materials. It utilizes the fact that the refractive index change is a function of the material composition. By replacing the lighter ions of the glass or LiNbO$_3$ with a heavier ion, the refractive index is increased. Therefore, a waveguide region is formed. Only a glass waveguide will be described here due to the similarity in the proton exchange technique in LiNbO$_3$ waveguide fabrication. Glass is usually composed of SiO$_2$ and B$_2$O$_3$ with some proportions of oxides, such as Na$_2$O, K$_2$O [17]. The first ion-exchanged glass waveguide was fabricated by exchanging Tl$^+$ ions in SiO$_2$ containing Na$^+$ and K$^+$ oxides at an elevated temperature. Since then, many cations such as Cs$^+$, Rb$^+$, Li$^+$, Ag$^+$ have been exchanged with sodium ions in glass for waveguide fabrication. Table 5.3 summarizes various physical parameters, exchange conditions, and the resulting index change for the

Figure 5.20 Planar waves propagating in planar medium (fan-out: 1:16).

Table 5.3
Results of Ion-Exchanges Waveguides For Various Ions with Na^+ as the Host Ion

Ion	Radius (A)	Polarizability (A minus 3)	Glass	Melt	Temperature (°C)	Δn
Tl^+	1.49	5.2	Boro-silicate	$TiNO_3 + KNO_3 + NaNO_3$	530	0.001–0.1
K^+	1.33	1.33	Soda-lime	KNO_3	365	0.008
Ag^+	1.26	2.4	Alumino-silicate	$AgNO_3$	225–270	0.13
Li^+	0.65	0.03	Soda-lime	$Li_2SO_4 + K_2SO_4$	520–620	0.012
Rb^+	1.49	1.98	Soda-lime	$RbNO_3$	520	0.015
Cs^+	1.65	3.34	Soda-lime	$CsNO_3$	520	0.03

Source: [17].

previously mentioned ions exchanged with Na^+. Typically, Tl^+ and Ag^+ ion-exchange methods are used for multimode waveguide fabrication due to the large index change ($\Delta n \geq 0.1$). For single-mode waveguides, K^+ with smaller index change (on the order of 10^{-3}) is used.

In the ion-exchange process, the net index variation depends on three major factors: ionic polarizability, molar volume, and stress induced by the substitution. The dominant factor will depend on the individual ion exchange. In the case of Ag^+-Na^+ ion exchange, the ionic polarizability factor is the dominant mechanism, as the polarizability of silver ion is much larger than the sodium ion in the glass. Thus, the exchanged region has a higher index than its surrounds.

A multimode planar waveguide, formed by exchanging Ag^+ ions with the Na^+ ions, is demonstrated in Figure 5.21. The light is coupled onto the waveguide by a prism using an evanescent field coupling technique. The bright streak of light is the scattering of the guided light. Glass waveguides have the advantage of easy and low-cost fabrication, extremely low loss (<0.1 dB/cm), and the feasibility of large area optical interconnect implementation. However, a shortcoming of glass waveguides is the lack of electro-optic (EO) effects, which excludes the implementation of functional devices on this material system. Glass waveguides have approximately the same index of refraction as do fiber waveguides. Therefore, insertion loss due to the index mismatch between the fiber and the waveguide is minimal. Glass is useful in passive, medium-distance (board-to-board) optical interconnects, multiplexing, and demultiplexing applications.

The ion-exchange process can be enhanced by the application of an electric field in the direction of the diffusion [27]. However, in the pure thermal diffusion process (without an external electric field), the waveguide formed is a graded index type. The electric field induced exchange process creates a step-index waveguide at a high field due to the domination of ion movement by the electric field instead of diffusion. The

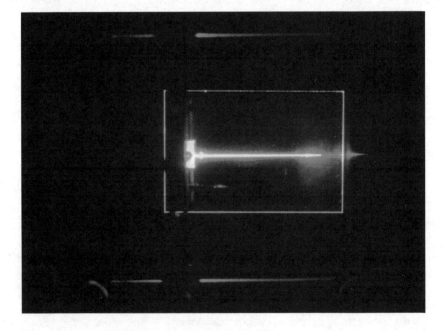

Figure 5.21 Ag^+ ion-exchange multi-mode waveguide using prism coupling.

exchange ion density becomes constant within the depth $E\mu t$, where t is the time for the ion exchange, μ is the ion mobility depending upon temperature, and E is the electric field.

The fabrication setup for the ion-exchange process is shown in Figure 5.22. The salt containing the exchange ion (e.g., $AgNO_3$ or KNO_3) is held in a crucible that is kept under constant temperature in a furnace. Polished glass substrates with channel waveguide aluminum mask openings (developed by photolithography) are immersed in the molten salt bath for a few hours at the appropriate temperature (250°C and 370°C for Ag^+ and K^+, respectively). In the Ag^+ ion-exchanged waveguide, the surface index change, Δn, is roughly on the order of 10^{-2}, and the thickness of the waveguide is approximately 2 μm.

5.2.6.2 Ti Indiffusion LiNbO$_3$ Waveguides

LiNbO$_3$ has relatively large electro-optic and acoustic-optic coefficients, so many functional devices can be fabricated. Thus, there is strong motivation to fabricate low-loss optical waveguides using LiNbO$_3$ as the substrate material. There are several methods for fabricating this type of optical waveguide, such as proton exchange, sputtering, epitaxial growth, outdiffusion, and indiffusion. The metal indiffusion technique is perhaps the only method that does not sacrifice the EO effect and has almost isotropic index change for both the ordinary and extraordinary propagating waves. V, Ni, Cu, and Ti have been diffused in LiNbO$_3$ to form optical waveguides. Of these, Ti indiffused waveguides exhibit the best optical characterization with no noticeable effect on the EO coefficient of the crystal [2,20,28].

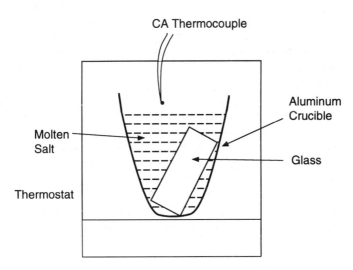

Figure 5.22 The fabrication setup for glass ion-exchange waveguide fabrication.

Usually, LiNbO$_3$ substrate's optical axis should be specified according to the design of the functional device. When $z(x)$-cut LiNbO$_3$ substrate is mentioned, it means that the wafer is cut normal to the $z(x)$ axis of the crystal. The y-cut LiNbO$_3$ are unpopular due to inferior device properties such as low coupling efficiency to fibers because of a strongly elliptical near-field pattern, and the strong tendency to exhibit outdiffused waveguides. However, y-cut LiNbO$_3$ is useful in acousto-optic devices. The first step in the waveguide fabrication procedure is to evaporate a thin layer (200 to 800Å) of Li$_2$O to form a waveguide pattern using a liftoff technique. The sample is then heated in a flow of moistured Ar or O$_2$ gas to temperatures around 850° to 1,000°C for 1 hour to prevent the oxidation of the metal [22]. A schematic of the fabrication setup is shown in Figure 5.23 with the gas flow time.

The reason for using the moistured gas is to prevent the outdiffusion of Li$_2$O. This outdiffusion increases the extraordinary index across the entire face of the crystal. Therefore, it decreases the confinement of light by the local indiffusion of Ti in the channel waveguide. The sample is left in the oven for about 4 hours for diffusion to completely occur. In the case of diffusion times that are long compared to the time required for the metal film to completely enter the crystal, the concentration profile approaches that of the Gaussian function. For a short diffusion time, where the metal has not completely entered the crystal, the index profile approximates that of a complementary error function (ERFC). Like the long diffusion time case where the metal is completely diffused, the index profile will be intermediate between the Gaussian and ERFC profiles [20]. Figure 5.24 shows a typical Gaussian profile of the effective index as a function of the depth of the waveguide. In this case, the Ti film thickness is 450Å. After diffusion, the waveguide supports various TM and TE modes. By measuring the effective index of the TM modes, delta n_e and the $1/e$ thickness of the waveguide are 0.022 and 2.6 μm, respectively.

The electro-optic effect of the LiNbO$_3$ crystal is degraded by the photorefractive effect at a shorter wavelength (~0.8 μm). However, no significant degradation is observed in the 1.3 μm range to about 5 mW. Therefore, Ti:LiNbO$_3$ waveguide devices are most useful in telecommunication wavelengths of 1.3 to 1.6 μm [2].

The combination of a strong EO effect with low-loss waveguides makes the versatility of LiNbO$_3$ integrated optic material system extremely competitive. The development of functional devices such as modulators and switches has already reached the commercial market place. This material system offers the most potential in hybrid systems. Unfortunately, however, the inability to integrate detectors, laser sources, amplifiers, and other essential electronic circuitry into the LiNbO$_3$ material precludes its usefulness for total optical integrated circuits at the present time. There are major efforts being made to fabricate optical waveguide amplifiers and lasers by doping rare-earth elements such as Er^{+++} into LiNbO$_3$ [29]. This development, however, is still in its infancy.

5.2.6.3 GaAs-Based Waveguides

The advantage of GaAs- [30] or InP-based [16] materials over other material systems is the possibility of monolithic integration of all functional components such as the source,

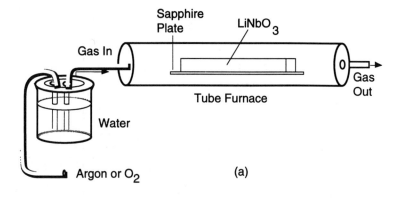

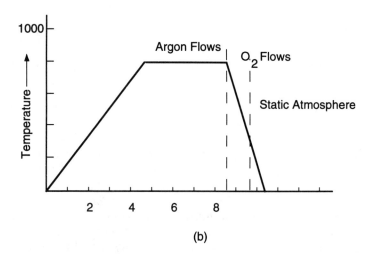

Figure 5.23 Proton-exchange waveguide fabrication: (a) the furnace with gas flow system; (b) temperature evolution and diffusion gas flow as a function of time.

the modulator and switch, detector, and high-speed electronic driving circuits on a single chip (namely, optoelectronic integrated circuit (OEIC) [31]). The propagation loss in semiconductor-based optical channel waveguides is generally much larger than in its counterpart in LiNbO$_3$ and glass. Therefore, reduction of the waveguide loss needs to be implemented for the material system to be practical. The propagation loss is mainly due to four different mechanisms: free-carrier absorption, waveguide wall scattering, bending loss, and coupling loss.

The two most popular types of GaAs-based channel waveguides are embedded and ridged waveguide. The embedded type is usually accomplished through ion implantation

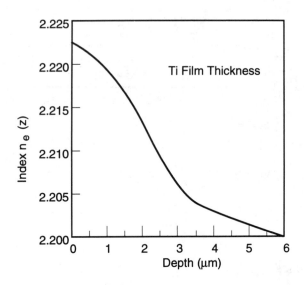

Figure 5.24 Index profile of the Ti-diffused waveguide in Z-cut LiNbO$_3$ with Ti film thickness of 450Å, $\Delta n_e = 0.022$, and $1/e$ thickness = 2.6 μm.

by high-energy protons (~300 keV). The bombardment causes defect centers for free-carrier trapping. This reduces the contribution of those free carriers to the refractive index through the negative plasma effect. Therefore, the index of refraction is increased in the ion implantation region with respect to other areas. Thus, optical channel waveguides are formed. However, the evanescent field of the propagating wave penetrates into the free-carrier region, which introduces loss due to free-carrier absorption.

Many integrated optical waveguides have ridged geometries. A cross-sectional view of this commonly used waveguide structure is shown in Figure 5.25.

The planar waveguide is fabricated from an n$^+$ GaAs substrate forming good ohmic contact. A layer (~2 μm) of very low doping density AlGaAs is epitaxially grown by either metal organic chemical vapor deposition or molecular beam epitaxy. This layer acts as the lower cladding layer of the optical waveguide. Figure 5.26 shows the index of refraction as a function of the photon wavelength and energy with x as the Al concentration in Al$_x$Ga$_{1-x}$As [32].

A layer of GaAs with low free-carrier concentration density (<10^{14}/cm^3) is grown to form the waveguide. A rule of thumb is that carrier concentrations should be kept as low as possible to minimize free-carrier absorption. The lateral confinement of the waveguide is made by dry (ion milling) or wet (chemical) etching the GaAs layer into a ridge shape. During the etching process, imperfections in the waveguide wall cause scattering loss. Therefore, extra fine photolithography and dry etching techniques should be used. Strong waveguide confinement is needed to ensure minimal penetration by the evanescent wave into the absorbing substrate and the rough surface of the waveguide wall.

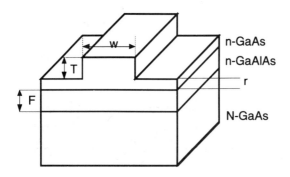

Figure 5.25 Ridge GaAs optical wavelength for GaAs/AlGaAs system.

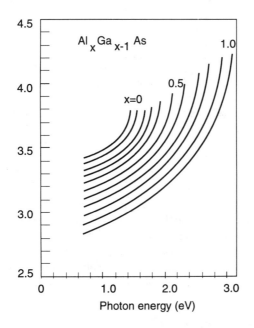

Figure 5.26 Refractive indices of $Al_xGa_{1-x}As$ as a function of the photon energy and waveguide with x-composition increments of 0.1.

It has been found that one of the best mode confinements in the vertical direction has a waveguide thickness, T, of 4.5 mm and cladding thickness, r, of 1.0 µm [30]. Figure 5.27 shows the normalized thickness of the fundamental TE-like mode V_o as a function of the etched step height, T, with various waveguide widths. To satisfy the single-mode condition, V_o must be less than 1.87, shown in Figure 5.27. For example, a strong confine-

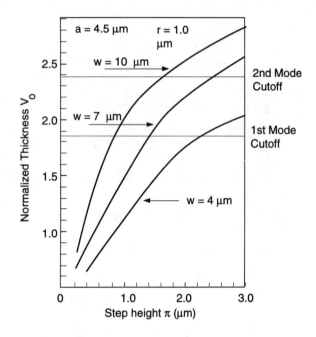

Figure 5.27 Theoretical calculation of the normalized thickness of the fundamental TE-like mode in ridge waveguide as a function of the ridge step height for various ridge widths: waveguide thickness of $a = 4.5$ μm; cladding thickness of $r = 1.0$ μm.

ment single-mode waveguide can be designed with dimensions of $W = 7.0$ μm, $T = 1.2$ μm, $a = 4.5$ μm, and $r = 1.0$ μm.

Clearly, the superiority of GaAs and InP material systems is in their feasibility for the monolithic integration of photonic devices. These semiconductor material systems have bright future potential in the areas of optical interconnects, optical computing, signal processing, and communications. However, the high costs of the GaAs and InP material and fabrication are major concerns in the commercialization community. The cost will be lowered considerably, however, as demand increases and the the technology matures.

5.2.6.4 Polymer-Based Waveguides

In recent years, polymer-based integrated-optic material systems have become increasingly important in optical interconnects. Some advantages of polymer systems are ease of fabrication, a standard lithographic technique suitable for mass production, low temperature processing, large EO effects for active devices, and cost-effective manufacturing [19].

There are many mechanisms in polymer waveguide formation, such as photo-induced bleaching, photopolymerization, and photo-crosslinking. The optical channel waveguide can be patterned through photo masks or direct laser writing techniques using appropriate wavelengths of light (coherent or not). The index change typically ranges from 0.001 to 0.2. Some of the common host polymer materials are polymethyl methacrylate (PMMA), polyimide, polyurethane, polyester, and even gelatin [11,33–34]. These host polymers are often doped with reactive photo chromophores for waveguide patterning. For example, a photosensitive nitrone chromophore ((4-N, N-dimethylaminophenyl)-N-phenyl nitrone or DMAPN) can be used as the photo reactive agent in PMMA. Upon exposure to UV light, the refractive index of DMAPN/PMMA films can be varied from that of PMMA (1.49) to greater than 1.57 through the photo-induced chemical change. This effect is ideal for channel waveguide fabrication by contact mask exposure techniques.

Polymer-based optical waveguides are very important in large area optical interconnects such as optical backplanes. The nature of spin-and-dip coating fabrication and the cost effectiveness of polymer materials give it a special advantage over other materials like $LiNbO_3$ and GaAs. The fact that the EO effect can be implemented by doping and attaching nonlinear dyes to the polymer matrix and backbones introduces functional device capability into this material system. However, the EO effect requires the removal of the centrosymmetry by electric poling. Temperature and long-term stability problems of the polymer must be solved before the system can become a practical reality.

5.2.6.5 Other Waveguide Materials

There are many other waveguide material systems that are important in the area of integrated optics. Each of these materials has its own unique features, and some of these materials have excellent electro-optic, acousto-optic, and magneto-optic effects. These materials include ZnO, heavy metal oxides such as Ta_2O_5 and Nb_2O_5, Si_3N_4, and Yttrium Iron Garnet (YIG: $Y_3Fe_5O_{12}$) films. Unfortunately, some inherent disadvantages such as high propagation loss (>5 dB/cm) may prevent their popular use [2].

For instance, ZnO has excellent piezoelectric properties as well as large EO and nonlinear effects [35]. ZnO can be deposited easily as a film by sputtering techniques. Due to the polycrystalline structure of the film, propagation loss is about 5 to 10 dB/cm on glass substrate. However, it has been shown that the loss can be decreased to 0.01 dB/cm by laser annealing technique on SiO_2/Si substrate.

YIG films are important materials for magneto-optic devices [36,37]. They are grown on Gadolinium Gallium Garnet (GGG) substrates by VPE, LPE, and sputtering techniques. Waveguides with transmission loss on the order of 1 to 10 dB/cm have been observed.

5.3 WAVEGUIDE COUPLERS

All integrated optics applications require a successful light-sources-to-waveguide coupling scheme. Unfortunately, this critical issue is also one of the most difficult problems in integrated optics, particularly for single-mode waveguide applications. This is largely caused by the extremely tight mechanical tolerances that must be satisfied, in spite of and in combination with many disturbing effects. These effects might include such events as the light source wavelength shift due to temperature changes and/or high-speed modulation. For example, a 1-μm long transversal misalignment of a 5-μm width single-mode waveguide can create a 20% coupling loss (i.e., 1 dB loss). This evaluation holds for two types of light sources: light emitting diodes (LEDs) and laser diodes (LDs). For our purposes, however, LDs are more important than LEDs because they are the preferable source candidates for high-speed and single-mode applications.

There are two major coupling schemes: transversal coupling and longitudinal coupling. Longitudinal coupling is preferable in laboratory conditions because it enables easy mechanical adjustment [38]. It is particularly useful for single-mode slab waveguide geometries in which 70% to 100% coupling efficiency can be achieved. Unfortunately, this coupling scheme is bulky and highly sensitive to source wavelength shift. Moreover, the solid-state packaging required for the majority of its interconnect applications eliminates the possibility of mechanical adjustment. Still, a number of promising solutions have recently been proposed. These solutions include microprism couplers and grating couplers in conjunction with surface-emitting LDs.

Transversal coupling is much more promising for future practical applications. It can be used for both single-mode and multimode waveguides and for both LDs and LEDs (including edge emitting LEDs (ELEDs)). The major problem with transversal coupling is related to unwanted Fresnel reflections from coupler interfaces that can create additional optical feedback, thus severely degrading the LD performance. Fortunately, this problem has been comprehensively analyzed and solved for LD-to-fiber coupling schemes for both the multimode and the single-mode coupling cases [39–41]. For this reason, use of an indirect coupling geometry is very attractive (i.e., an LD-to-fiber-to-waveguide approach).

Any theoretical analysis of waveguide coupling is complicated, particularly for the single-mode waveguide case. For single-mode waveguides, a knowledge of specific eigenmode field distributions (or, eigenfunctions) is necessary. This is beyond the scope of this chapter. Therefore, we have included an approximate analysis based on the Liouville (brightness) theorem. This method provides accurate results in the multimode waveguide case and the necessary conditions for high coupling efficiency (>50%) in the single-mode waveguide case. In addition, this method provides a precise distinction between the single-mode and multimode cases in terms of spatial coherence related to an optical analog of Heisenberg's uncertainty principle.

There are two major coupling schemes: transversal coupling (Fig. 5.28) and longitudinal coupling (Fig. 5.29). Slanted transversal geometry (Fig. 5.28(b)) can sometimes be preferable in order to minimize LDs optical feedback. Longitudinal coupling, either the prism type (Fig. 5.29(a)) or the grating type (Fig. 5.29(b)) was very popular in the beginning stages of integrated optics development, but in recent practical coupling schemes, the transversal coupling dominates.

5.3.1 Coupling of Laser Diode Beam Into Fiber

The schematic of LD fiber coupling is presented in Figure 5.30. The problem of semiconductor source coupling (LD, LED, ELED) into fiber has been comprehensively discussed

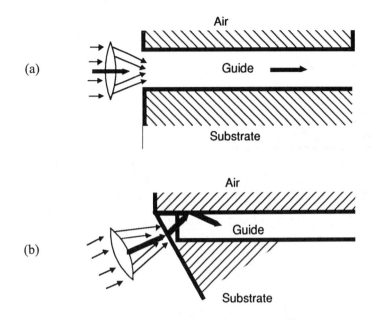

Figure 5.28 Transverse couplers: (a) axial geometry and (b) slanted geometry.

by many authors (too numerous to mention here), and solved on a custom-made basis. However, the problem of coupling a linear array of light sources into a linear array of fibers is still in the infancy stage. The reason for this is that the statistical location of fiber core centers within the cladding does not exactly coincide with the geometrical fiber centers in the fiber array. This is especially true in the single-mode case. Typical tolerances for 10/125-μm single-mode fibers are ± 2 μm for the cladding diameter and ± 1 μm for concentricity. However, it is possible to obtain fibers with tolerances of ± 0.5 μm. Using special alignment techniques, it is possible to obtain an average deviation of less than 1 μm for the fiber core center in respect to the array geometrical center. This is equivalent to about a 1-dB coupling loss. Moreover, to adjust the phase-space volumes of LD and fiber, lensed fibers are usually used. In such a case, the focal length of the fiber lens should be adjusted to optimize the trade-off between either high coupling efficiency (~90%) and very narrow lateral tolerance (± 1 μm at the 3-dB point) or lower efficiency (~20%) and broader lateral tolerance (± 3 μm at the 3-dB point).

The solution for this problem is shown in Figure 5.31 using a 250 μm pitch parallel four-arrayed optical coupling system between the LD or photodetector (PD) array and multimode ribbon fiber (GI 50/125 μm). The LD array chip was placed in a cavity formed on a submount. The submount consists of two Si (silicon) wafers and SiO_2 film attached by Si direct bonding technology, with self-aligning ribbon fiber held in an MT connector [40]. The measured coupling efficiency for a 100-μm distance between both arrays was 40% or 4 dB with an optical reflection ratio below -40 dB. It was degraded down to -10 dB for 30- to 40- μm transversal deviation.

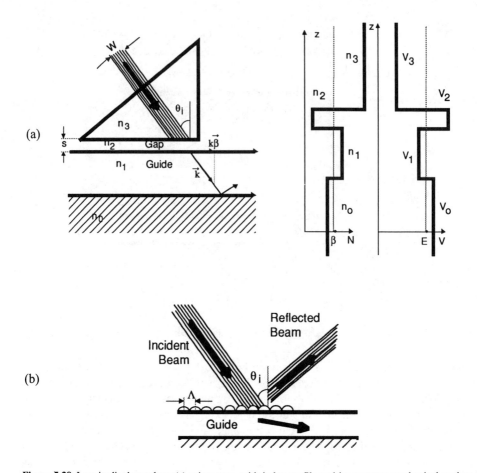

Figure 5.29 Longitudinal couplers: (a) prism-type, with index profile and its quantum mechanical analog;.(b) grating-type.

Within a few years, the coupling of LD arrays into fiber arrays should achieve a maturity similar to that for single-LD-into-fiber coupling for either the single-mode or the multimode case. However, for safety reasons, it is reasonable to assume that a 1 to 2 dB coupling loss should be accepted in future designs of the single-mode system's power budget. This loss should not be a problem for short distance (< 100m) intraprocessor optical interconnects with LDs as light sources.

5.3.2 Fiber-Waveguide Coupling

Fiber-waveguide coupling for optical interconnects is important for two reasons. The first reason is that some integrated optical devices (e.g., modulators and switches) can be part

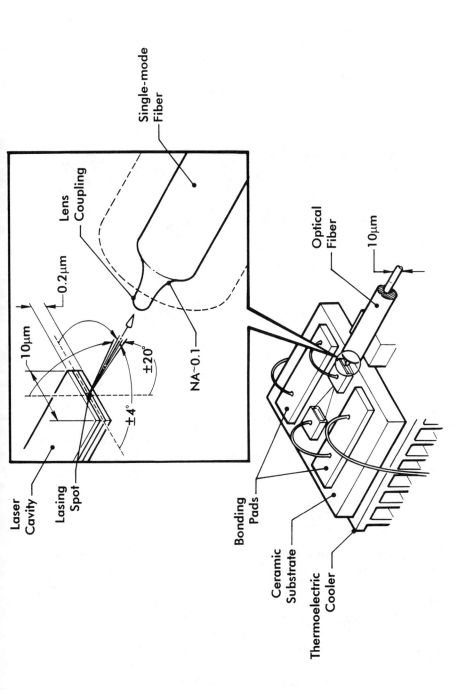

Figure 5.30 Laser diode to fiber coupling. The coupling area (expanded view) contains optical imaging elements: spherical lenses, cylindrical lenses, ball lenses, holographic optical elements, optical fiber tapers, and so on.

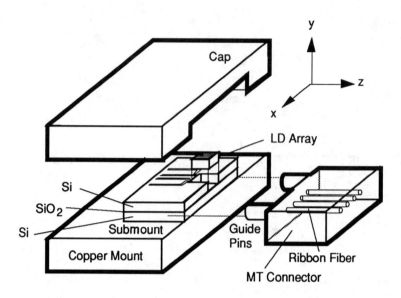

Figure 5.31 Optical coupling system between LD array and multi-mode fiber ribbon array.

of fiber optic communication systems. For example, high-speed (>10 GHz) integrated optic modulators are usually required to transmit an optical beam through a fiber modulated by a microwave signal. Second, LD-waveguide coupling can also be realized indirectly (i.e., in two steps: LD-to-fiber and fiber-to-waveguide).

In the case of single-mode fiber-waveguide coupling, the phase-space volumes for a fiber beam and a waveguide beam occupy the same elementary cell $\Omega_o = (2\pi)^2$, independent of their specific single-mode structure. In particular, the single-mode fiber beam in a near-field approximation (i.e., just behind the fiber) occupies the phase-space volume, W_f, in the form:

$$\Omega_f = [\pi k_o^2 (NA)_f^2]\left(\frac{\pi}{4}W_f^2\right) \cong (2\pi)^2 \tag{5.36}$$

where W_f is the beam spot size (or diameter), which is approximately equal to the fiber core diameter, d, (4–10 μm); and $(NA)_f$ is the single-mode beam numerical aperture.

It is sufficient in the single-mode coupling problem to compare the sizes of the beams, W_f, W_x, and W_y, because their angular sizes are automatically determined by Heisenberg relations. The same elementary principle holds for the Gaussian beam which approximates single-mode waveguide beams, $W_w = W_f = W_o$. Therefore, a suitable lens system can always be designed to obtain $W_f = W_x = W_y$ for almost 100% coupling efficiency. Here, "almost" means that the waveguide beam profile can be adjusted as quadratic but not circular. Unfortunately, fabrication of single-mode lensed fibers for improving coupling efficiency is difficult, especially for a highly anisotropic waveguide beam profile. Therefore, we must accept some discrepancy between the fiber and the waveguide beams' profiles by using a fiber-waveguide butt coupling procedure. Then, the coupling efficiency drops. Based on a Gaussian approximation of single-mode waveguide beams in Figure 5.32, the coupling loss was plotted as a function of laser mode spot size, W_f, horizontal waveguide mode spot size, W_x, and vertical mode spot size, W_y. Coupling loss depends only on ratios W_x/W_f and W_y/W_f when the beams are centered. In this case, we have

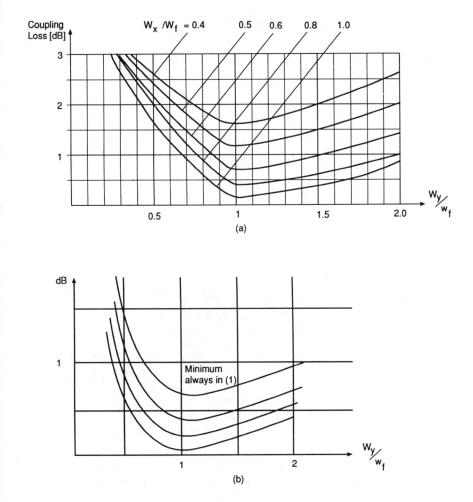

Figure 5.32 (a) Approximate single-mode fiber/waveguide coupling loss (excluding fresnel loss) as a function of: laser mode spot size, W_f, horizontal waveguide mode spot size, W_x, and vertical waveguide mode spot size, W_y; (b) blow up of the minimum loss that deviates around 1.

ignored the asymmetricity of the waveguide beam profile due to possible nonsymmetrical waveguide geometry. This creates an 0.1- to 0.2-dB error.

The coupling loss curves illustrated in Figure 5.32 are invariant with respect to permutation between W_x and W_y. As an example, consider the GaAs single-mode ridge channel waveguide cross-section geometry, illustrated in Figure 5.33 with $W_x = 10$ μm, $W_y = 3$ μm, and $W_f = 8$ μm. Thus, $W_x/W_f = 1.25$ and $W_y/W_f = 0.38$. Using the permutation rule mentioned previously, we obtain a coupling loss of 1.7 dB shown in Figure 5.32(a). Without antireflection coatings we need to add approximately 2 dB for Fresnel reflection loss.

The eigenvalues and eigenfunctions for both fiber and channel waveguides need to be solved numerically to maximize the overlap integral characterizing the mode field matching. This approach was used for titanium-diffused LiNbO$_3$ channel waveguides [42]. In this waveguide, a single-mode fiber with a very small core diameter ($d = 5$ μm) was chosen. The modal field distributions of both the fiber and waveguide were compared in a free-space near-field zone to maximize the overlap integral and, in turn, the mode-coupling efficiency. To make this maximization successful, the mode eigenequations for both the fiber and the waveguide were solved numerically to obtain a proper theoretical

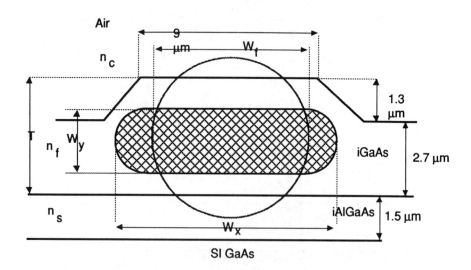

Figure 5.33 GaAs Single-mode ridge channel waveguide cross section with vertical thickness T-4 μm, including: cover with refractive index, n_c (Air); film with refractive index, n_f, (intrinsic (i) GaAs); substrate with refractive index, n_s (i AlGaAs); and under substrate (semi-insulating (SI) GaAs). The channel waveguide single-mode spot area (hatched) computed on the basis of eigenfunction analysis has horizontal size, $W_x = 10$ μm, and vertical size, $W_y = 3$ μm. The single mode fiber beam spot near-field area with size $W_f = 8$ μm (representing step index single-mode 8/125 μm fiber) is also included. This cross section is a fragment of channel waveguide linear array with 50 μm pitch.

match. Then, the diffusion process for waveguide fabrication was tuned to fit with the theoretical predictions (Fig. 5.34) for a z-plate cut.

According to Figure 5.34, the optimum temperature is about 1,000°C. After minimizing the Fresnel by matching the adhesive, a coupling loss in a 1-dB range has been reported. The coupling system transforms the light from LDs into a single-mode 8/125 μm fiber array, then into a single-mode channel waveguide array, then again to a 8/125 μm fiber array, and finally into a PD array. This is illustrated in Figure 5.35. The cross section of the GaAs ridge waveguide being used in this coupling system is illustrated in Figure 5.36. Using the data from Figure 5.33, a moderate 3-dB coupling loss between single-mode fiber and waveguide can be achieved as illustrated in Figure 5.32.

5.3.3 Other Transversal Coupling Schemes

Other interesting transversal coupling schemes include waveguide-to-waveguide transversal coupling, direct LD-to-waveguide coupling, and waveguide-to-PD transversal coupling. All of them can be analyzed using the design rules presented in the previous section. Direct LD-to-waveguide coupling is preferable if it is feasible to have direct access to LD fabrication technology (see Fig. 5.36). This problem is critical, especially for the single-mode case. Otherwise, we need to rely on off-the-shelf LDs, which have lasing area sizes that are not adjusted to this coupling scheme. Waveguide-to-PD transversal butt coupling is not a problem because the PD area is typically larger than the waveguide cross section. Another approach is to use indirect PD coupling as in Figure 5.35.

5.4 WAVEGUIDE LENSES

Waveguide lenses are one of the most important integrated optics components because they can perform various essential functions such as focusing, expanding, imaging, and planar waveguide Fourier transforms [43]. Thus, the search for various possibilities and structures began long ago in the infancy of integrated optics. A few methods of fabricating waveguide lenses have been developed. One way is to build a two-dimensional analog of the classical optic lens by changing the index of refraction of a lens-shaped area as shown in Figure 5.37, Luneberg and Geodesic lenses, within a uniform planar waveguide. For example, this can be done either by diffusion (e.g., Ti:indiffusion in $LiNbO_3$) or direct deposit of a higher index material directly on top of the planar waveguide [44]. However, these methods involve numerous defects such as scattering and mode conversion at the lens boundary, large aberration, and relatively large F-numbers. Although Luneberg lenses can solve the problems of scattering mode conversion and aberration by graded index distribution and the rotational symmetry design, small F-numbers are still difficult to obtain because the change in the effective index of the waveguide is rather small. In addition, a precision fabrication procedure is needed due to the sensitivity of the focal length by the variation of the deposited high-index film. Geodesic lenses can be used to

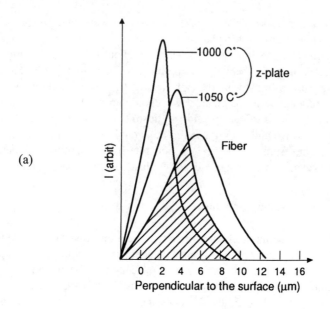

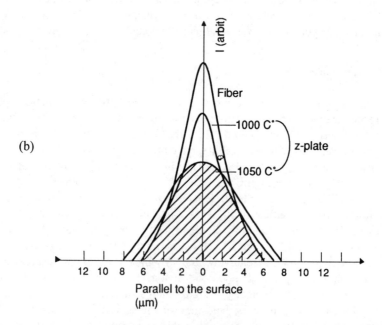

Figure 5.34 Overlap integral for 1050°C diffusion temperature for a single-mode coupling. Between 5-μm diameter fiber and LiNbO$_3$ channel waveguide with width 2 dy = 6.4 μm and depth dz = 3.6 μm for hybrid mode and z-plate cut: (a) perpendicular to the surface; (b) parallel to the surface.

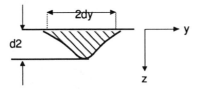

Figure 5.34(c) Waveguide cross section.

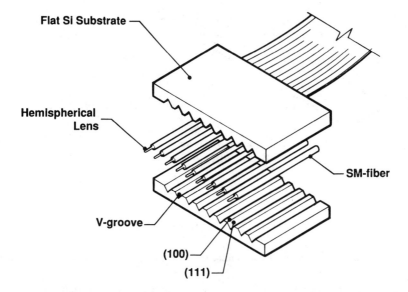

Figure 5.35(a) Fiber/waveguide array coupling system, for coupling between fiber ribbon (from both sides) and integrated optical circuit: general view, with opening illustrating single-mode GaAs ridge waveguide (see Figure 5.33) coupled to single-mode lensed fiber.

alleviate this problem. When the planar waveguide is deformed so that a curved surface is formed, the optical rays at different parts of the deformation travel different optical path lengths. The lens function can be realized by appropriately shaping the curved surface. However, this requires a time-consuming precision mechanical grinding process that makes high production rates and cost effectiveness impossible.

Recently, there has been increasing interest in waveguide lenses based on light diffraction mechanisms in periodic structures such as Fresnel lens and grating lenses. They can be fabricated and mass produced by planar fabrication techniques such as holographic interference and photo and E-beam lithographs. At the same time, they offer

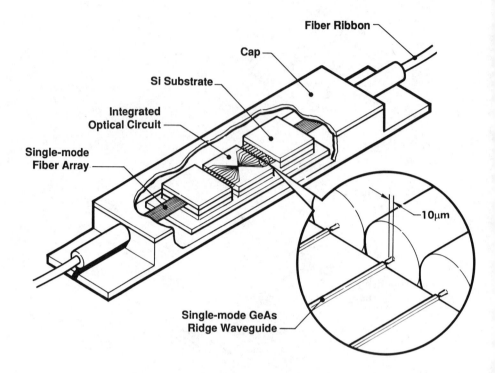

Figure 5.35(b) Fiber/waveguide array coupling system, for coupling between fiber ribbon (from both sides) and integrated optical circuit: the silicon V-groove fiber array.

diffraction limited performance [43]. Thus, they constitute a very important element in integrated optics.

Waveguide Fresnel lenses consist of periodic grating structure that causes a spatial phase difference between the input and the output wavefronts. The grating periodic structure gives a wavefront conversion by spatially modulating the grating. Assuming that the phase distribution functions of the input and output wavefronts are denoted by ϕ_i and ϕ_o, respectively, the difference in the phase in the guided wave structure can be written as:

$$\Delta\phi = \phi_o - \phi_i \tag{5.37}$$

The desired wavefront conversion is achieved by given a phase modulation equal to $\Delta\phi$ to the input wavefront. The grating for such phase modulation consists of grating lines described by:

$$\Delta\phi = 2m\pi \tag{5.38}$$

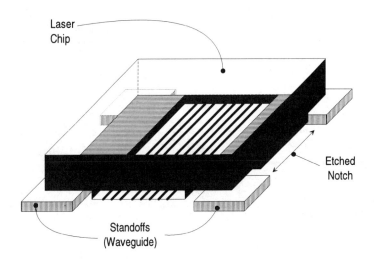

Figure 5.36 Direct LD-waveguide transversal coupling to 50 μm-wide multi-mode channel waveguides.

where m = integer.

The phase difference $\Delta\phi$ for a planar waveguide converging wave follows the expression:

$$\Delta\phi(x) = Kn_{eff}(f - \sqrt{x^2 + f^2}) \tag{5.39}$$

where f is the focal length, n_{eff} is the effective index of the waveguide, and x is the direction of the spatial periodic grating modulation. Figure 5.38 shows two typical waveguide Fresnel lens configurations. Fresnel lenses follow the phase modulation like their three-dimensional counterpart:

$$\phi_F(x) = \Delta\phi(x) + 2m\pi \tag{5.40}$$

for $x_m < |x| < x_{m+1}$, $\Delta\phi(x_m) = -2m\pi$ which is obtained by segmenting the modulation into Fresnel zones so that ϕ_F has amplitude 2π.

Under the thin lens approximation, the phase shift is given by $K\Delta nL$. Therefore, the phase of the wavefront for a specific wavelength can be controlled by the variations of Δn and L. If Δn is varied as a function of x direction, as shown in Figure 5.38(a), it is called the GRIN Fresnel lens and is described by:

$$\Delta n(x) = \Delta n_{max}(\phi_F(x)/2\pi + 1) \tag{5.41}$$

where the lens thickness, L, is held constant. Figure 5.38(b) shows the configuration of the gradient-thickness (GRTH) Fresnel lens. The thickness of the lens has the following functional form:

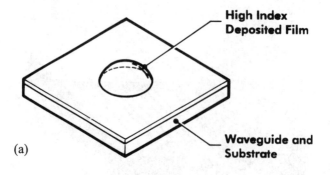

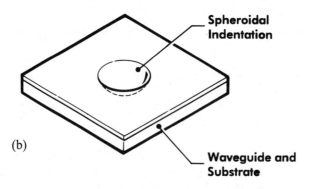

Figure 5.37 (a) Luneberg and (b) geodesic planar waveguide lens substrates.

$$L(x) = L_{max}(\phi_F(x)/2\pi + 1) \tag{5.42}$$

where Δn is held constant.

To have the 2π phase modulation, in either case, the modulation amplitude must be optimized so that $K\Delta n_{max}L = 2\pi$ or $K\Delta nL_{max} = 2\pi$ is satisfied. The binary approximation of the phase modulation results in the step-index Fresnel zone lens. The maximum efficiency of 90%, limited only by diffraction, can be obtained in GRIN and GRTH lenses, while the maximum efficiency of an step-index Fresnel zone lens is only 35%.

Another type of waveguide lens has been designed by spatially changing the K-vector as a function of distance to the central axis using a chirped Bragg grating configuration. The architecture of the chirped Bragg grating waveguide lens is shown in Figure 5.39.

The chirped Bragg grating lenses give the index modulation:

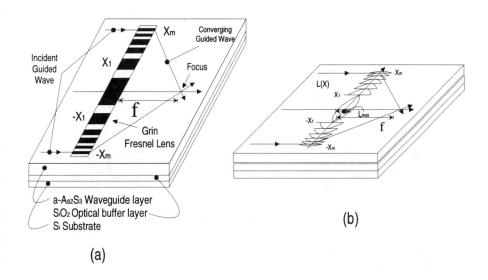

Figure 5.38 Waveguide fresnel lenses: (a) graded-index (GRIN) Fresnel lens; (b) gradient-thickness (GRTH) Fresnel lens.

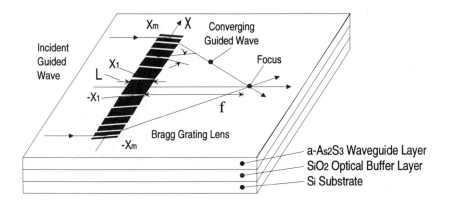

Figure 5.39 Configuration for waveguide chirped Bragg grating lens.

$$\Delta n(x) = \Delta n \, \cos[\Delta \phi(x)] = \Delta n \, \cos\{Kn_e[Kn_e(f - \sqrt{x^2 + f^2})]\} \qquad (5.43)$$

to the mode index on the lens aperture. As required by any device based on grating deflection, the Q parameter needs to be greater than 10 to reach the Bragg region in order to have high efficiency. The grating lines need to be gradiently slanted following the expression:

$$\Psi(x) = \frac{1}{2}\tan^{-1}(x/f) \cong x/2f \qquad (5.44)$$

so that the Bragg condition is satisfied over the entire aperture. The condition for maximum efficiency is:

$$\kappa L = \pi \Delta n L/\lambda = \pi/2 \qquad (5.45)$$

The on-axis focusing chirped Bragg grating waveguide lens configuration is actually difficult to make due to the fact that the on-axis grating cannot satisfied high-efficiency condition. Therefore, the off-axis chirped Bragg grating waveguide lens is usually used for high-efficiency waveguide focusing. Lenses of 1- to 3-μm aperture and F/3 to 10 have been fabricated in glass waveguides by etching and cladding, in $As_2S_3/SiO_2/Si$ waveguides by direct E-beam writing technique, and in waveguides by etching, cladding, and proton-exchange techniques. Nearly diffraction-limited focusing and efficiencies of 50% to 90% have been obtained.

By combining the various possibilities of waveguide phase front conversion with free-space radiation mode-coupling of gratings, different devices can be fabricated using the combination. For example, Figure 5.40 shows a grating pattern that converts a guided wave into a focused free-space radiation [45].

The grating pattern of such a focusing grating coupler can be specified as adding another dimensional in the y direction. From this case follows:

$$\Delta\phi(x,y) = (2\pi/\lambda)[n_e y + \sqrt{x^2 + (y - f\sin\theta)^2 + (f\cos\theta)^2}] = 2m\pi + \text{const} \qquad (5.46)$$

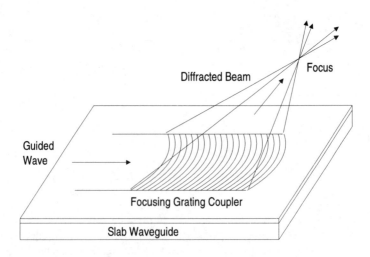

Figure 5.40 Focusing grating coupler configuration and focus spot and the intensity profile.

where, with reference figure 5.40, $n_e \equiv$ waveguide refractive index, $f \equiv$ focal length of grating coupler, $\theta \equiv$ off-orthogonal angle of propagating beam, and $\phi \equiv$ phase difference. The radiation decay factor should satisfy the condition to realize high-efficiency and the minimum focus spot width simultaneously. Other examples of the application of output coupling grating are shown in Figure 5.41 [46].

A wavelength demultiplexing application is realized by using the wavelength selectivity of the grating coupler. By cascading various periodic gratings as shown in Figure 5.42(a) or chirped grating, shown in Figure 5.42(b), the incoming guided wave with different wavelengths are demultiplexed. The demultiplexing need not be only restricted to free-space mode. A series of in-plane Bragg diffraction gratings can also be cascaded as shown in Figure 5.42. The system approaches the wafer-scale integration in which the demultiplexing function and detector arrays for the various wavelength are monolithically integrated on a single wafer [47].

5.5 ACTIVE GUIDED WAVE DEVICES

In an integrated optic system, it is desirable to control the properties of the guided wave. This is done by applying an external field to the functional devices. The external field may change the absorption, phase, or mode characteristics of the guided wave. This enables the external signal to be encoded in the guided wave and transmitted to the desired destination by switching and subsequently processing. It is impossible to cover all the active devices as the mechanisms range from electro-optic [48], acousto-optic [49], magneto-optic [37], to thermo-optic [50]. Thus, only representative device structures will be given.

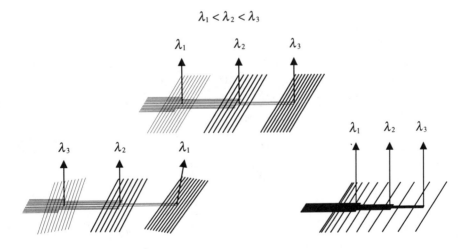

Figure 5.41 Wavelength demultiplexer using wavelength selectivity of grating coupler: migrating array; chirped grating.

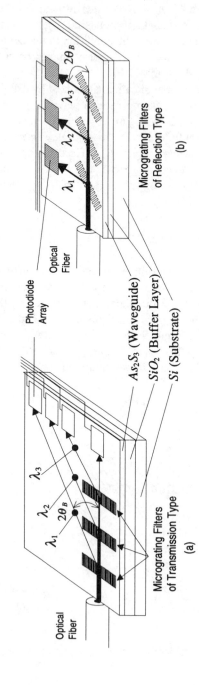

Figure 5.42 Wafer scale integration of WDM receiver terminal using: (a) transmission gratings (b) reflection grating types.

5.5.1 Electro-Optic Waveguide Devices

Several material systems are important in integrated optic devices. LiNbO$_3$ has become very useful for fabricating integrated optic devices because of its large electro-optic coefficient and its ability to be easily fabricated into low-loss optical waveguides [51,52]. The major disadvantage of LiNbO$_3$ is its inability to be integrated with semiconductor diode lasers and detectors to form a total integrated system. However, there have been some initial successes at developing LiNbO$_3$-based waveguide amplifiers and lasers by doping rare-earth ions into its matrix [29]. One competing material technology is a semiconductor-based integrated optic system. Although its electro-optic effect is smaller than LiNbO$_3$'s, it is possible to integrate the semiconductor-based material with semiconductor lasers, detectors, and other high-speed devices such as tunnel diodes [53] and heterojunction bipolar transistors (HBT) [54]. In addition, the semiconductor-based material offers more flexibility in devices using the effect of free-carrier induced changes in the index of refraction and electro absorption in multiquantum well (MQW) structures [55]. These effects have been proved to be larger than electro-optic effects so that devices of small sizes are possible. Therefore, a higher device density in the chip is promising. Polymer [57,58] and silicon-gel [59] systems have recently become serious contenders in the integrated optic arena. Polymer benefits from its inexpensiveness, ease of fabrication, and higher speed operation. However, the temperature and time stability of the EO effect still need further development. The general approach in the design of EO devices is similar in all material systems. The basic EO devices will be emphasized instead of the device structures. Figure 5.43 represents the four basic types of EO devices: phase, mode control, directional coupling, and cutoff types.

5.5.1.1 Phase Modulators

Using the electro-optic effect, it is possible to control the phase retardation of optical waves by means of an electric field without any reduction in transparency to light. Voltage applied over a pair of electrodes placed alongside or over the channel area creates an internal electric field that modifies the refractive index of the waveguide experienced by the propagating optical wave. The applied field direction depends on the crystal orientation and electrode placement. A typical electrode configuration for Ti:LiNbO$_3$ waveguide is shown in Figure 5.44. For the electrode placed on either side of the waveguide, a horizontal electric field is used to modulate the refractive index as shown in Figure 5.44(a). When a vertical electric field is used, one of the electrodes is placed on top of the waveguide. Care must be taken to ensure the minimization of such an arrangement. Usually, a buffer layer is used to isolate the metallic electrode from the optical waveguide, especially in the case of a TM (vertical) polarized wave. The orientation of the crystal is chosen so that the largest electro-optic coefficient is with the applied electric field. For LiNbO$_3$, Γ_{33} is the largest coefficient and should be parallel to the electric field. Therefore, for the

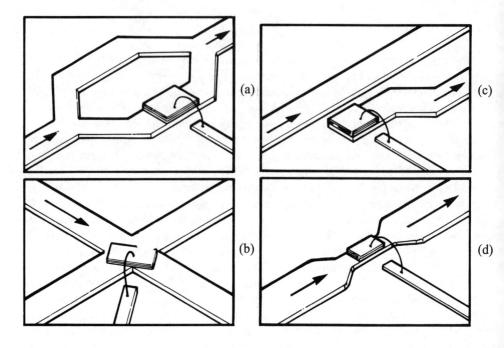

Figure 5.43 Four basic types of EO devices: (a) phase modulation via Mach-Zehnder; (b) mode control of x-switch; (c) directional coupling modulator; and (d) cut-off modulator.

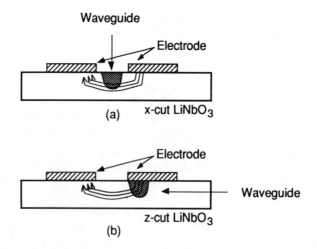

Figure 5.44 A typical electrode configuration for Ti:LiNbO$_3$ waveguides: (a) with an x-cut crystal (b) with a z-cut crystal.

electrode structure in Figure 5.44(a) and TE mode operation, an x-cut crystal is used. A z-cut crystal is arranged in the Figure 5.44(b) structure for the TM mode configuration.

There are two types of modulating electrode: lumped and traveling wave types. When the length of the electrode is short compared to the RF modulating wavelength, the electrode can be considered as a lumped electrode structure. The modulation bandwidth of the lumped electrode modulator is either limited by the optical (or electrical) transit time or the RC time constant of the structure, whichever is smaller. The capacitance per unit length of an electrode can be solved by conformal mapping techniques and is expressed by [48]:

$$C = \epsilon_{eff} \frac{K'(r_s)}{K(r_s)} \quad (5.47)$$

where $r_s = (G+1)/2w$ and $\epsilon_{eff} = (\epsilon_o/2)(1 + \epsilon_s/\epsilon_o)$, w is the electrode width, ϵ_s is the substrate dielectric permittivity, K is the complete elliptical integral of the first kind, and $K'(r_s) = K(\sqrt{1 - r_s^2})$.

The 3-dB bandwidth of a lumped electrode modulator can be approximated (neglecting inductance) by:

$$\Delta f_o \cong \frac{1}{2\pi R_s CL} \quad (5.48)$$

where C is the capacitance per unit length, L is the electrode length, and R_s is the total shunt resistance seen by the capacitor. When the applied RF field changes appreciably during the transit time of the optical wave, the electrical transit time bandwidth is:

$$\Delta f_t = \frac{c}{\pi \sqrt{\epsilon_{eff}} L} \quad (5.49)$$

where c is the speed of light, ϵ_{eff} is the effective permittivity of the RF modulating signal, and L is the length of the electrode. The bandwidth length product of such an electrode structure is in the order of 1 to 2.

To circumvent the electrode charging time limitation on the speed of the modulation, a traveling wave electrode structure can be used. The 3-dB bandwidth of the traveling wave electrode can be expressed as [60]:

$$\Delta f_{t.w.} = \frac{1.4c}{\pi |n_o - n_m| L} \quad (5.50)$$

where n_o is the optical refractive index, n_m is the RF refractive index, and L is the length of the electrode. The bandwidth of the modulator is inversely proportional to the different indices of refraction between the optical and the microwaves, rather than limiting by the RC constant and transit time effects. Traveling wave electrodes can be considered to be microwave transmission lines that have a characteristic impedance of the source, cable, and termination for formation and reflection of minimal standing wave. The modulation principle can be easily visualized. The optical wave and microwave are traveling at (ideally) the same speed so that the traveling optical wave sees the same electric traveling field. Due to the intrinsic material dispersion of refractive index, the phase between the microwave and the optical wave is changed. This results in the reduction of the modulation efficiency and a complete phase cancellation. For $LiNbO_3$, the bandwidth length product is limited to roughly 9 GHz-cm due to the large dispersion of the refractive index [61]. For GaAs, the bandwidth length product is greater than 60 GHz-cm [62]. The small dispersion of polymer that provides a bandwidth length of 150 GHz-cm [63] resulted in the recent achievement of a high-frequency EO polymer modulator with a 3-dB bandwidth greater than 40 GHz [64].

Figure 5.45 gives some of the commonly used traveling wave electrode structures such as microstrip, coplanar, coplanar strip, and asymmetric coplanar strip transmission lines [65,66]. All these transmission lines, with the exception of the coplanar strip line, are easily designed to have a characteristic impedance of 50Ω. A transmission line that

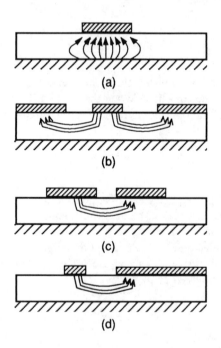

Figure 5.45 Commonly used traveling wave electrode structures (a) microstrip; (b) coplanar waveguide; (c) coplanar strip; (d) asymmetric coplanar strip.

does not have the conventional 50Ω design will require the same impedance and will also suffer some reflection loss due to the transition from the RF launcher to the transmission line caused by the impedance mismatch.

A simple, high-frequency phase modulator is shown in Figure 5.46. The physical operating principle can be easily understood by the following visualization: Assuming the optical wave input has a sinusoidal function, the RF electric field increases and decreases the index of refraction as a function of RF phase. In other words, it could advance and retard the speed of the traveling optical wave in the waveguide. At the output of the optical waveguide, one sees more and less optical cycles in the same period of time. This translates into different optical frequencies. This is a fundamental characteristic of a phase modulator. It can distribute energy into the sidebands $\omega \pm n\omega_{RF}$, where ω is the frequency of the optical wave, ω_{RF} is the RF modulating frequency, and n is an integer.

For high-dispersion materials like LiNbO$_3$ and GaAs, a different electrode structure is required to go beyond the 20 to 30 GHz range. A popular method is to use a periodic electrode structure as shown in Figure 5.47. A periodic phase reversal modulating scheme is used to circumvent the walkoff effect due to microwave and optical wave dispersion [60]. The electrode modulates the optical wave until it is out of phase. Then, a delay line is inserted into the electrode to bring the two waves in phase again. Results as high as 60 GHz were achieved in phase modulators using periodic antenna coupled electrode structure LiNbO$_3$ [67].

Phase modulators are easily turned into intensity modulators by using an interferometer [68]. The intensity modulator can be constructed by combining two y-junction splitters back to back as shown in Figure 5.48. By applying a voltage to the electrode, the phase shift of one arm can be made to be 180 degrees out of the phase with the other arm. When those two out-of-phase waves are recombined at the output junction, the second order mode with antisymmetric structure is excited. Due to the single-mode design of the

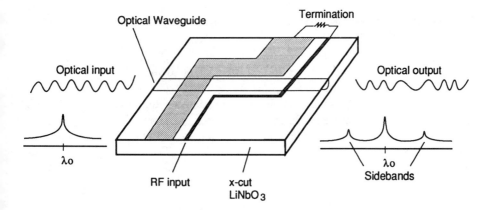

Figure 5.46 A simple high-frequency phase modulator.

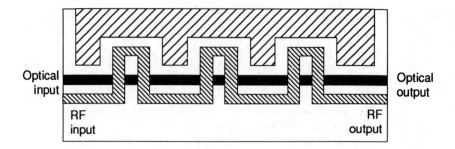

Figure 5.47 A periodic electrode structure.

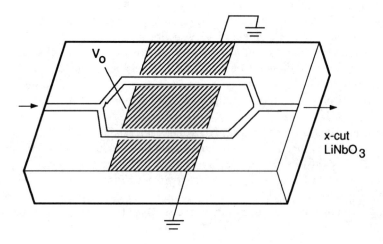

Figure 5.48 Structure corresponding to a Mach-Zehnder push/pull interferometer modulator that provides more-efficient modulation.

waveguide, the higher order modes are scattered around the substrate, resulting in a minimum intensity transmission along the waveguide. If the phase difference is 0 or a multiple of 360°, the two waves combine in phase at the output y-junction and continue to propagate with undiminished intensity. The structure shown in Figure 5.48, which corresponds to a Mach-Zehnder interferometer modulator operating in a push-pull configuration and provides a more efficient modulation. The efficiency of modulation can be expressed as:

$$\eta = \cos^2\left(\frac{\pi}{2} \cdot \frac{v}{v_\pi}\right) \tag{5.51}$$

where equal power split by the y-junction is assumed. Here, v is the modulating voltage and v_π is the voltage required to cause a π (180 degrees) phase shift.

5.5.1.2 Directional Coupler Switch/Modulator

Optical directional coupler waveguides consist of two waveguides placed very close together for a length L [69]. The overlap of the evanescent fields of the two waveguides causes light energy to exchange between them with a coupling coefficient per unit length κ, where κ is a function of the waveguide dimensions and parameters, the spacing between them, and the wavelength of the light. Complete light transfer is accomplished when the waveguides are fully phase matched, which means $\Delta\beta = \beta_1 - \beta_2 = 0$. In addition, the interaction length has to satisfy the condition $L = (2\nu + 1)l$, where ν is an integer, and l is the coupling length. A switch can be constructed by inducing a phase mismatch $\Delta\beta$ between the waveguides as shown in Figure 5.49(a). The phase mismatch can be induced electrically by fabricating the directional coupler on an electro-optic material. Optical switching of the directional coupler is also possible through the free-carrier induced changes in the index of refraction and through semiconductor electron-hole pair generation. This assumes no free-carrier absorption.

However, it is virtually impossible to construct a directional coupler switch/modulator with sufficient tolerance to eliminate the crosstalk completely. An option is to use an alternating delta b phase reversal directional coupler as shown in Figure 5.49(b). The crossover and the straight through states can be controlled electrically for any length, L, that is longer than the complete crossover length, l. There is always a voltage for the light to cross over completely and to go straight through. It consists of two electrically controlled directional couplers, each having the length $l/2$ connected back-to-back with reverse polarity. The condition required for the complete crossover state can be expressed as:

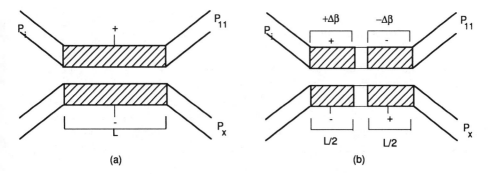

Figure 5.49 Directional coupler modulator/switch: constructed by (a) inducing a phase mismatch, $\Delta\beta$, between the waveguides; (b) using an alternating $\Delta\beta$ phase reversal directional coupler.

$$\frac{(2\kappa)^2}{(2\kappa^2) + (\Delta\beta)^2}\sin^2\frac{1}{4}\sqrt{(2\kappa^2) + (\Delta\beta)^2} = \frac{1}{2} \equiv \sin^2\frac{\pi}{4} \qquad (5.52)$$

And the straight through // conditions are met with the criterion:

$$\left(\frac{L}{l}\right)^2 + \left(\frac{\Delta\beta L}{\pi}\right)^2 = (4\nu)^2 \qquad (5.53)$$

This condition corresponds to concentric circles with radius 4, 8, 12, etc., in the switching diagram (shown in Fig. 5.50(a)). The diagram indicates that the $\Delta\beta$ reversal can achieve a total crossover for a series of L/l values (e.g., $L/l = 1-3$). The diagram also gives the condition for switching that can be controlled electrically.

Consequently, an additional section of alternating $\Delta\beta$ can be added to the directional coupler. The analysis and result are quite similar to the two-section $\Delta\beta$ reversal configuration. However, three or more sections do offer a low-voltage adjustment for the crossover state when the L/l is large. Figure 5.50(b) shows the switching diagram for a two-section alternating $\Delta\beta$ directional coupler. The $\otimes$ sign marks the cross-state conditions, and the $\ominus$ sign marks the bar-state conditions.

The directional coupler can also be used as an optical bandpass filter. By using two waveguides with different $\beta(\beta_a \neq \beta_b)$, the waveguide dispersion causes the phase matching condition ($\beta_a = \beta_b$) to be satisfied only at one particular wavelength. The full width at half-maximum (FWHM) of the optical filter is determined by the degree of asymmetry of the directional coupler. The center wavelength of the bandpass filter can be tuned by a voltage or optical signal depending on the waveguide material, which is either electro-optic or optical active. However, the occurrence of large sidelobes limits the wavelength spacing between filtering bands. The side lobes can be reduced to -25 dB by using a tapered directional coupler in which the spacing between the waveguides are varied [70].

5.5.1.3 Index Distributed Switch/Modulator

Another promising modulator that involves a simple electrode structure is a TIR switch/modulator. The simple structure, as shown in Figure 5.51, uses the electro-optically induced index change. This causes the light from the input Pi to be totally internally reflected into the upper waveguide to form $P//$. When there is no voltage applied, the light passes straight through the junction to the lower waveguide. The value of voltage required for switching can be approximated by [71]:

$$v \geq \frac{d}{n^2\Gamma}\left|\frac{\theta_i}{2}\right|^2 \qquad (5.54)$$

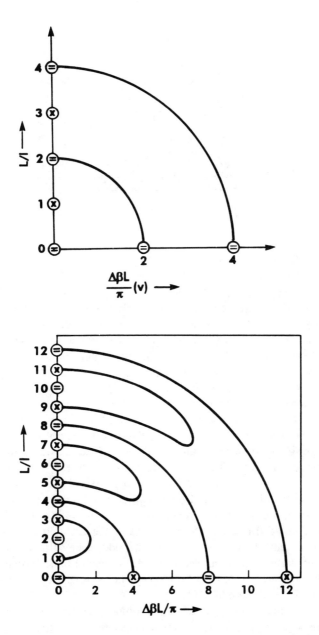

Figure 5.50 Switching diagrams for: (a) switched directional coupler; (b) two section alternating $\Delta\beta$ directional coupler.

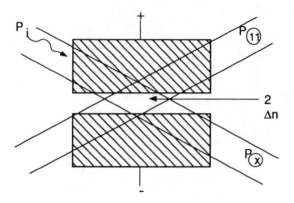

Figure 5.51 A simple total internal reflection (TIR) switch.

where θ_i is the half-angle between the waveguides, d is the gap between electrodes, n is the effective index of the waveguide, and Γ is the electro-optic coefficient. To reduce the switching voltage, a small θ_i is preferable. However, evanescent field coupling between waveguides outside of the electrode region can be undesirable to the switch.

The simple approach of total internal reflection is not always adequate to describe all effects of the switch. A more rigorous approach using the beam propagation method (BPM) [72] has been shown to give very accurate results. Figure 5.52 shows the evolution of optical wave propagating along the x-switch structure. The fundamental mode is excited at one waveguide. This wave then generates a superposition of even (TE_{00}) and odd (TE_{01}) modes in the intersection region. The relative phases at the end of the interaction region determine the power distribution in the output ports. The phases of the mode in the interaction region can be controlled via an electro-optic effect.

Figure 5.53 shows the mode distribution when the voltage is 0V and 25V. This corresponds to the switching of power from the lower waveguide to the upper waveguide. The effect of $2\Delta n$ in the intersection region is to reduce loss because of stronger light confinement. The reduced loss may be explained by the fact that the Δn crossing is a single mode in the intersection region. The excited odd mode suffers more loss because it is closer to the cutoff condition. The principle of electro-optic switching is that the electro-optically induced changes of the propagation constants of the even modes are much larger than those of the odd modes.

The same principle can be applied to a two-mode interference switch/modulator in which a two-moded waveguide section is coupled to single-mode input and output waveguides by y-junctions as shown in Figure 5.54(a). The crossing $\otimes$ and the parallel // switching voltage are shown in Figure 5.54(b).

One point that needs to be emphasized here is that most of the modulation schemes mentioned need not be limited to the electro-optic effect. As it well known in semiconductor

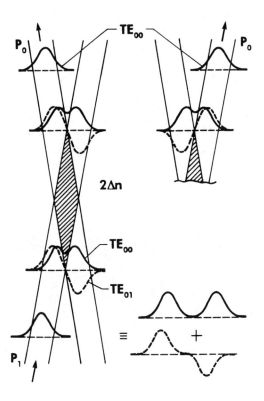

Figure 5.52 Optical field evolution in an x-switch structure [72].

laser diodes, the refractive index change induced by carrier injection is roughly two orders of magnitude greater than that of electro-optic effect [73]. Thus, smaller interaction and higher density devices can be realized. A number of these modulators, including TIR and Mach-Zehnder switch/modulators, have been constructed using semiconductor materials such as Si [74], GaAs [54], and InP [55], with current and optical injection methods.

Another class of modulator can be built by fabricating the waveguide with a lateral dimension just below the cutoff condition of the fundamental mode [75]. A pair of electrodes is placed properly depending on the cut of the crystal. This increases the refractive index of the waveguide when an electric field is applied. Instead of scattering into the substrate, the waveguide now supports the fundamental mode of the incoming wave. Due to the small dimensions of the waveguide, a very high density cutoff modulator can be constructed. However, the disadvantages of the cutoff modulators are inherently high insertion loss due to the dimension proximity to the cutoff condition and the crosstalk between waveguide channels caused by the scattered light into the substrate.

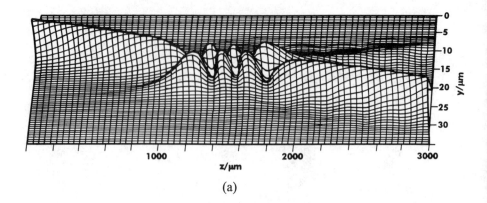

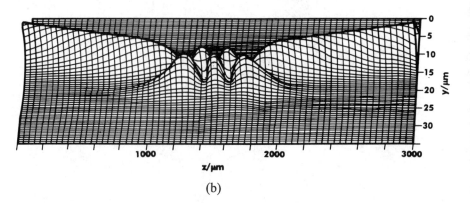

Figure 5.53 Calculated optical field distributions in the x-switch (angle = 0.6°) for two different modulation voltages: (a) $v = 0$V; (b) $v = 25$V [72].

Up to this point, all the modulators and switches have used the on-diagonal elements of the electro-optic tensor. The off-diagonal element can also be used, mainly for TE/TM mode conversion [76]. By integrating a metallic overlay waveguide polarizer, the TE/TM mode converter can be easily turned into an intensity modulator. Furthermore, TE/TM mode conversion is a very strong function of the light wavelength for highly birefringent crystals such as LiNbO$_3$. Thus, it can be used as a wavelength multiplexing/demultiplexing element. A schematic of such a mode converter is shown in Figure 5.55.

In the case of LiNbO$_3$, the off-diagonal element $\Gamma_{51} = 28$ *pm/v* is used by choosing a z-cut x-propagating orientation of the crystal. The TE/TM mode-coupling coefficient is given by:

$$\kappa = \Gamma\left(\frac{\pi}{2\lambda}\right) n_s^3 \Gamma_{51}\left(\frac{v}{d}\right) \tag{5.55}$$

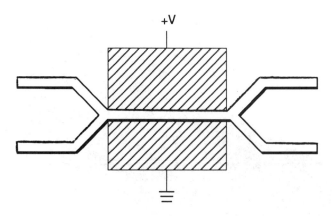

Figure 5.54(a) Structure for two-mode interference switch.

where v is the applied voltage, d is the interdigit electrode gap, n is the substrate refractive index, λ is the wavelength, and Γ is the overlap integral of the electric field and the TE and TM mode field. Period Λ of the electrode needs to satisfy the phase matching condition.

$$\frac{2\pi}{\Lambda}|N_{TE} - N_{TM}| = \frac{2\pi}{\lambda} \tag{5.56}$$

where N_{TE} and N_{TM} are the refractive indices of TE and TM mode respectively. The conversion efficiency is given by:

$$\eta_{TE/TM} = \frac{\sin^2\{\kappa L[1 + (\delta/\kappa)^2]^{1/2}\}}{1 + (\delta/\kappa)^2} \tag{5.57}$$

where $\delta = (1/2)\Delta\beta$ and L is the interaction length. Complete conversion occurs at $\delta = 0$ and $\kappa L = (1/2)n\pi$, n being an odd integer. The approximate FWHM of the mode converter is:

$$\frac{\Delta\lambda}{\lambda} \cong \frac{\Lambda}{L} = \frac{1}{N} \tag{5.58}$$

where N is the number of electrode periods. A typical wavelength response of the mode converter is shown in Figure 5.56.

5.5.1.4 Electroabsorption Modulator

III-V semiconductors such as GaAs are very versatile material systems for integrated optics. Many active integrated-optic devices using the conventional EO coefficient have

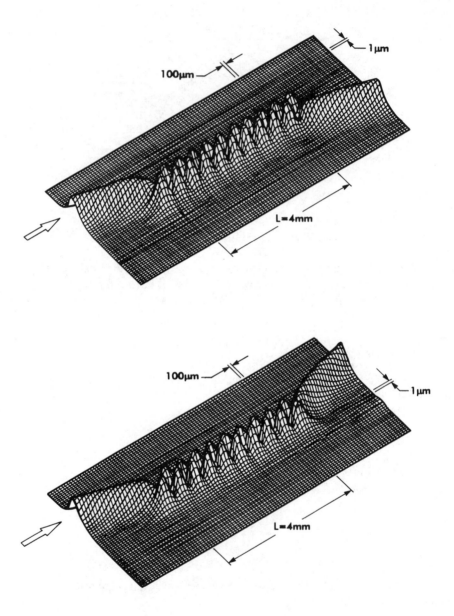

Figure 5.54(b) Beam propagation method simulation for the switching at on (top) and off (bottom) voltages [72].

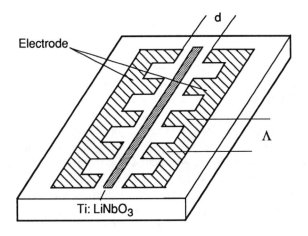

Figure 5.55 Structure of a mode converter.

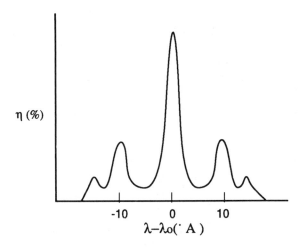

Figure 5.56 Typical characteristic of a mode converter.

been realized [77–79]. However, another form of modulation can be obtained by varying the absorption coefficient electrically. The most promising technology involves the quantum-level effects in semiconductors. With the advancement in the material growth technology by methods of molecular beam epitaxy and metal organic chemical vapor deposition, layers of semiconductors such as GaAs, InP and their alloys can be controlled down to a single atomic layer, which is in the order of a few angstroms. By sandwiching a low bandgap material such as GaAs between two layers of high bandgap material such as AlGaAs, a quantum well in the macroscopic level is formed. Electrons and holes in the material tend to become confined in the region of the low bandgap material where the potential energy is low. Light is absorbed when the photon creates an electron-hole pair bound together in an excitonic state. Figure 5.57(a) shows the energy band diagram of a quantum well where the exciton energy is the sum of the GaAs bandgap and the zero point energies of the electron and the hole, with a correction of the exciton binding energy. When an electric field is applied across the quantum well, the well is tilted, and the potential energy seen by the electron and the hole becomes what is shown in Figure 5.57(b). The reduced exciton energy gap increases the absorption coefficient for light that is just below the unbiased energy gap. By switching on and off an electric field, the transparency of the quantum well can be modulated at that particular wavelength of light. This is called the quantum-confined Stark effect (QCSE). This electroabsorption effect is approximately 50 times larger than that of bulk semiconductors. A typical quantum well waveguide modulator has the structure shown in Figure 5.58 [56]. TE and TM waves have different absorption spectra in the function of photon energy and wavelength as shown in Figure 5.59. The on/off ratio, R, of the waveguide modulator is given by:

$$R = e^{\Gamma \Delta \alpha L} \tag{5.59}$$

where $\Delta \alpha$ is the change in the absorption coefficient, L is the interaction length, and Γ is the overlap between the quantum wells and the optical mode of the waveguide. Typical values of Γ are on the order of 2–10. The quantum well material need not be limited to the GaAs/AlGaAs system that operates in the neighborhood of 0.85 μm. For fiber communication, longer wavelength operation at 1.3 and 1.55 μm is desirable due to minimum propagation loss and dispersion at these wavelengths. There are presently three promising material systems in this wavelength range: GaInAs/AlInAs, GaInAs/InP, and

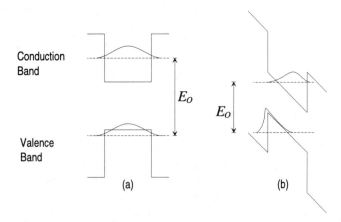

Figure 5.57 The electroabsorption effect in a multiquantum well: (a) $\epsilon = 0$; (b) $\epsilon \neq 0$.

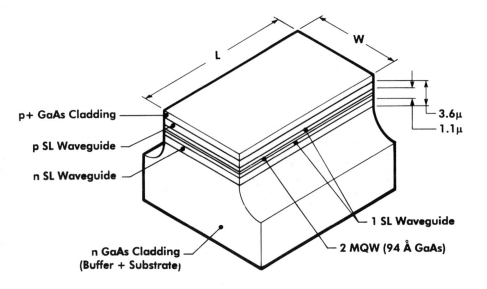

Figure 5.58 A typical quantum well waveguide structure.

GaSb/AlGaSb. GaInAs/AlInAs and GaInAs/InP waveguide devices can be grown on an InP substrate which is conveniently transparent to the operating wavelength. GaInAs/InP has been widely used in lasers and detectors. Thus, integration of the waveguide modulators with these components to form a total integrated system can be achieved.

5.5.2 Acousto-Optic Modulators

Various methods exist in optical guided wave control besides the electro-optic effects. One important modulation scheme is to use the acoustic wave to modify the refractive index characteristics of the guiding material. This is called the acousto-optic effect. However, the speed of these devices cannot be compared to the ultrafast electronic response of the electro-optic devices. On the other hand, acousto-optic devices do offer some unique operations, such as real-time spectral analysis of wide band RF signals, correlation, convolution, pulse compression, and matched filtering of RF signals [49,80].

A basic acousto-optic Bragg modulator configuration is shown in Figure 5.60. An interdigital finger electrode array is used to induce the surface acoustic wave (SAW). As the light enters the SAW region, it experiences a higher (lower) value of the refractive index at the maximum compression (rarefaction) to create an index grating. As far as the light is concerned, it sees a stationary variation in the index of refraction due to the slow speed of the SAW. The incident light is therefore Bragg scattered into different diffraction orders. Y-cut LiNbO$_3$ is commonly used as SAW modulator material due to its strong

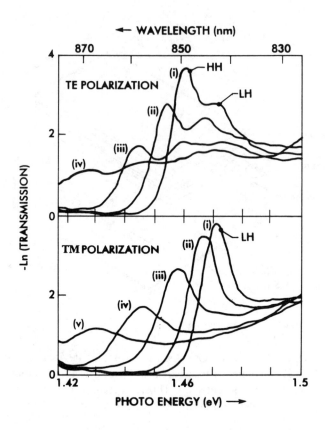

Figure 5.59 TE and TM absorption spectra as a function of photon energy and wavelength.

piezoelectric characteristics that easily excite SAW and its maturity in optical waveguide technology.

The diffraction of the optical wave by the SAW can be either the Raman-Nath type or the Bragg type. The acousto-optic parameter, Q, determines the working region of the device [49].

$$Q \equiv \frac{2\pi\lambda_o L}{n\Lambda^2} \qquad (5.60)$$

When $Q \leq 0.3$, the device is working as a Raman-Nath type, which consists of several side orders of diffraction. As for the case of $Q \geq 4\pi$, the device enters the Bragg region, which consists of mainly one order diffraction. The Bragg type diffraction is considered in most SAW devices because of its larger modulation bandwidth and dynamic range. The incidence and diffraction angles with respect to the SAW wavefront, θ_m and θ_n are:

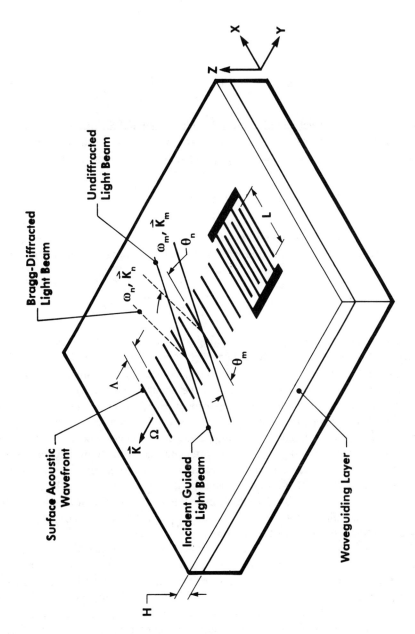

Figure 5.60 A basic acousto-optic Bragg modulator structure.

$$\sin \theta_m = \frac{\lambda_o}{2n_m\Lambda}\left[1 + \frac{\Lambda}{\lambda_o^2}(n_m^2 - n_n^2)\right] \qquad (5.61)$$

and

$$\sin \theta_n = \frac{\lambda_o}{2n_n\Lambda}\left[1 - \frac{\Lambda}{\lambda_o^2}(n_m^2 - n_n^2)\right] \qquad (5.62)$$

where n_m and n_n are the effective indices of the undiffracted and diffracted light, λ_o is the optical wavelength in free space, and Λ is the SAW wavelength. In the case of $n_n = n_m$, the undiffracted and the diffracted lights are propagating in the same mode.

$$\sin \theta_n = \sin \theta_m = \frac{\lambda_o}{2n_m\Lambda} \qquad (5.63)$$

which is the well-known Bragg condition.

The efficiency for light diffraction via SAW under the Bragg condition is given by:

$$\eta = \sin^2\left\{\left(\frac{\pi}{\sqrt{2}\lambda_o}\right)M_{2mn}^{1/2}C_{mn}(f)\left(\frac{L}{\cos \theta_m \cos \theta_n}\right)^{1/2}P_a^{1/2}\right\} \qquad (5.64)$$

where $C_{mn}(f)$ is the frequency-dependent coupling coefficient, which relies on the waveguide materials and dimensions, P_a is the total power flow of the SAW, and M_{2mn} is defined as $(n_m^3 n_n^3 P^2/\rho V^3)$, where P is the relevant photoelastic constant, ρ, V designates the density and the acoustic propagation velocity, and L is the acoustic aperture. For constant frequency, the efficiency is approximately proportional to P_a in the low-efficiency region ($\eta \leq 0.5$). The SAW power is a product of the RF driving power and the conversion efficiency (maximum of -3 dB) of the interdigital finger transducer. The bandwidth of the SAW modulator is determined by the frequency dependence of the transducer conversion efficiency, the SAW confinement, and the phase-matching requirement. The general conclusion is that the diffraction efficiency and the bandwidth product of the modulator are rather limited. Thus, a large bandwidth is obtainable at a reduced diffraction efficiency and vice versa. In addition, both the total acoustic requirement and RF drive powers are inversely proportional to the acoustic aperture.

It is desirable to construct a SAW modulator with a large oandwidth and a high diffraction efficiency. There are currently five popular schemes for wide-bandwidth modulators with different interaction and transducer combinations: isotropic Bragg diffraction with a multiple tilted transducers of staggered center frequency; isotropic Bragg diffraction with a phased-array transducer; isotropic Bragg diffraction with a phased-array transducer that combines the tilted transducers with staggered center frequency; isotropic Bragg diffraction with a tilted-finger chirp transducer; and optimized anisotropic Bragg diffraction with multiple transducers of staggered center frequency or a parallel-finger chirp transducer. Detailed descriptions of each configuration can be found in [49].

The transducer configurations are shown in Figures 5.61 through 5.64 [49]. Figure 5.61(a) shows the multiple tilted array transducer with staggered center frequencies. They are tilted in angle to satisfy the Bragg conditions at each center frequency to enable wide frequency operation. A 680 MHz bandwidth has been achieved using a four-stage array with center frequencies of 380, 520, 703, and 950 MHz [49]. A curved transducer, as shown in Figure 5.61(b), can be constructed if the finger electrodes of a large array of small aperture tilted transducers are joined side-by-side to form a single stage wide bandwidth transducer.

However, the impedance matching is difficult and the RF power loss is larger due to electrode resistance. Isotropic Bragg diffraction with phased-array transducer is shown in Figure 5.62. The wide bandwidth characteristic involves a frequency-controlled SAW beam steering by arranging a series of isotropic transducers in a stepped configuration. The steps introduce a variable phase shift between adjacent SAWs, resulting in the scanning of the acoustic wavefront. The SAW beam steering property enables the resultant SAW

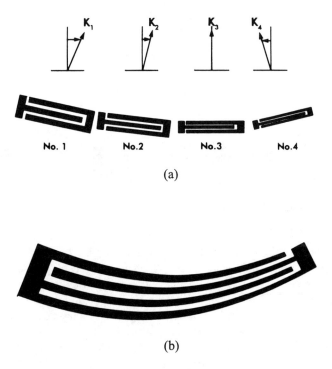

Figure 5.61 (a) Multiple tilted SAW transducers of staggered center frequency; (b) curved transducer evolved from a large number of tilted transducers of staggered center frequency.

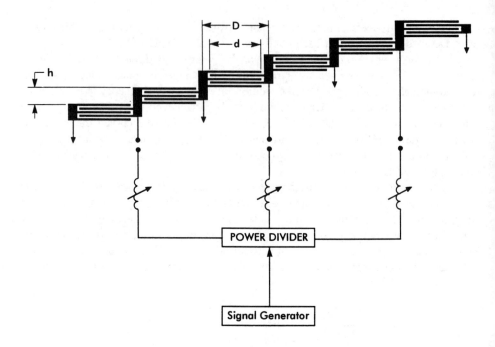

Figure 5.62 Phased-array transducer with matching and driver circuits.

Figure 5.63 Multiple tilted phased-array transducers.

with wide acoustic aperture to satisfy the Bragg condition for a relatively wide frequency range. By integrating the tilted transducer (which is capable of providing very large bandwidth at reasonable diffraction efficiency with the phased-array transducer), very large device bandwidth and very high diffraction efficiency can be achieved as shown in Figure 5.63. However, this transducer configuration involves a high degree of complexity in transducer design, fabrication, and RF driving circuits.

Figure 5.64 shows another wide bandwidth SAW modulator, using the SAW beam steering property. The SAW beam steering is achieved by gradually tilting the finger electrodes. The disadvantage with this configuration is the low SAW excitation efficiency

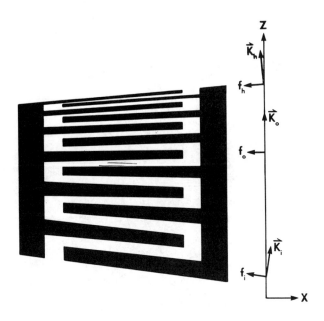

Figure 5.64 A tilted-finger chirp transducer evolved from a large number of tilted transducers of staggered center frequency.

due to impedance mismatch of the transducer. A 1-GHz bandwidth has been achieved with this configuration at center frequencies of 500 and 1,000 MHz.

Large bandwidth can be achieved by using the anisotropic Bragg diffraction. The wave vector of the diffracted light is orthogonal or nearly orthogonal to the SAW with TE/TM mode conversion in an anisotropic crystal. The optimum phase-matching condition can be maintained for a wide range of acoustic frequency without acoustic beam steering. The Bragg bandwidth for optimized anisotropic diffraction is inversely proportional to the square root of the acoustic aperture and is considerably larger than that for isotropic diffraction. Consequently, both high diffraction efficiency and large Bragg bandwidth can be realized simultaneously through the anisotropic phase matching condition as shown in Figure 5.65.

With the advancement in integrated optic devices, such as waveguide lenses, integrated laser sources, and photodetector arrays, system integration using SAW devices on a common substrate is becoming a reality. An example of such an integrated system is shown in Figure 5.66, in which a SAW time-integrating correlator can be constructed using GaAs-based material as opposed to the hybrid configuration. Many unique applications in wide bandwidth multichannel communications and signal processing have been realized using SAW modulators. They constitute a major component in the advancement of integrated optics.

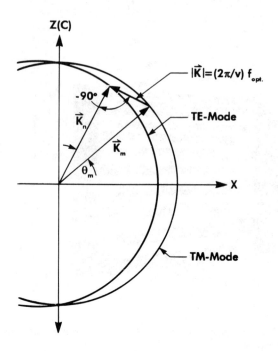

Figure 5.65 Phase-matched diagram indicating wavevectors of incident light, Bragg diffracted light and SAW in a y-cut LiNbO$_3$ waveguide with optimized anisotropic case.

5.5.3 Magneto-Optic Modulators

Similar to acousto-optic waveguide modulators using SAW, magnetostatic waves (MSW) can be used to interact with the guided optical wave for guided wave control. In contrast to the SAW devices, MSW modulators offer the possibility of a large time-bandwidth product in the tens of GHz of the microwave spectrum [36,37]. However, the diffraction efficiency is much lower than its acousto-optic and electro-optic counterparts.

Guided wave MSW modulators are usually implemented in thin YIG film waveguide on GGG substrate. Like the SAW devices, an electrode transducer is needed to induce the MSW in YIG film. There are three types of MSWs that can be excited depending on the direction of the biasing dc magnetic field, namely, magnetostatic forward volume waves (MSFVW), magnetostatic backward volume waves (MSBVW), and magnetostatic surface waves (MSSW). Those MSW can be efficiently excited via the electromagnetic field in a microstrip transducer to induce a magnetic perturbation on the YIG film. The wavelengths of the excited MSW on the YIG film with typical transducer frequency of 1 GHz are in the order of μm. Optical interaction with the MSW can be collinear where the wave vectors are in the same direction. It can also be in a noncollinear configuration in which the optical guided waves are propagating in a orthogonal or near-orthogonal

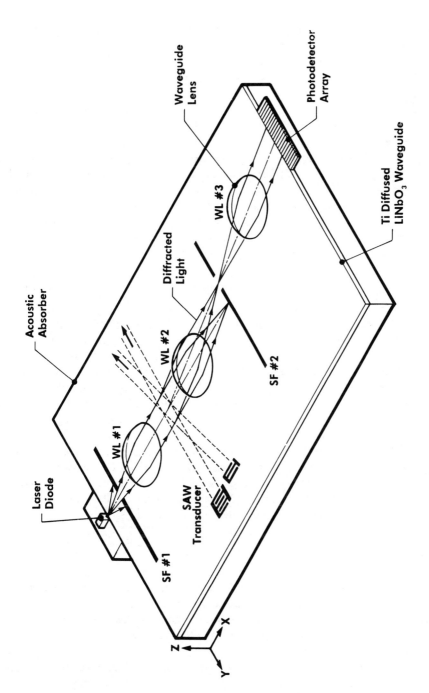

Figure 5.66 An acousto-optic time-integrating correlator using hybrid optical waveguide structure.

direction with the MSW. Similar to the SAW devices, the noncollinear configuration is expected to offer more versatility than the collinear interaction.

In the case of noncollinear interaction in MSW, the RF magnetizations and the dc magnetic field induce changes in the permittivity tensor delta epsilon that has benefited from the Faraday effect and Cotton-Mouton effect. TE_{01} and TM_{01} modes in the waveguide are coupled through the induced changes in the permittivity tensor. As a result, three additional waves are excited in the waveguide, namely the diffracted TM_{01} and TE_{01} modes and the undiffracted TE_{01} mode.

A typical schematic magneto-optic Bragg deflector using MSFVW is shown in Figure 5.67 [49]. An incident guided optical wave is excited in either TM or TE mode propagating along the x-axis. A dc magnetic field in the z-direction and a microstrip transducer are used to excite a y-propagating MSFVW. Through the dynamic Faraday and Cotton-Mouton effect, coupling results are achieved between the incident light in the mode and the diffracted light in the mode, and vice versa. The appropriate phase matching condition can be expressed in the same way as the SAW devices.

$$\beta_{TE_o}^{(d)} = \beta_{TM_o}^{(u)} - K \qquad \omega_d = \omega_u - \Omega \qquad (5.65)$$

where $\beta_{TE_o}^{(d)}$, $\beta_{TM_o}^{(u)}$, and K are the wave vectors of the TM_{01} mode undiffracted light, the TE_{01} mode diffracted light, and the MSFVW, respectively, while ω_d, ω_u, and Ω are the corresponding frequencies. The frequency of the MSFVW can be tuned by varying the dc magnetic field. In the experiment shown in Figure 5.67, the frequency is tuned from 2.0 to 7.0 GHz by varying the magnetic field from 2,000 to 3,500 Oe. Another way of to tune is to vary the frequency of the RF excitation wave and keep the CD magnetic field constant. The diffracted light is spatially separated by the phase matching condition and imaged on an infrared camera.

The applications of the MSW modulators are similar to their SAW device counterparts, but they are working at a higher frequency region (GHz). One drawback is the requirement of a powerful electromagnet to provide the biasing dc field and the tuning of the MSW frequency. This adds bulkiness to the total device, which may cause inconvenience relative to the compact packaging issues.

5.6 CONCLUSIONS

This chapter has described basic principles of integrated optics, including planar and channel waveguides, passive planar devices, and active devices. However, due to space limitations, many important subjects have not been discussed. This chapter provides the readers with some basic understanding of integrated optical waveguide and functional devices (planar micro-optics) fitting in the context of optical interconnects. From the concepts described in this chapter, the reader may apply the same principles to any upcoming new integrated-optic material systems. In addition, this chapter enables the

221

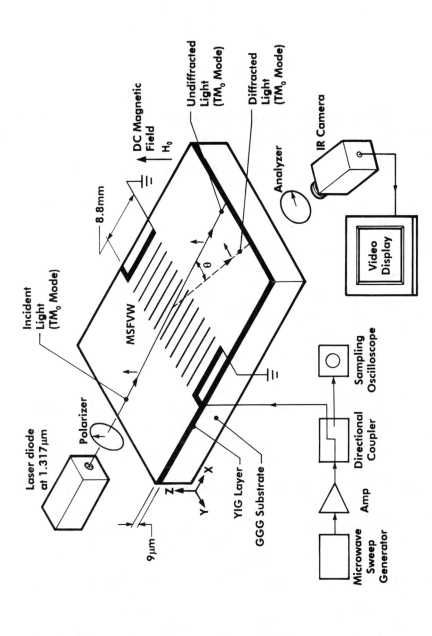

Figure 5.67 Experimental arrangement for scanning of guided-light beam in YIG-GGG waveguide using magnetostatic forward volume waves.

reader to estimate the order of magnitude of competitive effects in integrated optics. This will provide the basis for distinguishing first-order effects from higher order effects and thereby help the user to avoid focusing on the ant while the elephant is missed.

REFERENCES

[1] Taylor, H. F. and A. Yariv, "Guided wave optics," *IEEE Proc.*, Vol. 62, 1974, pp. 1044–1060.
[2] Nishihara, H., M. Haruna, and T. Suhara, *Optical Integrated Circuits*, McGraw-Hill Book Company, New York, 1985.
[3] Marcuse, D., *Light Transmission Optics*, Van Nostrend, 1982.
[4] Tamir, T., *Integrated Optics*, Springer Verlag, 1975.
[5] Marcuse D., *Theory of Dielectric Waveguides*, Academic Press, 1974.
[6] Tien, P. K., "Integrated optics and new wave phenomenon in optical waveguides," *Rev. Mod. Phys.*, Vol. 49, 1977, pp. 361–381.
[7] Kogelnik, H., "An introduction to integrated optics," *IEEE Trans. Microwave Theory Tech.*, Vol. MTT-23, 1975, pp. 2–16,.
[8] Marcatili, E. A. J., "Dielectric rectangular waveguide and directional coupler for integrated optics," *Bell Syst. Tech. J.*, Vol. 48, 1969, pp. 2071–2102.
[9] Marcuse, D., "TE modes of graded-index slab waveguides," *IEEE J. Quantum Electron.*, Vol. QE-9, 1973, pp. 1000–1006.
[10] Shiff, L. I., *Quantum Mechanics*, McGraw-Hill, 1968.
[11] Chen, R. T., M. Wang, and T. Jannson, "Polymer microstructure waveguide on BeO and Al2O3 substrates for optical interconnection," *Appl. Phys. Lett.*, Vol. 56, 1990, pp. 709.
[12] Veselka J. J. and G. A. Bogart, "Low-loss TM-pass polarizer fabricated by proton exchange for Z-cut Ti:LiNbO$_3$ waveguides," *Electron. Lett.*, Vol. 23, 1987, pp. 37–38.
[13] Hocker, G. Benjamin, W. K. Burns, "Modes in diffused optical waveguides of arbitrary index profile," *IEEE J. Quantum Electron.*, Vol. QE-11, No. 6, 1975, pp. 270–276.
[14] Chen, R. T., M. R. Wang, and T. Jannson, "Intraplane guided wave massive fanout optical interconnects," *Appl. Phys. Lett.*, Vol. 57, 1990, p. 2071.
[15] Goell, J. E. and R. D. Standley, "Sputtered glass waveguide for integrated optical circuits," *Bell Syst. Tech. J.*, Vol. 48, 1969, pp. 3445–3448.
[16] Koren, U., B. I. Miller, T. L. Koch, G. D. Boyd, R. J. Capik, and C. E. Soccolich, "Low loss InGaAs/InP multiple quantum well waveguides," *Appl. Phys. Lett.*, Vol. 49, No. 23, 1986, pp. 1602–1604.
[17] Ramaswamy, R. V. and R. Srivastava, "Ion-exchanged glass waveguides: A review," *J. Lightwave Tech.*, Vol. 6, No. 6, 1988, pp. 984–1002.
[18] Stutius, W. and W. Streifer, "Silicon nitride films on silicon for optical waveguides," *Appl. Optics*, Vol. 16, No. 12, 1977, pp. 3218–3222.
[19] McDonach, A. and M. Copeland, "Polymeric guided wave optics," *SPIE Proc.*, Vol. 1177, 1989, p. 67.
[20] Schmidt, R. V. and I. P. Kaminow, "Metal-diffused optical waveguides in LiNbO$_3$," *Appl. Phys. Lett.*, Vol. 25, 1974, pp.458–460.
[21] Findakly, T. and C. L. Chen, "Diffused optical waveguides with exponential profile: Effects of metal-clad and dielectric overlay," *Appl. Opt.*, Vol. 17, 1978, pp. 469–473.
[22] Chen, B. and A. C. Pastor, "Eliminmation of Li$_2$O out-diffusion waveguide in LiNbO$_3$," *Appl. Phys. Lett.*, Vol. 30, 1977, pp. 570–571.
[23] Jackel, J. L., C. E. Rice, and J. J. Veselka, "Proton exchange for high-index waveguides in LiNbO$_3$," *Appl. Phys. Lett.*, Vol. 47, 1982, pp. 607–608.
[24] Goodwin, M. and C. Stewart, "Proton-exchanged optical waveguides in Y-cut lithium niobate," *Electron. Lett.*, Vol. 19, 1983, pp. 223–225.

[25] Jackel, J. L., V. Ramaswamy, and S. P. Lyman, "Elimination of out-diffused surface guiding in titanium-diffused LiNbO$_3$," *Appl. Phys. Lett.*, Vol. 38, No. 7, 1981, pp. 509–511.
[26] Jackel, J. L., C. E. Rice, and J. J. Veselka, "Proton exchange for high-index waveguides in LiNbO$_3$," *Appl. Phys. Lett.*, Vol. 41, No. 7, 1982, pp. 607–608.
[27] Lilienhof, J. J., E. Voges, D. Ritter, and B. Pantschew, "Field-induced index profiles of multimode ion-exchanged strip waveguides," *IEEE J. Quantum Electron.*, Vol. QE-18, 1982, p. 1877.
[28] Hocker, B. and W. K. Burns, "Mode dispersion in diffused channel waveguides by Effective Index Method," *Appl. Opt.*, Vol. 16, 1977, p. 113.
[29] Gill, D. M., A. Judy, L. McCaughan, and J. C. Wright, "Method for the local incorporation of Er into LiNbO$_3$ guided wave optic devices by Ti co-diffusion," *Appl. Phys. Lett.*, vol. 60, no. 9, 1992, pp. 1067–1069.
[30] Inoue, H., K. Hiruma, K. Ishida, T. Asai, and H. Matsumura, "Low loss GaAs optical waveguides," *J. Lightwave Tech.*, Vol. LT-3, No. 6, 1985 pp. 1270–1276.
[31] Forrest, S.R. "Optoelectronic integrated circuits," *Proc. of the IEEE*, Vol. 75, No. 11, 1987, pp. 1488–1497.
[32] Adachi, S., "GaAs, AlAs, and Al$_x$Ga$_{1-x}$As: Material parameters for use in research and device applications," *J. Appl. Phys.*, Vol. 58, No. 3, 1985, pp. R1–R29.
[33] Shi, Y., W. H. Steier, L. Yu, M. Chen, and L. R. Dalton, "Large stable photoinduced refractive index change in a nonlinear optical polyester polymer with disperse red side groups," *Appl. Phys. Lett.*, Vol. 58, No. 11, 1991, pp. 1131–1133.
[34] Thackara, J. I., G. F. Lipscomb, M. A. Stiller, A. J. Ticknor, and R. Lytel, "Poled electro-optic waveguide formation in thin-film organic media," *Appl. Phys. Lett.*, Vol. 52, No. 13, 1988, pp. 1031–1033.
[35] Dutta, S. and H. E. Jackson, "Scattering loss reduction in ZnO optical waveguides by laser annealing," *Appl. Phys. Lett.*, Vol. 39, No. 3, 1981, pp. 206–2208.
[36] Fisher, A. D., J. N. Lee, E. S. Gaynor, and A. B. Tveten, "Optical guided-wave interactions with mangetostatic waves at microwave frequencies," *Appl. Phys. Lett.*, Vol. 41, No. 9, 1982, pp. 779–781.
[37] Tsai, C. S. and D. Young, "Wideband scanning of a guided-light beam and spectrum analysis using magnetostatic waves in an Yttrium Iron Garnet - Gadolinium Gallium Garnet waveguide," *Appl. Phys. Lett.*, Vol. 54, No. 3, 1989, pp. 196–198.
[38] Ulrich, R., "Theory of the prism-film coupler by plane-wave analysis," *J. Opt. Soc. Am.*, Vol. 60, 1970, pp. 1337–1350.
[39] Hunsperger, R. G., A. Yariv, and A. Lee, "Parallel end-butt coupling for optical integrated circuits," *Appl. Optics*, Vol. 16, No. 4, 1977, pp. 1026–1032.
[40] Murphy, E. J. Murphy and T. C. Rice, "Self-alignment technique for fiber attachment to guided wave devices," *IEEE Quantum Electron.*, Vol. QE-22, No. 6, 1986, pp. 928–932.
[41] Sarawatari, M. and K. Nawata, "Semiconductor laser to single-mode fiber coupler," *Appl. Opt.*, Vol. 18, No. 11, 1979, pp. 1847–1856.
[42] Fukuma, M. and J. Noda, *Appl. Opt.*, Vol. 19, 1980, p. 591.
[43] Suhara, T. and H. Nishihara, "Integrated optics components and devices using periodic structures," *IEEE J. Quantum Electron.*, Vol. QE-22, No. 6, 1986, pp. 845–867.
[44] Chang, W. S. C. and P. R. Ashley, "Fresnel lenses in optical waveguides," *J. Quantum Electron.*, Vol. QE-16, No. 7, 1980, pp. 744–745.
[45] Suhara, T., S. Ura, H. Nishihara, and J. Koyama, "An integrated-optic disc pickup device," *Int. Conf. Integrated Opt. and Opt, Fiber Commun., Tech. Dig.*, Venezia, Italy, Oct. 1–4, 1985, pp. 117–120.
[46] Winzer, G., "Wavelength multiplexing components-A review of single-mode devices and their applications," J. Lightwave Tech., Vol. LT-2, 1984, pp. 369–378.
[47] Spear-Zino, J. D., R. R. Rice, J. K. Powers, D. A. Bryan, D. G. Hall, E. A. Dalke, and W. R. Reed, "Multiwavelength monolithic integrated fiber optics terminal: An update," *Proc. SPIE*, Vol. 239, 1980, pp. 293–298.
[48] Alferness, R. C., "Waveguide electooptic modulators," *IEEE Trans. Microwave Theory and Tech.*, Vol. MTT-30, No. 8, 1982, pp. 1121–1137.

[49] Tsai, C. S., "Guided-wave acoustooptic Bragg modulators for wideband integrated-optic communications and signal processing," *IEEE Trans. Circuits Syst.*, Vol. CAS-26, 1979, pp. 1072–1098.
[50] Haruna, M. and J. Koyama, "Thermooptic deflection and switching in glass," *Appl. Opt.*, Vol. 21, No. 19, 1982, pp. 3461–3465.
[51] Voges, E. and A. Neyer, "Integrated-optic devices on LiNbO$_3$ for optical communication," *J. Lightwave Tech.*, Vol. LT-5, No. 9, 1987, pp. 1229–1238.
[52] Thylen, L., "Integrated optics in LiNbO$_3$: Recent developments in devices for telecommunications," *J. Lightwave Tech.*, Vol. 6, No. 6, 1988, pp. 847–861.
[53] Grave, I., S. C. Kan, G. Griffel, S. W. Wu, A. Sa'ar, and A. Yariv, "Monolithic integration of a resonant tunneling diode and a quantum well semiconductor laser," *Appl. Phys. Lett.*, Vol. 58, 1991, pp. 110–113.
[54] Toda, K., Y.Okada, "Bipolar transistor carrier-injected optical modulator/switch: proposal and analysis," *IEEE Electron. Device Lett.*, EDL-7, 1986, p. 605.
[55] Ishida, K., H. Nakamura, and H. Matsumura, "InGaAsP/InP optical switches using carrier induced refractive index change," *Appl. Phys. Lett.*, Vol. 50, No. 3, 1987, pp. 141–142.
[56] Wood, T. H., "Multiple quantum well (MQW) waveguide modulators," *J. Lightwave Tech.*, Vol. 6, No. 6, 1988, pp. 743–757.
[57] Haas, D., H. Yoon, H. T. Man, G. Cross, S. Mann, and N. Parsons, "Polymeric electro-optic waveguide modulator; materials and fabrication," *SPIE* Vol. 1147, 1989, pp. 222–232.
[58] Small, R. D., K. D. Singer, J. E. Sohn, M. G. Kuzyk, and S. J. Lalama, "Thin film processing of polymers for nonlinear optics," *SPIE* Vol. 682, 1986, pp. 160–169.
[59] Zaugg, T., B. Fabes, L. Weisenbach, and B. Zelinski, "Waveguide formation by laser irridiation of sol-gel coatings," *Proc. SPIE*, Vol. 1590, 1991, pp. 26–35.
[60] Schaffner, J., "Analysis of a millimeter wave integrated electro-optic modulator with a periodic electrode," *Proc. SPIE*, Vol. 1217, 1990, pp. 101–110.
[61] Becker, R. A., "Traveling-wave electroptic modulator with maximum bandwidth-length product," *Appl. Phys. Lett.*, Vol. 45, 1984, pp. 1168–1170.
[62] Walker, R., "High-speed electrooptic modulation in GaAs/GaAlAs waveguide devices," *J. Lightwave Tech.*, Vol. LT-5, No. 10, 1987, pp. 1444–1453.
[63] Girton, D. G., S. L. Kwiatkowski, G. F. Lipscomb, and R. S. Lytel, "20 GHz electro-optic polymer Mach-Zehnder modulator," *Appl. Phys. Lett.*, Vol. 58, No. 16, 1991, pp. 1730–1732.
[64] Teng, C. C., "Traveling-wave polymeric optical intensity modulator with more than 40 GHz of 3-dB electrical bandwidth," *Appl. Phys. Lett.*, Vol. 60, No. 13, 1992, pp. 1538–1540.
[65] Marcuse, D., "Optimal electrode design for integrated optics modulators," *IEEE J. Quantum Electron.*, Vol. QE-18, No. 3, 1982, pp. 393–399.
[66] Simons, R. N. and G. E. Ponchak, "Modeling of some coplanar waveguide discontinuities," *IEEE Trans. Microwave and Tech.*, Vol. MTT-36, No. 12, 1988, pp. 1796–1803.
[67] Bridges, W. B., F. T. Sheehy, and J. H. Schaffner, "Wave-coupled LiNbO$_3$ electrooptic modulator for microwave and millimeter-wave modulation," *IEEE Phot. Tech. Lett.*, Vol. 3, No. 2, 1991, pp. 133–135.
[68] Kaminow, I. P., "Optical waveguide modulators," *IEEE Tran. Microwave Theory Tech.*, Vol. MTT-23, 1975, pp. 57–70.
[69] Kogelnik, H. and R. V. Schmidt, "Switched directional couplers with alternating ," *IEEE J. Quantum Electron.*, Vol. QE-12, No. 7, 1976, pp. 396–401.
[70] Alferness, R. C., "Optical directional couplers with weighted coupling," *Appl. Phys. Lett.*, Vol. 35, No. 3, 1979, pp. 260–262.
[71] Tsai, C. S., B. Kim, and F. R. El-Akkari, "Optical channel waveguide switch and coupler using total internal reflection," *IEEE J. Quantum Electron.*, QE-14, 1978, pp. 513–517.
[72] Neyer, A., W. Nevenkamp, L. Thylen, and B. Lagerstrom, "A beam progapation method analysis of active and passive waveguide crossings," *J. Lightwave Tech.*, Vol. LT-3, No.3, 1985, pp. 635–6422.
[73] Dutta, N. K.Dutta, N.A. Olsson, and W.T.Tsang, "Carrier-induced refractive index change in AlGaAs quantum well lasers," *Appl. Phys. Lett.* Vol. 45, 1984, p. 836.

[74] Lorenzo, J. P. and R. A. Soref, "1.3 μm electro-optic silicon switch," *Appl. Phys. Lett.*, Vol. 51, No. 1, 1987, pp. 6–8.
[75] Neyer, A. and W. Sohler, "High-speed cutoff modulator using a Ti-diffused $LiNbO_3$ channel waveguide," *Appl. Phys. Lett.*, Vol.35, 1979, pp. 256–258.
[76] Alferness, R. C., "Efficient waveguide electro-optic TETM mode converter/wavelength filter," *Appl. Phys. Lett.*, Vol. 36, No. 7, 1980, pp. 513–515.
[77] Donnelly, J. P., N. L. Demeo, G. A. Ferrante, and K. B. Nichols, "A high-frequency GaAs optical guided-wave electrooptic interferometric modulator," *IEEE J. Quantum electron.*, Vol. QE-21, No. 1, 1985, pp. 18–21.
[78] Walker, R. G., "High-speed electrooptic modulation in GaAs/GaAlAs waveguide devices," *J. Lightwave Tech.*, Vol. LT-5, No. 10, 1987, pp. 1444–1453.
[79] Tada, K. and N. Suzuki, "Electrooptic coefficient of InP," *Japan J. Appl. Phys.*, Vol. 19, 1980, pp. 2295–2297.
[80] Taylor, H. F., "Application of guided-wave optics in signal processing and sensing," *Proc. of the IEEE*, Vol. 75, No. 11, 1987, pp. 1524–1535.

Chapter 6

Optoelectronic Interconnects and Electronic Packaging

Lynn D. Hutcheson

6.1 INTRODUCTION

When designing an optoelectronic interconnect, it must be remembered that the performance depends on the optical components as well as the electronics. To see a significant improvement in the performance of an optoelectronic integrated circuit (OEIC) over a hybrid circuit, the processing sequence must not compromise either the electronics or the optoelectronic components (e.g., lasers and detectors). This means that the circuit must accommodate the materials and processes needed for the electronics as well as those of the optoelectronics. This is where the similarities in the development of integrated optoelectronic components differ greatly from those in the silicon integrated circuits (IC). The drive to increase the level of complexity of silicon ICs is based on considerations of cost, performance, density, and ease of packaging. All parts of the silicon ICs are made with the same materials and processing sequence. For OEICs, however, the materials and processes for the optoelectronics and the electronics are vastly different and the process for integration is more complex still.

The obvious question then is: Why bother integrating the devices? There is no single answer to this question. It is actually application dependent. There are many applications where the performance required cannot be achieved by electrical interconnects and can be achieved only by optics. The major reasons for integrating the devices, then, fall into three areas: lower parasitics means higher performance, optical integration increases density, and fewer parts eases the packaging task. Another potential reason for integration is cost. However, it is not clear whether there will be a reduction in cost for producing optoelectronic ICs over an IC connected to a discrete optical device, particularly if the OEIC has a medium-scale integration (MSI) level of circuitry. This is because the integration of the optical components adds processing steps to the electronics, thereby increasing the cost per circuit and decreasing the circuit yield. In addition, the crystal growth on a substrate for an integrated optoelectronic circuit costs just as much as 1 for a wafer to be made into discrete parts, but the integrated wafer will produce relatively few OEICs, making each OEIC more expensive. So instead of trying to justify the components from a cost standpoint, they should be examined on the basis of what can be done with this technology that cannot be done any other way.

In this chapter, the components that make up an IOC—materials, electronics, and optoelectronics—will be described. The parameters that are important to a designer of interconnects (e.g., bandwidth, power, density, and bit error rate (BER)) are described. Other operating characteristics, such as temperature sensitivity, are discussed to provide an appreciation of what must be considered when designing optical transmitters and optical receivers. Finally, a few examples are given of the expected performances of OEICs and their impact on the system.

6.2 DESIGN AND IMPLEMENTATION CONSIDERATIONS

One implementation for optoelectronic interconnects is an optical network for processor-to-processor communications [1]. In a distributed network, multiple processors communicate with each other on an electrical bus. An electrical bus is typically 32 to 40 bits wide. In a local area network (LAN), the processors can be separated by up to 1 km, and transmission is generally asynchronous. This means that the bits are not transmitted at set times known a priori to both the sender and receiver. Instead, the receiver must synchronize itself with the incoming signal. Synchronization is accomplished by coding the data in a way that guarantees a certain number of transitions in a given time period. The transitions are detected and generate a signal that controls the frequency and phase of a local oscillator.

One common type of LAN is the ring network. In the ring, each node transmits to one other node and receives from one other node. Therefore, each line is a point-to-point interconnect. The nodes are typically connected by a parallel electrical bus consisting of perhaps 40 coaxial or twisted pairs of cables. Optical interconnects are being developed to greatly reduce the complexity and size of such a bus. A schematic of the components [1] in an optical bus interface unit (BIU) for a ring bus is shown in Figure 6.1. The information enters the BIU in parallel from the microprocessor. The parallel lines are converted to serial data for transmission by the multiplexer and coder. At the receiver end, the signal is detected and amplified, and the timing is recovered using phase-locked loop and decoder.

The complexity of the node, excluding the media access controller, is typically several thousand gates. The bandwidth of the optical signal is determined by the number and bandwidth of the parallel lines and the overhead due to the coding. As an example, 40 lines at 50 MHz each are multiplexed using a code that produces 20% overhead. The

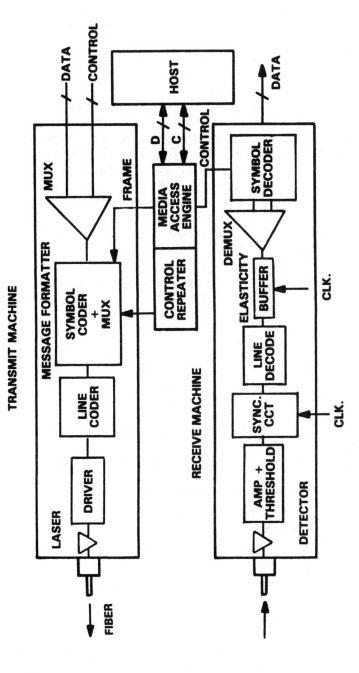

Figure 6.1 The components within an optical bus interface unit for a ring bus.

serial transmission rate would be 2.4 GHz. A rough order of magnitude for the power required for the optical node is shown in Table 6.1. The assumptions are that enhancement-mode metal-semiconductor field-effect transistors (MESFET) [2] are used for the GaAs electronics and that the parallel outputs from the silicon IC to the GaAs are CMOS compatible, unmatched with 20 pF of capacitance, and 50% duty cycle. The power required for a comparable electrical node that transmits 40 individual 50 MHz lines is given in Table 6.2, which clearly shows the optical approach provides a significant power savings.

One measure of an optical communication system performance is a BER. For the example, a 2.4 GHz data rate will be used, and a BER of 10^{-15} or better will be needed to meet a reasonable system performance. Typical telecommunication systems require 10^{-9} BER. However, computing systems require much better performance because a single bit having an error can have significant consequences at the end of a long calculation. As shown in Figure 6.2, the BER is strongly dependent on the power at the receiver. A plot calculated from Smith's data [3] of the received optical power versus log (BER) is shown. As can be seen from the figure, a small change in the received power changes the BER by orders of magnitude. This is true for all communications systems. Therefore, a design of at least -35 dBm optical power must strike the optical detector. For most optical interconnect applications other than telecommunications, the maximum fiber length will be approximately 1 km. Assuming a 2 dB/km transmission loss at 850 nm operating

Table 6.1
Power Required for an Optical Node

Components	Power Requirements	Power (mW)
400 gates	0.5 mW / Gate	200
1600 gates	0.05 mW / Gate	80
Laser/driver	2V at 50 mA	100
Detector/amplifier	2V at 30 mA	60
PLL		100
40 parallel I/O to host	10 mW each	400
Total		840

Table 6.2
Power Required for a Comparable Electrical Node

Components	Power Requirements	Power (mW)
40 gates (coding)	0.05 mW/gate	20
40 I/O on to the network (3.3V swing into 50Ω 50% duty)	100 mW each	4,000
40 I/O to the host (3.3V swing into 20 pF)	10 mW each	400
Total		4,420

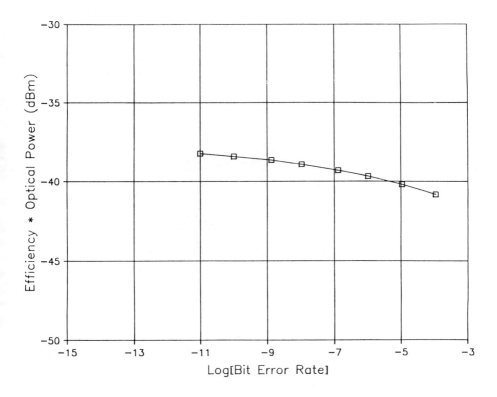

Figure 6.2 Bit error rate versus optical power.

wavelength and a 10dB laser-to-fiber coupling loss (10% coupling efficiency), the required optical power must exceed −23 dBm. Because most lasers emit powers of 10 dBm or more, the error rate specification can easily be met.

Another example in which to consider optoelectronic interconnects is in the CPU-to-memory link of a GaAs high-speed processor. Optical interconnects can provide a speed advantage in this chip-to-chip application. In this case, speed is the critical factor. In all processors, accessing a high-speed memory is one of the greatest bottlenecks to throughput. A schematic for an electronic implementation of the CPU-to-memory is shown in Figure 6.3. Ideally, the CPU will request data from the RAM by transmitting the address of the bits to be addressed. At the RAM, the address bits are decoded, and the access bits appear at the bit line outputs. The data are then transmitted back to the CPU, where they are latched at the input buffers. The minimum amount of time from the clock pulse, which starts the address on its way toward the memory until the returning data can be latched at the buffers, determines the fastest clock speed at which the processor can operate. This time can be estimated for the electrical implementation as well as the optical

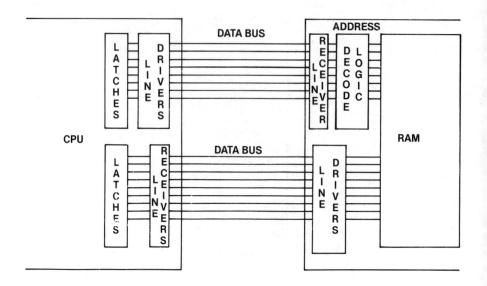

Figure 6.3 The connection of a CPU chip to memory.

by estimating the delays encountered at each stage during the retrieval of data from the memory.

In both cases, it is assumed that the latch in the final stage of the GaAs electronics is capable of driving approximately 1.0 mA into either the laser driver or the output line drivers. For the sake of comparison, assume that the line driver and the laser driver can sink 5 mA of current. This would require an enhanced-mode MESFET approximately 200 μm wide [1]. Such a field-effect transistor (FET) would have a gate capacitance of 0.4 pF. The delay encountered for the latch to raise the input of the driver to 0.5V can be estimated from:

$$t = C\left(\frac{\Delta V}{I}\right) = 0.4 \text{ pF}\left(\frac{0.5\text{V}}{1.0\text{mA}}\right) = 200 \text{ ps} \tag{6.1}$$

First, the electrical case will be analyzed. Assume that the CPU is surface mounted on a multichip package approximately 1 cm from a memory with a 1 ns access time. The capacitance of the pads plus the interconnect lines is estimated to be 2.0 pF. The time required to cause a voltage swing of DV on the line can be estimated from:

$$t = C\left(\frac{\Delta V}{I}\right) = 200 \text{ ps} \tag{6.2}$$

The total delay for the data request is estimated from this delay and the other delays shown in Table 6.3. Therefore, the electrical interconnect will limit the maximum clock rate to 500 MHz.

In an optical link, the output line drivers are replaced by a laser and laser driver, and the line receivers are replaced by a detector and amplifier. For a laser, C is approximately 1 pF, and ΔV will be determined by the series resistor, R_s. $\Delta V = IR_s = 5$ mA $\times$ 10Ω = 50 mV. Thus,

$$t = C\left(\frac{\Delta V}{I}\right) = 0.01 \text{ ns} \tag{6.3}$$

The 5-mA pulse into a prebiased laser will produce a change in the output power, P, given by:

$$P = \left(\frac{h\nu}{q}\right)\Delta I = 2.75 \text{ mW} \tag{6.4}$$

In a short point-to-point interconnect, the losses in the optical coupling would be only about 10 dB. Therefore, −7 dBm of optical power would fall on the detector. This is orders of magnitude more than required to meet the 10^{-15} BER requirement. The extra power, however, can be used to reduce the number of gain stages in the amplifier. The amount of optical power (−7 dBm) would provide a large enough electrical signal at the detector so that only two small stages of amplification would be required. The delay through the amplifier would be only about 0.15 ns. The total data request for an optical link is shown in Table 6.4. Therefore, one can see that the optical interconnect would provide a cycle time that is shorter than the electrical interconnect.

A major improvement for the optical interconnect would be gained if the laser would require a pulse current of only 1 mA and the optical coupling were improved. The

Table 6.3
Total Delay for an Electrical Data Request

Delay Mechanism	Delay Times (ns)
Buffer on output latch	0.2
Rise time of 1 line	0.2
Media	0.05
RAM access time	1.0
Buffer	0.2
Rise time	0.2
Media	0.05
Settling time	0.1
Total	2.0

Table 6.4
Total Delay for an Optical Data Request

Delay Mechanism	Delay Times (ns)
Buffer on latch output	0.2
Laser driver	0.01
Media	0.05
Detector/amplifier	0.15
RAM access time	1.0
Output buffer	0.2
Media	0.01
Detector/amplifier	0.05
Settling time	0.1
Total	1.92

buffer delay would be reduced to nearly zero, but the improved coupling would prevent the detector/amplifier delay from degrading. The total time would be reduced to about 1.3 ns. The reduction of current into the driver would result in a power saving for the link. The electrical link, on the other hand, would get slower. The rise time would increase to approximately 1.0 ns, making the total data request delay 2.8 ns.

Moving our attention inside a processor, optical links are being investigated for the computer backplane to reduce the number and volume of cables. A schematic of how two boards in the processor would communicate is shown in Figure 6.4. All the chips on the board are silicon except for an I/O chip, which is GaAs and contains the optical and high-speed electrical functions. Most processors are synchronous, meaning that a single clock controls all operations. Therefore, the interconnect has the time between two clock pulses to send a bit, have it propagate to the receiver board, and be valid and ready for latching into the input registers. In a 32-bit machine, each electrical bus contains between 32 and 40 parallel lines. Assuming each line operates at 50 MHz, then impedance matching will probably be required.

For a board-to-board interconnect, the parallel electrical lines are multiplexed into a single optical channel. The question then is: How fast must the multiplexer and demultiplexer operate? The answer is determined by calculating the time available for multiplexing and dividing it into the number of parallel lines. Assume 32 lines to be transmitted in 20 ns (50 MHz system clock) across 1m of fiber. The clock period for the high-speed serial rate of the multiplexer is designated by T. The time delays associated with transmitting the data from board-to-board are given in Table 6.5. Therefore, the multiplexer speed is calculated from the following equation:

$$20 \text{ ns} - 7.5 \text{ ns} = (32 + 2)T \qquad 1/T = 2.7 \text{ GHz} \qquad (6.5)$$

This is obviously a fast multiplexer. The speed, however, is a result of the need to have information valid in 20 ns. If the system could be designed with asynchronous interconnect timing, the speed of the multiplexer would be a more manageable [1].

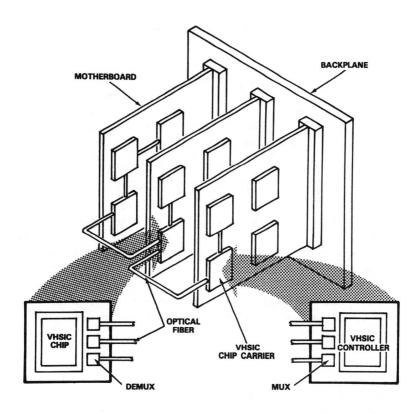

Figure 6.4 The interconnections of boards in a computer back-plane using optical interconnects.

Table 6.5
Time Delays for Board-to-Board Data Transmission

Delay Mechanism	Delay Time (ns)
Settling time of silicon output buffer	1 ns
Latching of parallel data in MUX	T
Time to transmit last bit	32 T
Latency of laser/driver	0.3 ns
Delay through fiber (2×10^8 m/s)	5 ns
Latency of detector/amplifier	0.7 ns
Latch in D-MUX	T
Settling time at silicon buffer input	0.5 ns

Note: T = clock period for the high-speed serial rate of the multiplexer.

$$f = 32/20 \text{ ns} = 1.6 \text{ GHz} \tag{6.6}$$

Optical interconnects may also find an application at the chip-to-chip level. High-speed multiplexing and demultiplexing can be used to greatly decrease the interconnect complexity of VLSI-level parts. The advantages are twofold. First, the number of signal lines coming out of a package will be greatly reduced by multiplexing. Second, the elimination of the capacitive loading due to electrical routing on the circuit board should make it possible to achieve higher I/O rates from the VLSI chips.

6.3 MATERIAL AND PROCESSING REQUIREMENTS

Optoelectronic ICs for optical interconnects begin with either a GaAs or InP substrate and require the growth of epitaxial layers to fabricate the optoelectronic components. For this discussion, we will consider only the GaAs material system because this material is more advanced for high-speed electronics. These materials must provide a method for the electronics to be compatible with the optoelectronics. In production, this means that the substrate must be at least 3 inches in diameter and the epitaxial growth system must be able to handle the 3-inch substrate. Further requirements are placed on both the substrate and the epitaxial layers by the devices that are to be fabricated.

The requirements of the starting GaAs substrate are dependent on the electronics and the optoelectronics. The electronic technologies that are fabricated using selective ion implantation require uniform, high-resistivity substrates that maintain their properties after the implant anneal steps. The circuits that require epitaxial layers are not as dependent on the electrical properties of the substrate, but they do depend on the density of defects. Optoelectronic components, especially lasers, are very susceptible to defects in the substrate, which propagate up through the active region. These defects have proved to be one of the major causes of short-lived lasers.

There are two basic methods for growing GaAs substrates: horizontal Bridgeman (HB) and liquid-encapsulated Czochralsky (LEC). Most people today choose to use 3-in LEC material that has resistivities in the range of 1×10^8 Ω-cm. These resistive properties are maintained fairly well during implant annealing. The major problem with the quality of the substrates is the defect density. Present 3-in round LEC wafers have etch pit densities less than $10^3/cm^2$. Nonuniformities in the threshold voltage of FETs and defects in the epitaxial growth have been attributed to these defects. However, "zero-defect" LEC material is becoming more available from a number of companies. One method for achieving zero defects is to dope the wafer with indium, which ties up any defects and thus prevents them from propagating. Because the indium is isoelectronic, it does not cause the same electrical problems as chrome during implant anneals. It is possible that the success of integrated optoelectronics depends on the cost and availability of zero-defect material. This is because the laser lifetime may not be long enough on the high-defect standard LEC material. Therefore, low-defect density is a must.

The requirements for the growth of the epitaxial layers on the substrate are also dependent on the type of component to be fabricated. The electronics technologies requiring epitaxial growth generally need low background carrier concentrations (around 1×10^{13} cm^3) and the ability to control doping accurately. Lasers are not as dependent on electrical properties but need high photoluminescence efficiency. Waveguide structures need low carrier concentration (low capacitance) and excellent morphology (low scattering). Quantum-well lasers and MODFET-type electronics also require extremely sharp interfaces (on the order of a few angstroms) between layers of GaAs and AlGaAs.

There are two epitaxial growth techniques that can be considered for use in the production of integrated optoelectronic circuits: molecular beam epitaxy (MBE) and metal organic chemical vapor deposition (MOCVD). Liquid phase epitaxy (LPE) has been used for years to manufacture high-quality lasers and light emitting diodes. The substrates used in LPE, however, are limited in size to approximately 2 in^2, which is not compatible with GaAs electronics. MBE [4] has been the standard technique for the development of MODFET structures. The system can achieve the low background carrier concentrations, layer thicknesses, and sharp interfaces needed for low-threshold lasers [5] and optoelectronic circuits [6]. There are two drawbacks to MBE as a production process for IOCs. They are growth rate and defect density. MOCVD has a higher growth rate; the layers can be made with very low defect levels; and the photoluminescent efficiency is extremely high. MOCVD is currently being used in production by a number of laser diode manufacturers.

6.4 ELECTRONICS FOR INTEGRATION

For optoelectronic ICs to meet the needs of high-speed processors, it is imperative that the OEICs be fabricated in a way that guarantees an adequate supply of chips. This will occur only if the OEICs are fabricated using a standard GaAs production process to which the extra steps for the optoelectronics have been added. It is very doubtful that a new GaAs electronics process will be developed solely for the purpose of allowing optoelectronic components to be integrated.

Figure 5.5 shows the cross sections of the most common transistors fabricated in GaAs. Figure 5.5(a) is the depletion-mode MESFET (metal-semiconductor field-effect transistor). Figure 5.5(b) shows the enhancement-mode MESFET, and Figure 5.5(c) is the junction FET (JFET). Each of these technologies is fabricated using undoped, semi-insulating (10^8 Ω-cm) GaAs as a starting substrate. The channel and contact regions are formed by selective ion implantations that are activated using a high-temperature annealing step. The most common GaAs OEICs are fabricated using depletion-mode MESFETs.

6.5 INTEGRATED OPTOELECTRONIC DETECTORS AND RECEIVERS

Semiconductor optical detectors are two-terminal devices that convert optical inputs into electrical carriers. By connecting the detector to an appropriate circuit, the electrical

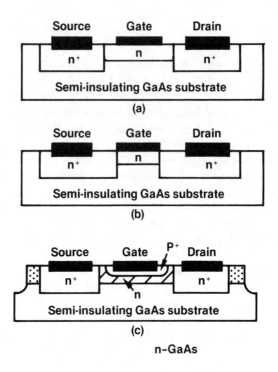

Figure 6.5 GaAs field effect transistors: (a) depletion-mode MESFET; (b) self-aligned gate enhancement-mode MESFET; (c) junction FET.

carriers are collected and the signal is amplified to levels adequate to drive a digital IC. The detector and amplifier must be designed as a unit. The integration of the detector with the amplifier provides a significant improvement in bandwidth and sensitivity over a hybrid circuit [1]. The reason for the improvement is that the capacitance at the connection of the detector to the amplifier can be made as low as 0.2 pF for a monolithic circuit as opposed to >0.5 pF for a discrete detector/amplifier pair. The significance of this can be seen in Figure 6.6 where the detected optical power versus bit rate for several values of input capacitance is plotted for a BER = 10^{-9} [7]. The best experimental results for a hybrid receiver are shown in Figure 6.6 as closed circles [8] and are consistent with a total capacitance between 0.6 and 0.8 pF. At 1 Gbp/s, there is a 5 dBm increase in receiver sensitivity between a hybrid and an integrated receiver. Alternately, this threefold or fourfold decrease in capacitance would permit the detector/amplifier to operate at nearly three or four times the bit rate with no degradation in accuracy. This figure also shows that the effect of capacitance is even more dramatic at higher bit rates. This is a strong argument for integration.

Figure 6.6 also shows that avalanche photodiodes (APD) have much higher sensitivity than pin structures. However, APDs require maintaining gain control at high reverse

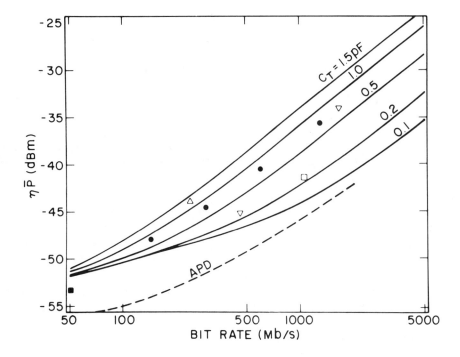

Figure 6.6 Sensitivity of 10^{-9} BER versus bit rate as a function of node capacitance.

biases (typically 60V or more) over a wide range of temperatures due to the temperature sensitivity of the avalanche gain. This can significantly increase complexity and cost.

If a photoconductor is to be integrated on a circuit with more than a few electronic components, it must meet a number of criteria. First, the detector must be compatible with the electronics processing. Therefore, it must be processed on a production line and be compatible with the substrates used for the electronics. For GaAs, this means a 3-in round semi-insulating substrate. Second, the material for the detector cannot interfere with the electronics. Thus, if epitaxial material growth is needed, it must be excluded from the regions in which the electronics will be fabricated, and the transition to the epitaxial region must be smooth enough to permit fine-line photolithography. Third, any process needed to fabricate the detector cannot degrade the performance of the electronics. For example, a very high temperature step may cause unacceptable surface damage. Fourth, the detector and electronics must be adequately electrically isolated from each other on the substrate. Finally, the detector must meet specifications after integration.

Although most of these sound obvious, they are not easily accomplished. If the devices are fabricated on semi-insulating substrates, a method must be found to make electrical contact to both sides of the detector other than via the substrate. The photoconductor and back-to-back Schottky detectors have both contacts at the surface. However, for both the *p-i-n* and avalanche detector, one contact is below the surface. A process step must be added to make contact to this area. One method for solving this problem is shown in Figure 6.7, where the layers for the *p-i-n* detector are grown in a well [9]. After the epitaxial layers are removed from the area in which the electronics are to be fabricated, the *n*+ region is exposed at the edge of the well, as shown in Figure 6.7, permitting easy access for contacting. Growing in a well is one step closer toward a monolithic receiver.

The first and simplest detector/amplifiers reported were *p-i-n*/FETs fabricated at Bell Labs in InGaAs/InP [10]. Both the detector and FET were fabricated in epitaxial layers grown by LPE. One such example is shown in Figure 6.8 where a transimpedance amplifier is integrated with a *p-i-n* detector [11]. The layers for the detector and amplifier were selectively grown in the two regions by a two-step MOCVD process. As in the structure shown in Figure 6.7 the three *p-i-n* layers were grown first in a pre-etched 7-μm deep well. Next the FET layers were grown, forming a nearly planar surface for photolithography. The *p-n* junction was formed by a Zn diffusion into the lightly doped GaAs absorption region. The technique of selectively growing the *p-i-n* and FET layers allows for the independent optimization of both circuit segments. For example, high-transconductance FETs require thin (<0.5 μm) *n*-type channels with high impurity concentration. The transimpedance amplifier consists of six GaAs MESFETs. The amplifier consists of two stages with a transimpedance of 1 kΩ. The output impedance is 50Ω, and the circuit has a 400 ps rise and fall time.

The metal semiconductor metal (MSM) detector (which is an interdigitated set of electrodes) has probably received the most attention for integrated optical receivers. The reason is its ease of fabrication, low capacitance, high-speed, and compatibility with FET IC processes. Several demonstrations [12–15] of OEIC receivers have taken place both in GaAs and InP material systems. Another advantage of the MSM detector is its flexibility to change parameters (e.g., finger spacing and bandwidth) by a simple adjustment of the photolithographic mask. Even though one might think that MSM detectors do not compare favorably with standard *p-i-n* detectors, Figure 6.9 shows a comparison [14] between *p-i-n* and MSM detectors. It is clear that MSM detectors can be made to operate at very high speeds. Speeds as high as 40 GHz have been demonstrated [13].

A early example of an integrated MSM receiver fabricated by Honeywell researchers is a GaAs receiver [12] consisting of a detector, amplifier, and a 1-to-4 GaAs demultiplexer operating at 1 GHz clock rates. The detector is an interdigitated back-to-back Schottky diode with 1-μm lines and 3-μm spaces fabricated directly on the semi-insulating substrate. The circuit consists of depletion-mode MESFETs fabricated using selective ion implantation on a 3-in processing line. This is truly a production compatible part with GaAs IC production techniques.

6.6 INTEGRATED OPTOELECTRONIC TRANSMITTERS

The fabrication of semiconductor laser diodes requires crystal growth and processing steps similar to IC manufacturing to define the electrical and optical cavity in the two dimensions

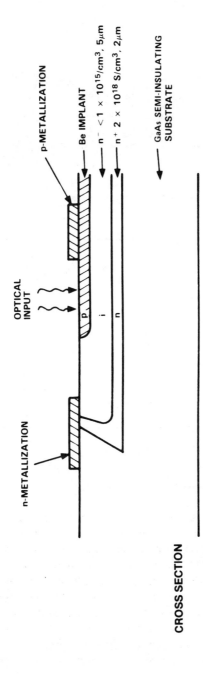

Figure 6.7 A *p-i-n* detector grown in a well on a semi-insulating substrate.

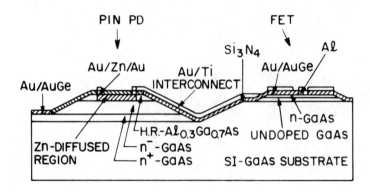

Figure 6.8 A monolithic optical receiver fabricated in GaAs/AlGaAs.

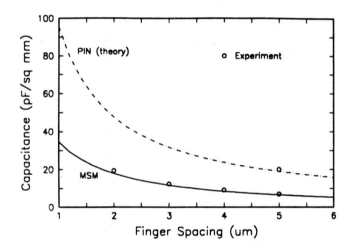

Figure 6.9 MSM versus *p-i-n* capacitances.

perpendicular to the direction of light propagation. The length of the cavity is defined by partial mirrors that are formed by cleaving the semiconductor along parallel crystal planes. The integration of the laser diode with the associated electronics is much more difficult from a materials and processing standpoint than the integration of the detector with its associated electronics. There are three major reasons for this: (1) lasers require a multilayered heterostructure up to 7-μm thick; (2) they need two high-quality parallel mirrors separated by a distance on the order of 200 μm; and (3) a method is needed that can achieve electrical and optical confinement in the lateral dimension.

The requirements for monolithically integrating a laser diode with high-speed electronics fall into three categories: compatibility, required components, and performance.

The decisions that must be made to obtain a working component will be application dependent. In general, the materials and processing for the laser and the electronics will be different. In addition, one of the major requirements for monolithic integration is that the epitaxial material grown for the laser covers only specific regions of the substrate while the rest of the surface consists of high-quality planar material for the fabrication of the electronics. As a second requirement, the epitaxial layers for the laser must be grown on a semi-insulating substrate. It is also necessary that the resulting substrate, after processing of the laser, be highly planar to permit fine-line lithography for the electronics. Finally, it is necessary to isolate the laser from the electronics both electrically and optically.

The first monolithic laser/electronics demonstration was reported by researchers at the California Institute of Technology [16]. This component consisted of an AlGaAs laser that was integrated with a GaAs Gunn oscillator. Since that time, there have been numerous demonstrations of different laser and electronic devices in both GaAs and InP material systems. The demonstrations were limited to single lasers, grown by LPE, integrated with a single transistor. The laser mirrors were formed by cleaving. These parts demonstrated the feasibility of integrating lasers with electronics, but they were far from being practical.

In more recent years, advances have been made in the areas of crystal growth, circuit design, and complexity that bring these components closer to production. As discussed in the earlier section on material growth, both MBE and MOCVD are capable of growing on 3-in round substrates that are standard for current production of GaAs ICs.

Demonstrations have also been reported that show an increase in the complexity of the electronics, an on-chip mirror, and power monitor. Figure 6.10 shows an optoelectronic IC with the laser grown in a well etched in the semi-insulating substrate and has a multiquantum-well active region [17]. The back facet was formed by reactive ion etching

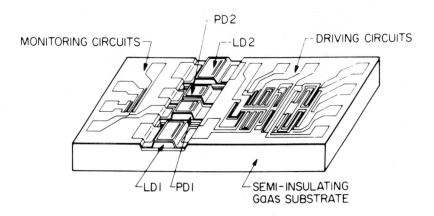

Figure 6.10 Integrated optoelectronic laser transmitter fabricated with a quantum well laser and selective ion implantation.

and resulted in a laser threshold of 40 mA. The electronics were formed by selective ion implantation into the semi-insulating substrate. The circuit design uses input buffers as well as a differential drive. It was demonstrated at modulation rates up to 2 Gb/s using nonreturn-to-zero (NRZ) format.

A more sophisticated and complex demonstration consisted of a TJS laser integrated with a 4-to-1 multiplexer [18]. The chip was approximately 1.8 by 1.8 mm^2. The 4-to-1 multiplexer (MUX) and its associated circuitry is formed by selective ion implantation and contains 36 NOR gates (approximately 150 D-mode MESFETs). The TJS laser was grown by LPE in an etched well and the rear facet was formed with an undercut mirror process. This particular chip was tested at speeds up to 160 MHz.

Integrating lasers on the same substrate as electronics has been demonstrated at a number of laboratories. However, it is far more difficult than integrating detectors on the same substrate. The primary reason is the epitaxial growth and stringent material requirements for the lasers. The additional processing steps required to fabricate a high-quality laser significantly reduce yields and thus increase cost. Using a hybrid approach is more cost effective. This involves fabricating the high-performance GaAs electronics on its own chip and the laser or laser arrays on a separate chip [15] and carefully packaging the two chips with fiber pigtails as shown in Figure 6.11. The figure shows the two chips, with the bottom chip being the transmitter chip that contains four laser modulators, and bias control circuit shown in the top portion of the chip. In the lower-right portion of the picture is the phaselock loop clock synthesizer. The lower-left portion of the chip contains a 10-to-1 multiplexer and synchronization control circuit.

The laser array chip with fiber pigtails can been seen more clearly in Figure 6.12. The laser array consists of four individually addressable, AlGaAs, single quantum well, graded index, ridge waveguide lasers on 250-μm centers. The average threshold and differential quantum efficiency are 11 mA and 0.55 mW/mA, respectively. The fibers are mounted in a silicon V-groove alignment fixture that is notched to provide clearance for the wirebounds on the top surface of the laser array chip. The width of the V-grooves are dimensioned so that when the fibers are seated in the bottom of the grooves, the axes of the fibers are level with the polished silicon surface. When the notched silicon alignment fixture containing the fibers is flipped and the protrusions are placed in intimate contact with the top surface of the laser array, the axes of the fibers are automatically aligned with respect to the lasers in the vertical direction and in the rotational direction about the fiber axes.

6.7 PACKAGING

One of the more difficult challenges of optical interconnects is the packaging of high-speed electronics with optoelectronic components and waveguides. Unique packaging problems arise when optical interconnects are used with high-speed electronics. The coupling between optoelectronic components and waveguides requires critical alignment

Figure 6.11 GaAs transmitter chip and laser submount assembly. Photograph courtesy of IBM.

tolerances of less than a micron. In addition, thermal conditions on the chip carriers must be carefully managed to keep the optoelectronic devices from drifting.

Conventional electronic packaging has electrical in/electrical out, whereas optoelectronic packaging requires optical in/electrical out, or electrical in/optical out, or optical in/optical out. Packaging technology having discrete optoelectronic components (lasers and detectors) with fiber in/out is well established. Integrating these components on a common substrate presents unique challenges.

One technique that has been developed and demonstrated [19–21] at a number of companies is a technique that aligns several fibers to a multiple detector array. Detectors typically require the light to impinge on the device from the surface. The fiber/detector

Figure 6.12 Four channel AlGaAs laser array and optical fiber assembly. Photograph courtesy of IBM.

coupling approach shown in Figure 6.13 uses a planar packaging technique that maintains the two-dimensional integrity of conventional packaging technology. V-grooves are etched in silicon to provide an excellent alignment fixture because the spacing between the V-grooves can be delineated exactly to match the detector spacing.

As shown in Figure 6.13, the exit end of the fiber is polished at an angle such that the light strikes the end of the fiber and is totally internally reflected through the bottom of the fiber. Because the fiber used is multimode (50-μm core diameter) the fiber end must be polished at 58° so that all of the modes are totally internally reflected. If the fiber is polished at 45°, the fiber end needs to be metalized to minimize the loss. However,

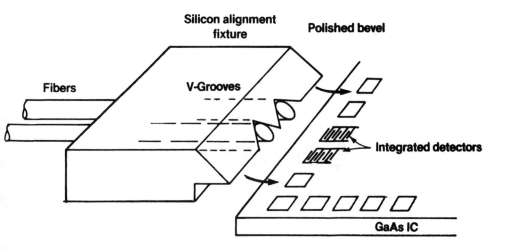

Figure 6.13 Planar technique for fiber to detector alignment.

a 45° polish does yield the smallest spot size. An increase in spot size of 15% to 30% can be anticipated [21] with a 58° polish, as can be seen in Figure 6.14.

A photograph of four fibers coupled to four MSM detectors is shown in Figure 6.15. The maximum efficiency is limited to about 70% using the polished V-groove technique technique due to Fresnel reflection from the surface of the GaAs. Fresnel reflective losses can be reduced by introducing an index matching fluid or antireflection coating on the surface of the detector.

Another technique allows vertical bonding of the fiber to the chip in which the fiber axis is normal to the semiconductor surface [22]. The advantage of this technique is that the fibers can be attached anywhere on the surface, which allows for higher fiber interconnect density. The drawback to this approach is that a three-dimensional package is required rather than the conventional two-dimensional package used in the previous technique.

The photodetectors were fabricated using standard photolithography in MOCVD grown epitaxial layers. The detectors are MSM, much the same as in the technique described previously, and photosensitive MESFETs. Fiber coupling cavities are etched from the backside of the wafer to the epitaxial detector sites as shown in Figure 6.16(a). The etched cavity is needed because GaAs is opaque. In this case, reactive ion etching is used to form the cavities because this etching technique permits higher packing densities than wet chemical etching. Efficient coupling from the fiber end-face to the detector is achieved by inserting an AlGaAs window/stop-etch layer having a very low etch rate such that etching stops at the required point and leaves a very smooth surface. A picture of a cross section of an etched well with an attached fiber is shown in Figure 6.16(b).

The method for aligning and fixing multiple fibers to a detector array is shown in Figure 6.17. Fibers are inserted in a fixture suitable for a two-dimensional fiber array.

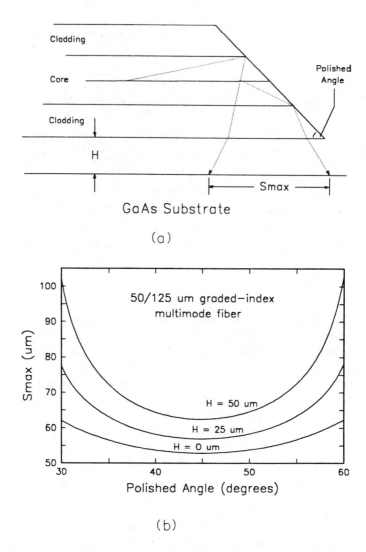

Figure 6.14 Schematic cross-section (a) analysis; (b) spot size formed at the detector surface.

The fibers are attached by either epoxy or soldering. In the case of soldering, the fibers are metalized and the fibers are soldered using conventional soldering techniques.

A packing density of 1,600/cm^2 can be achieved using this technique, assuming a 250-μm pitch. While this measure alone may not have much meaning, it gains significance when translated to information capacity. Assuming single channel bandwidths of 1 GHz and 100 MHz for optical and electrical respectively, the capacity for the two optical and two electrical I/O techniques is summarized [22] in Figure 6.18. The maximum

Figure 6.15 Beveled fiber array coupled to a four channel GaAs photoreceiver. Photo courtesy of IBM.

communication capacity is plotted as a function of chip size for each of the techniques. We can see that optics has a clear performance advantage.

Each of the two packaging approaches has its own advantages and disadvantages. The planar technique is most similar to conventional electronic packaging, which allows for using standard techniques for mounting the devices on boards. The advantage to using the vertical bonding technique is the flexibility to put the optical detectors at any position on the chip. This allows the flexibility to receive information at any position, giving the designer more flexibility in designing and laying out the chip or board. Additionally, more detectors can be used, which gives more inputs and outputs and yields the potential

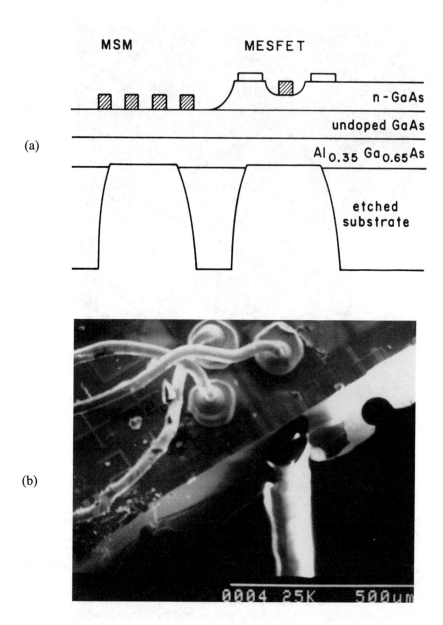

Figure 6.16 (a) GaAs fiber-optic coupler; (b) fiber coupled to the GaAs chip. Photographs courtesy of Columbia University.

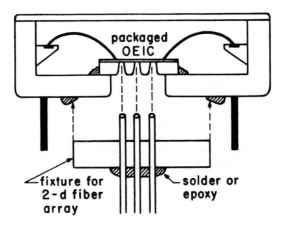

Figure 6.17 A packaged device accommodating backside fiber optic interconnect.

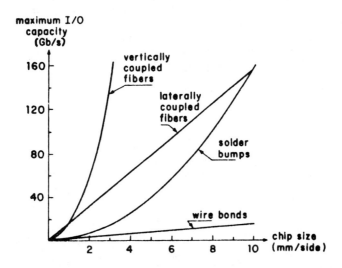

Figure 6.18 Maximum off-chip communication capacity for several interconnection techniques. A pitch of 250 μm is assumed for both planar and linear arrays.

for much higher information capacity. However, because this technique is not planar, an unconventional package needs to be implemented.

REFERENCES

[1] Hutcheson, L. D., ed., *Integrated Optical Circuits and Components: Design and Applications*, Marcel Dekker, New York, 1987.
[2] Hutcheson, L. D., *Proceedings SPIE*, Vol. 869, Cannes, France, November 1987.
[3] Smith, R. G. and S. D. Personick, *Semiconductor Devices for Optical Communication*, Springer-Verlag, New York, 1980.
[4] Cho, A. Y., "Thin Solid Films," Vol. 100, 1983, p. 291.
[5] Tsang, W. T., *Appl. Phys. Lett.*, Vol. 40, 1982, p. 217.
[6] Sanada, T., J. et al., *Appl. Phys. Lett.*, Vol. 44, 1984, p. 325.
[7] Forrest, S. R., *J. Light. Tech.*, Vol. LT-3, 1985, p. 1248, 1985.
[8] Smythe, P. P., et al., *Tech. Digest 10th European Conf. Opt. Fibre Comm.*, Stuttgart, Germany, paper 11B-5, 1984.
[9] Kolbas, R. M., et al., *Appl. Phys. Lett.*, Vol. 43, 1983, p. 821.
[10] Leheny, R. F., et al. *Elect. Lett.*, Vol. 16, 1980, p. 353.
[11] Makiuchi, M., et al., *Tech. Digest of IEDM*, 1984, p. 862.
[12] Ray, S. and M. B. Walton, *Proc. IEEE Microwave and Millimeter IC Symposium*, Baltimore, 1986.
[13] Sano, E., *IEEE Trans. Elect. Devices*, Vol. 37, 1990, p. 1964.
[14] Rogres, D. L., *IEEE J. Light. Tech.*, Vol. 9, 1991, p. 1635.
[15] Ewen, J. F., K. P. Jackson, R. J. S. Bates and E. B. Flint, *J. Light. Tech.*, Vol. 9, 1991, p. 1755.
[16] Lee, C. P., S. Margalit, I. Ury and A. Yariv, *Appl. Phys. Lett.*, Vol. 32, 1986, p.574.
[17] Nakano, H., S. Yamashita, T. Tanaka, N. Hirao and N. Naeda, *J. Light. Tech.*, Vol. LT-4, 1986, p. 574.
[18] Carney, J. K., M. J. Helix and R. M. Kolbas, *Tech. Digest GaAs IC Symposium*, Phoenix, 1983, p. 48.
[19] Hartman, D. H., M. K. Grace, and F. V. Richard, *IEEE J. Light. Tech.*, Vol. LT-4, 1986, p. 73.
[20] Haugen, P. R., S. Rychnovsky, A. Husain, and L. D. Hutcheson, *Opt. Eng.*, Vol. 25, 1986, p. 1076.
[21] Jackson, K. P., A. J. Moll, E. B. Flint, and M. F. Cina, *SPIE Proc.*, Vol. 994, 1988, p. 40.
[22] Ade, R. W. and E. R. Fossum, *SPIE Proc.*, Vol. 994, 1988, p. 33.

Section III
Applications of Optical Interconnects

Chapter 7

Polymer-Based Photonic Integrated Circuits and Their Applications in Optical Interconnects and Signal Processing

Ray T. Chen

7.1 INTRODUCTION

The development of advanced optical materials that can focus, modulate, multiplex, transmit, receive, demultiplex, and demodulate optical signals will be key to the realization of economical and reliable wide band (~THz) optoelectronic systems for optical signal processing and computing applications. To date, efforts have focused on the development of hybrid and monolithically integrated devices and systems in the $LiNbO_3$ and III-V material systems, respectively. A number of technology-related issues, however, currently impede further progress. In particular, both $LiNbO_3$ and III-V semiconductors are incapable of producing the large index modulations that are required to create multiplexed phase gratings, which constitute one of several important building blocks in very large scale optically interconnected systems. Second, the requirements of lattice-matching have severely restricted the number, size, and types of materials that can be grown on top of the III-V compound substrates and epilayers. Finally, device yields have been relatively low, while costs associated with the growth and processing of related microstructures have been high. Hence, the development of new, low-cost materials that can be processed into microstructural optical components, such as waveguides and gratings, will be invaluable to the future optoelectronic integration effort.

Due to the constraints of $LiNbO_3$ and III-V material systems, an array of guided wave devices research has been concentrated on polymer-based materials. Polymer molecules are formed by combining a myriad of monomers. Therefore, by definition, an infinite number of polymers can be formed. The polymeric materials suitable for guided wave device application in general and for optical interconnection in specific are the ones that demonstrate acceptable optical and mechanical properties. Both passive and active polymer-based guided wave devices have been demonstrated using different polymeric materials [1–20]. This chapter reports on the development and study of new polymer-based passive and active guided wave devices employing photolime gel which is a nonsynthesized superpolymer extracted from animal bones. It overcomes many of the problems associated with the fabrication of conventional microstructural thin-films, as described previously, and shows promise for becoming a new building block in photonic circuit systems. Polymer microstructure waveguides (PMSWs), which exhibit low loss (0.1 dB/cm)[21] and excellent optical quality (low defect number), have been formed on a variety of substrates, including GaAs, $LiNbO_3$, glass, Al, Al_2O_3, and BeO. Recent experimental efforts have also demonstrated PMSWs on Si, quartz, fused silica, phenolic PC board, Cu, Cr, AlN, and Kovar. Optical waveguides with slab guiding layers as large as 50 × 50 cm have been constructed. In addition, the ability to control the refractive index profile of the guiding region during the fabrication process has been demonstrated. As a result of this refractive index profile tuning capability, it is possible to fabricate low-loss polymer waveguide structures on any type of substrate, including semiconductors, conductors, insulators, and ceramics, regardless of the substrate refractive index and conductivity. A local sensitization technique, used in conjunction with the polymer films, has been developed to facilitate the formation of χ^2 nonlinear polymer, polymer waveguide amplifier and multiplexed holograms. The results that have been achieved thus far are summarized in Table 7.1, with the device features on $LiNbO_3$ and GaAs as references. Details of these polymer-based guided wave devices are addressed sequentially in this chapter.

7.2 FORMATION OF POLYMER MICROSTRUCTURE WAVEGUIDES

High-quality thin polymer films, which exhibit propagation losses of less than 0.1 dB/cm, can be formed from pure photo-lime gel polymers. When the gel is first formed, it exists, in aqueous solution, as a series of single polymer chains surrounded by adjacent water molecules. After standing for a period of time at temperatures below 30°C, solutions containing more than 1% photolime gel solidify, forming films that are rigid and, henceforth, rubber-like in their mechanical properties [21]. An optical thin film is formed from the pure photolime gel polymer by mixing solutions having various gel-to-water ratios and spinning the solutions on top of a substrate material. By changing the gel-to-water ratio or the spin speed of film coating, film thicknesses can be achieved that vary from less than 1 mm to greater than 100 mm in dimension. The optical transmission characteristics of a 10-mm thick gelatin polymer film, formed in the previously described fashion, are shown in Figure 7.1. It can be seen that the film is nearly 100% transparent from a wavelength of ~300 nm to greater than 2700 nm. These film properties were found to be relatively insensitive to temperature changes. In particular, recent experimental results indicate that the transmission and index properties are stable over the temperature range

Table 7.1
Features of Polymer-based Integrated Optical Devices

Features	Polymer-based Technology	GaAs	LiNbO$_3$
Planar waveguide	Yes	Yes	Yes
Channel waveguide	Yes	Yes	Yes
Electro-optic modulator	Yes*	Yes	Yes
Waveguide propagation loss	< 0.1 dB/cm	0.2 to 0.5 dB/cm	<0.1 dB/cm
OEIC size	Unlimited†	Limited†	Limited†
Formation of multiplexed grating	Yes‡	No	No
Channel waveguide packaging density (channels/cm)	High◇	High	High
Implementation on other substrates	Easy△	Difficult•	Difficult•
Large area multiple-guiding layer on single substrate	Yes	No	No
Waveguide lens	Yes	Yes	Yes
Dielectric constant dispersion	Low□	High	High
Potential modulation speed	> 100GHz§	~30 GHz	~30 GHz
Fabrication cost	Low	High	High
Moldability	Yes	No	No
Waveguide amplifier	Yes	Yes	Yes

Note: Results are from the Microelectronics Research Center of the University of Texas, Austin.
*Nonlinear polymer with g_{33} larger than LiNbO$_3$ and GaAs has been reported.
†Polymer can be implemented on any large substrate while GaAs- and LiNbO$_3$-based OEICs are limited by the crystal dimension.
‡High index modulation of same polymeric material allows us to multiplex hundreds of gratings on the same area for 1-to-many fanout (useful for high-speed clock signal distribution).
◇Up to 1250 channels/cm on polymer 500 channels/cm on GaAs and 333 channels/cm on LiNbO$_3$ were reported.
△Thin film coating.
•By definition GaAs- and LiNbO$_3$-OEICs are thick film devices which are difficult to transfer to other substrates.
□Polymer dielectric constant is controlled by electron oscillation which has very small dispersion from microwave to optical wave.
§Small dielectric constant dispersion gives very small walk-off between microwave and optical wave.

from −196°C to +180°C for dry films or holograms prebaked at +180°C [22]. Other tests performed on the polymer films have also indicated a high degree of immunity and radiation hardness to some nuclear and high-power microwave radiation sources [23]. In addition, films prepared in the previous manner have been shown to possess step-index profiles [24,25]. Their formation on absorptive, or higher index, substrate materials would, therefore, result in the creation of leaky mode waveguides having excessive propagation loss. The influence of substrate loss on guided wave propagation behavior was experimentally confirmed by depositing polymer films on top of Al$_2$O$_3$ and BeO ceramic substrates, respectively, and measuring the mode attenuation via a prism coupling technique. High losses, measured in excess of 40 dB/cm, make the step-index polymer films, as processed

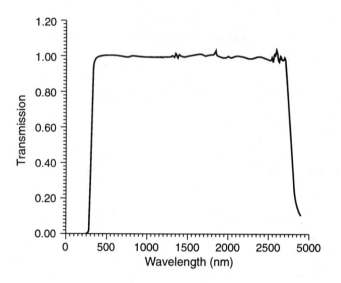

Figure 7.1 Transmission bandwidth of the polymer thin film.

previously, unsuitable for waveguiding applications. Ideally, the ability to change the index from a step to a graded-index (GRIN) profile with a higher surface index would enable polymer films to be used on low-loss, as well as extremely lossy, substrate materials. In this case, the lower index portion of the GRIN polymer film would function as a waveguide cladding layer. It would reduce the evanescent field overlap with the underlying substrate and provide tighter mode confinement closer to the polymer film surface. To achieve low-loss waveguiding in the gel films, a method of tuning the refractive index profile from a step-index to a GRIN profile was developed.

A combination of systematic wet and dry processing techniques was employed to perturb the mass density of the polymer and, hence, the polymer refractive index. The index of refraction of the newly perturbed polymer film can be qualitatively estimated using the Lorentz-Lorenz formulation [26], in terms of the average number of molecules per-unit volume that possess different mass densities. The process of index tuning is actually one of controlled absorption and dehydration. The film is first hardened in a fixer solution and then immersed in a de-ionized water bath. During this later step, water absorption causes the film to swell. The process is then reversed by dehydrating the polymer film in a temperature-controlled alcohol bath. To prevent film microcracking, the alcohol concentration is slowly and gradually increased during the final phase of processing. Qualitative data, not shown but obtained from scanning electron micrographs, indicate that the polymer mass density decreases monotonically towards the substrate surface. Various refractive index profiles for multimode PMSWs, created through the index tuning method, were experimentally measured using the prism coupling technique and analytically determined using the inverse Wentzel-Kramers Brillouin (IWKB) method

[27,28]. The results of these measurements and calculations are shown in Figure 7.2 for polymer films that have been deposited on top of Al_2O_3, Al, and GaAs substrates, respectively.

Refractive index profiles are calculated by finding solutions to the eigenvalue problem, based on a suitable application of the boundary conditions at the waveguide surface. The WKB approximation then yields, as a function of the depth parameter, h, the following integral expression for the polymer refractive index [26]:

$$\int_0^{h_q} \left(N^2(h) - N^2_{eff_q} \right)^{1/2} dh = \frac{4q-1}{8} \qquad q = 1, 2, \ldots \qquad (7.1)$$

Here, h is normalized to the free-space wavelength, l; h_q is defined by the relation $N(h_q) = N_{eff_q}$; $h_0 = 0$; $N_{eff_0} = N(0)$; and the integral is performed over the extent of the polymer film. This expression provides an accurate treatment in the present case because the effective index of the zeroth order mode, within the multimode waveguide, is very close to that of the waveguide surface index. As seen in Figure 7.2, a number of different tuned index profiles can be achieved through careful control of the polymer absorption and dehydration process. We note that the profile variations on the three different substrates result from differences in process parameters, rather than from substrate-related dependencies. In all cases, a reduction in the overlap between the guided-mode evanescent field and the underlying substrate, due to the presence of a GRIN profile, resulted in the observation of strong waveguiding and propagation losses of 0.1 dB/cm in the PMSWs. Low-loss waveguiding is also depicted in Figure 7.3 for polymer films deposited on Si, quartz, Au, and Al_2O_3, respectively. Similar results have been observed for polymers

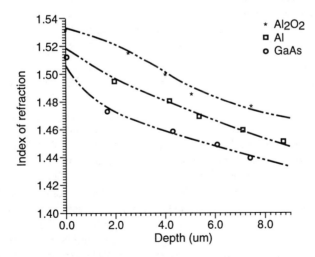

Figure 7.2 Graded index profiles of Al_2O_3, Al and GaAs determined by the IWKB method.

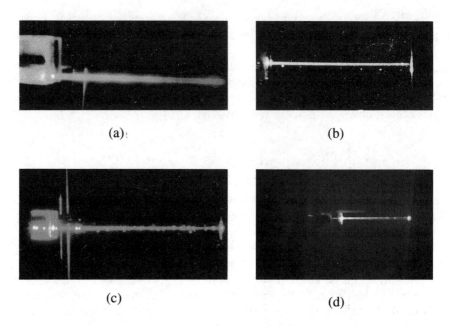

Figure 7.3 Graded index polymer microstructure waveguide on: (a) Si; (b) Quartz; (c) Au; and (d) Al_2O_3.

deposited onto other substrate materials, including Si, $LiNbO_3$, fused silica, Al, Cu, Cr, Kovar, BeO, AlN, and phenolic PC board, respectively. Without the use of special adhesion promoters or surface planarizing layers, the majority of polymer films displayed excellent deposition and adhesion properties on most of the substrates described previously. There are no observed effects on the substrate materials described previously, due to the presence of the polymer layer and the associated chemical processing techniques.

7.3 FORMATION OF LINEAR AND CURVED CHANNEL WAVEGUIDE ARRAY

In this section, we report the cross-link induced linear and channel waveguide arrays on GRIN photolime gel [21,30]. The GRIN characteristic of the polymer thin film allows us to implement such a channel waveguide on any substrate of interest. After the polymer film is spin-coated on a substrate, it is dipped into ammonia dichromate solution for sensitization. Formation of a channel waveguide is realized by cross-linking the polymer film through ultraviolet exposure. It is observed that the cross-linked area has a higher index of refraction than the unexposed area. The modulation index, due to the photo-induced cross-link, can be as high as 0.2 [31]. Consequently, the channel waveguide confinement and thus the packaging density (number of channels per centimeter) can be extremely high. Implementation of the waveguide pattern is realized either by laser beam direct writing or through a conventional lithographic process. The graph in Figure 7.4 was produced by computer simulation based on the effective index method [32]. It shows the optimal single-mode channel waveguide dimensions for an optical wavelength of 1.31mm. The cutoff dimension is shown with index modulation as a parameter. Note that the cutoff boundary defined here is for E_{12}^x [33], above which the channel waveguide becomes multimode. Marcatili's five-region method was also used for this purpose. The result (not shown) is very close to that of Figure 7.4. However, the cutoff dimension for E_{11}^x [29,30] determined by the effective index method is quite different from that determined by Marcatili's method. In any case, the waveguide cutoff dimension for E_{12}^x is well above the cutoff condition for. The discrepancy between these two methods for E_{12}^x is negligible.

The experimental results of a linear polymer channel waveguide array working at 0.63 mm and 1.31 mm were experimentally verified using the setup shown in Figure 7.5(a). A microprism was employed [34] to provide simultaneous coupling for multiple channels. The observed near-field patterns for 0.63 and 1.31 mm, using the setup shown in Figure 7.2(a), are displayed in Figure 7.5(b,c). A single-mode waveguide at 0.63mm was further confirmed by employing a Si charge coupled photo detector (CCPD) array to image the mode profile in both horizontal and vertical directions.

The packaging density of the waveguide device shown in Figures 7.5 and 7.6 is 1,250 channels/cm, which is approximately two orders of magnitude higher than that of electrical wiring for board-to-board interconnections. Confirmation of single-mode guiding at 0.63mm assures that the waveguide-mode for 1.31mm shown in Figure 7.5(c) is also single-mode [35].

To provide optical interconnects on the intra-multichip module (MCM), inter-MCM, and backplane levels, an optical bus may need to be curved to transmit optical signals to

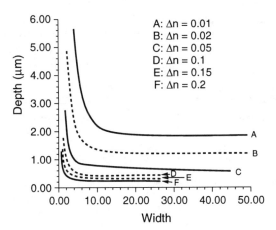

Figure 7.4 Single-mode polymer channel waveguide dimension based on the effective index method with index modulation as a parameter.

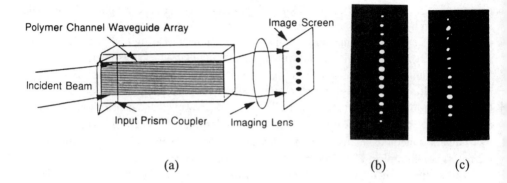

Figure 7.5 (a) Experimental setup for the Observation of polymer channel waveguide array (channel width = $2\mu m$) and near-field patterns of the channel waveguide array for wavelengths at (b) 0.63 μm and (c) 1.31 μm, respectively. The channel separation is $8\mu m$.

the addressed location (e.g., memory). To evaluate the feasibility of generating a curved polymer waveguide, channel waveguides with radii of curvature (ROC) of from 1 to 40 mm were made. Table 7.2 summarizes the parameters of the curved waveguides fabricated.

Large index modulation caused by photo-induced cross-linking provides a better waveguide confinement factor and thus a smaller ROC. Theoretically, the loss due to waveguide bending can be negligibly small if [36]:

$$\text{ROC} > 3N_{eff}^2 \lambda / \pi [N_{eff}^2 - N_c^2 + (\lambda/2b)^2]^{3/2} \qquad (7.2)$$

where N_{eff} is the guided wave effective index, N_c is the cladding layer index, and l and b are the optical wavelength and the width of the channel waveguide (see the inset of Fig. 7.4). Note that in deriving (7.1), the wave number of the guided wave along the x direction is assumed to be equal to $1/2\ b$. Figure 7.7(a–c) shows the experimental results of a curved polymer channel waveguide. The coupling angle of the input prism was set at the phase matching angle for E_{11}^x mode. No surface scattering can be observed from the image taken by a vidicon camera. The confirmation of curved waveguiding was made by scribing the curved channel waveguide surface (Figure 7.7(a)). A bright spot was observed. The bright

Table 7.2
Curved Waveguide Parameters Under Investigation

Channel Width (μm)	Curvature Radius (mm)	Degrees of Rotation	Channel Width (μm)	Curvature Radius (mm)	Degrees of Rotation
10	1	90,180	10	4	90,180
10	2	90,180	10	4.5	90,180
10	3	90,180	10	40	90,180

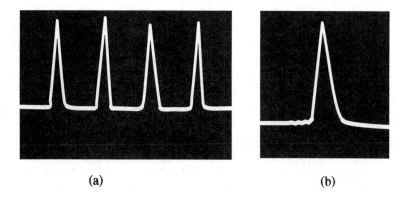

Figure 7.6 Mode profile of the single-mode channel waveguide array in the (a) horizontal (peak-to-peak separation is 8μm) and (b) vertical directions.

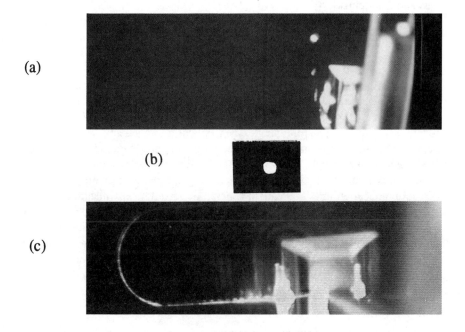

Figure 7.7 Experimental results of curved channel waveguides: (a) observation of a curved channel waveguide (ROC = 3mm) at the scribed waveguide surface (no surface scattering can be observed); (b) near-field pattern of the curved channel waveguide with the same ROC as Figure 7.7(a); (c) observation of the curved channel waveguide with a ROC equivalent to that of Figure 7.7(a) after surface treatment.

streak on the waveguide surface, which can be an indicator of loss [34], disappeared (Figure 7.7(a)) in our linear and curved channel waveguide devices. The photograph shown in Figure 7.7(b) is the linear field mode pattern at the output end of the curved channel waveguide using end-face imaging. Special surface treatment is needed to increase the loss and thus make surface scattering visible. Figure 7.7(c) shows a curved waveguide with the same pattern of Figure 7.7(a) after surface treatment.

The value of the index modulation plays an important role in minimizing the propagation loss for the curved region. The larger the index modulation, the better the waveguide confinement factor. the evanescent field decays drastically in the cladding region. The radiation loss due to velocity mismatch [36,37] is thus minimized. As far as the waveguide propagation loss is concerned, the measured loss in the neighborhood of 0.1 dB/cm has been consistently observed. Purification of the polymer thin film significantly reduced the volume scattering centers within the guiding medium.

The advancement in high-speed computers requires an interconnection technology capable of routing high-speed signals, especially clock distribution for a synchronous bus, such as NUbus [38], and high-speed data for a compelled asynchronous bus, such as VMEbus and Futurebus [38].

Electrical interconnects using either thin metal films or transmission lines are not efficient enough to provide highly parallel, high-speed (>1 GHz) connection for distances longer than 1cm. Optical interconnections based on polymer waveguides have been reported to provide two-dimensional and three-dimensional optical interconnects with 60 GHz modulation bandwidth and 22-dB signal-to-noise ratio [39]. The results presented herein give us a packaging density as high as 1,250 channels per cm with propagation loss as low as 0.1 dB/cm. For computing systems using hypercube, daisy chain, and star interconnect architectures, the single-layer optical interconnects reported herein are sufficient to provide the required interconnectivity. Implementation of fiber arrays for current backplane buses such as VMEbus and Futurebus is not practical because these bus architectures connect all processor and memory cardboards in parallel along a set of common communication lines that can, in general, be driven by any machine and listened to simultaneously by all machines (i.e., using low-efficiency coupling holograms). Thus, any machine on a bus can communicate directly with any other (subject to bus ownership protocols). As a result, implementation of a practical optical backplane bus requires a microlithographic process that is not applicable for fiber arrays. Furthermore, the GRIN property of the polymer waveguide facilitates the implementation of optical integrated circuits on any optoelectronic substrate.

7.4 FORMATION OF WAVEGUIDE HOLOGRAMS BY LOCAL SENSITIZATION WITH APPLICATIONS TO WDM

The formation of tuned index polymer waveguides, on virtually any kind of substrate, as described previously, provides the basis for the development of other passive and active optical devices that are suitable for signal routing and modulation. In particular, devices that use multiplexed holograms for wavelength multiplexing and demultiplexing operations are key components for advanced optical interconnection architectures. Based on the refractive index tuning capability within the polymer film, microlithographic techniques are used, in conjunction with a local sensitization process, to generate waveguide multiplexed holograms. After the polymer film is deposited and cured on top of a suitable substrate, a photoresist window is defined using standard photolithography. Local sensitization is then achieved by immersing the selectively-masked sample into a room-temperature ammonia dichromate solution [40]. The masking material is then removed, and within 2 hr after drying and stabilization of the sensitized region, the sample is ready for holographic recording and processing.

A similar process is used to form multiplexed holographic phase gratings in the PMSW films. The gratings are created by successively exposing holographic patterns within the selectively defined and sensitized regions of the waveguide. A two-beam interference recording method is used to define individual holographic gratings, each at a different recording angle and each having a sinusoidal phase modulation profile, such that:

$$K_i = 2k_{\lambda i} \sin(\alpha_i/2) \tag{7.3}$$

where k_{li} and K_i are defined as:

$$k_{\lambda i} = N_{\it{eff}_{\lambda i}} \frac{2\pi}{\lambda_i} \tag{7.4}$$

and

$$K_i = \frac{2\pi}{\Lambda_i} \tag{7.5}$$

where a_i is the angle of Bragg diffraction, L_i is the ith holographic grating period, K_i is the ith grating wave vector, and $N_{\it{eff}_{li}}$ is the waveguide-mode effective index at l_i. The resultant grating wave vector for each hologram lies within a plane that is parallel to the waveguide surface. A schematic of the locally sensitized polymer waveguide, containing a multiplexed holographic phase grating, is depicted in Figure 7.8.

The previously described technique was used to develop a four-channel wavelength division demultiplexer (WDDM) that operates at the center wavelengths of 632.8 nm (red), 611.9 nm (orange), 594.1 nm (yellow), and 543.0 nm (green), respectively. The device, consisting of a locally-sensitized single-mode PMSW and prepared on a Soda-Lime glass substrate, is shown in Figure 7.9. Only the TE_0 guided mode of the waveguide is excited via the prism coupler and used in the present device. Holographic recording angles, for each of the deflected wavelengths, were chosen, commensurate with the

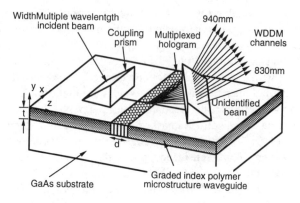

Figure 7.8 Locally sensitized polymer waveguide, containing a multiplexed holographic phase grating.

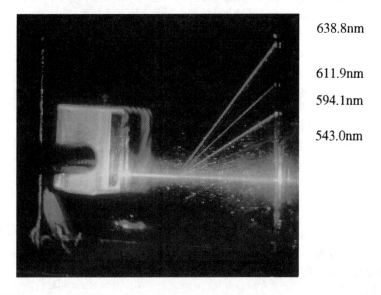

Figure 7.9 Four-channel polymer waveguide wavelength division demultiplexer working at 638.8 nm (red), 611.9 nm (orange), 594.1 nm (yellow), and 543.0 nm (green).

effective index dispersion relations of the polymer film. The Bragg diffraction angle of each resulting transmission phase grating was designed to satisfy the phase-matching condition for the signal wavelength of interest. Exposure parameters were adjusted during successive holographic recordings in an attempt to optimize diffraction efficiencies. Each grating was, therefore, designed to be capable of deflecting only one wavelength within a 4- to 10-nm spectral bandwidth. Device measurements yielded a crosstalk figure of less than −40 dB between adjacent channels and a corresponding diffraction efficiency of better than 50% at each wavelength. The angular and spectral sensitivities for the demultiplexer were theoretically determined to be within 0.2° to 0.4° and ~4 to 10 nm, respectively. It is expected that even higher diffraction efficiencies for each grating can be achieved if the grating modulation index and the interaction length are optimized accordingly. In addition, while excellent crosstalk figures were obtained for the present waveguide demultiplexer, it is conceivable that the presence of substrate radiation modes from each signal carrier, generated by the interaction with other existing gratings, will limit the overall device efficiency for closely spaced channels. These modes might be present as a consequence of random fluctuations in the grating modulation index and waveguide thickness or from variations in the tuned refractive index profile.

A 12-channel WDDM device was further demonstrated on a semi-insulating GaAs substrate [41]. The gratings were designed to operate at the diffraction angles of 10°, 15°, 20°, 25°, 30°, 35°, 40°, 45°, 50°, 55°, 60°, and 65° to selectively disperse signals at the center wavelengths of 830, 840, 850, 860, 870, 880, 890, 900, 910, 920, 930, and 940 nm, respectively. For each l_j, the corresponding recording angles were selected to generate a waveguide transmission hologram with the desired grating periodicity. The interaction length of the multiplexed waveguide hologram is 0.4 mm. The mode dots coupled out of the prism coupler are shown in Figure 7.10 with the corresponding wavelengths as indicated. The observation of these clean mode dots verified the quality of the polymer waveguide. A propagation loss in the neighborhood of 0.1 dB/cm has been routinely achieved in a class 100 clean-room environment.

The noise of the fanout channels of the WDDM device is mainly from the crosstalk of the signals from adjacent channels. An average crosstalk figure of −20.5 dB was measured with diffraction efficiency from 40% to 55% among these output channels. The

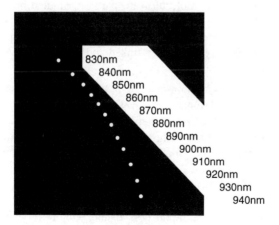

Figure 7.10 Twelve-channel WDDM on a semi-insulating GaAs substrate.

spectral width of the Ti:Al$_2$O$_3$ laser from Spectrophysics was also measured. A −3 dB bandwidth of ~4nm was found. The results suggest that the WDM devices of better than 4nm wavelength separation cannot be experimentally realized without significant channel crosstalk. Theoretically, our device structure is capable of operating at a channel-to-channel spacing as small as 1nm under the current design when a DFB laser diode is employed. The −20.5 dB crosstalk is primarily due to the wavelength spreading of the Ti:Al$_2$O$_3$ laser rather than the waveguide device itself.

As far as the throughput intensity is concerned, the 33% diffraction efficiency represents an output power as high as 50 mW. For a communication system involving the reported WDDM device, the system power budget will be determined by laser power, modulation speed, bit error rate, and detector sensitivity. Employing a PIN-FET as the demodulation scheme, we can theoretically use a ~0.5-mW semiconductor laser to obtain 1 Gb/s communication with a 21.5-dB signal-to-noise ratio. The power budget assumes 50% diffraction efficiency, 1-dB waveguide propagation loss, 3-dB waveguide coupling loss, 2-dB hologram excess loss, 4-dB fiber propagation loss, 5-dB system power margin, and room temperature operation condition with an amplifier noise figure equal to four. The current design allows us to provide 60-channel multiplexibility with the maximum value of index modulation set at 0.1.

7.5 χ^2 ELECTRO-OPTIC POLYMER

Gelatin is a class of biopolymer that consists of thousands of 10Å to 20Å amino-acids. Gelatin has been classified as a superpolymer [42] because of its extraordinary chemical and physical properties and its molecular structure. The remarkable fact is that biological gelatin has two extraordinary microstructural morphologies:

- Two-dimensional network: molecules are partially aligned and cross-linked with interchain hydrogen bonds to form a helical collagen-like sheet (Fig. 7.11);
- Thermoplastic −> thermosetting: the processible gelatin (thermoplastic form) can be readily transformed into a highly cross-linked and insoluble polymer (thermosetting form) using heat or UV radiation curing.

When gelatin is incorporated with ammonia dichromate, it becomes a widely used holographic recording material, dichromate gelatin (DCG). Holographic gelatin has attracted a great deal of interest because it has a variety of applications. Incorporation of different guest molecules significantly changes the electro-optic (EO) properties of the polymer matrix. Such a material characteristic motivates us to further develop χ^2 nonlinear polymer using a nonlinear dye and photolime gel combination.

The search for materials with fast response times (<1 ns) combined with a large EO effect ($Dn > 10^{-4}$) and which are simple to process has been of intense interest for some time. EO materials, such as liquid crystal and photorefractive and piezoelectric materials, offer large EO effects but exhibit a switching time that is too slow (50 ns to 500 ms) for practical applications. Also, the widely used inorganic crystalline (LiNbO$_3$)

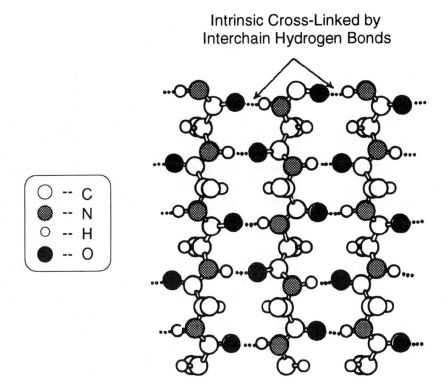

Figure 7.11 The sheet-like structure of photolime gel with intrinsic crosslinking.

thin film is very difficult to process. In contrast, EO polymers offer excellent properties, including large EO coefficients (Dn 5 × 10^{-4}), low-temperature processing (<200°C), and fast response times (<300 ps). In recent years, EO poled polymers have evoked international attention and effort. High glass transition temperature (T_g) polymers or cross-linked polymers have proved their effectiveness for stability enhancement. Unfortunately, their long-term EO stability is still a bottleneck for practical device applications [43]. While tremendous efforts have been put into the characterization of poled materials [44], device research on EO polymers is still in its infancy. For the most part, this is because the basic criterion for guided wave materials is that they have the capability to be integrated with other optoelectronic components. This aspect of most EO polymers has yet to be examined. Polymeric gelatin has been proved to be an excellent material for the fabrication of an array of guided wave optical elements, but no active EO device has been built in the gelatin media. The purpose of this section is to describe the achievement of assembling an EO modulator by the incorporation of active EO materials in the highly cross-linked gelatin matrix.

The gelatin used in this experiment is known as type-A gelatin, an acid-treated protein derived from animal tissues. It has an isoelectric point between pH 7 and pH 9. Gelatin is soluble in aqueous solutions and insoluble in most organic solvents, such as benzene, acetone, petroleum ether, and absolute alcohol. This low solubility limits the use of a large number of EO dyes, which are frequently water insoluble. In this experiment, 4-nitrophenol (NP) from Aldrich was selected as the active EO material. Nitrophenol has very high solubility in water (25 mg/ml) and can homogeneously disperse in gelatin matrix. The electron acceptor NO_2 group and electron donor OH group form a linear dipole moment across the p conjugated electronic system. The molecular nonlinear hyperpolarizability of NP in solution has been gauged to be about 1/3 that of methyl nitroaniline (MNA) by measuring the electric field induced second harmonic generation [43]. A solution containing polymer gelatin (30 g), nitrophenol (~20% to 30% by weight), and water (400 ml), was set in a water bath (60°C) until it became clear. The solution was filtrated and spin-coated on an aluminum (Al) glass substrate. The absorption peak of NP is at 304 nm. It has a very wide optical bandwidth, from 350 to 2,700 nm, which is much wider than most EO polymers. From the interference fringe, the thickness of the thin film was measured to be 10 mm. The thickness of polymer thin films can be varied from submicron to 100 mm by changing the ratio of water and gelatin or the speed of spin-coating. The dry thin film was covered with an optically transparent indium tin oxide (ITO) glass substrate. The Al and ITO electrodes served as poling fields as well as modulating fields. Although gelatin molecules are partially aligned microscopically, no macroscopic birefringence was found in the gelatin film before poling. The poling and curing processes were performed at the glass transition temperature (T_g), 60° to 70°C, for 30 min.

The experiment was performed in the reflection geometry [44] (see Fig. 7.12) with a 3-mW HeNe laser beam polarized at 45° to the plane of incidence so that the parallel

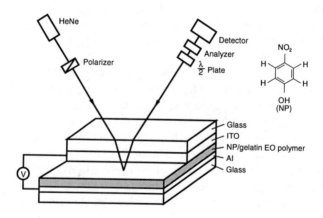

Figure 7.12 Electro-optic measurement setup.

and perpendicular components of the optical field were equal in amplitude. The beam reflected from the Al coating was propagated through a half-wave plate, an analyzer, and then into a photodiode. The analyzer was set at a cross-polarization angle with respect to the polarizer. The half-wave plate was used to generate $p/2$ phase retardation between the x and y polarizations of the laser beam. When the phase shift, y, is biased at $p/2$ with a 1/2 plate, the output intensity is approximately linearly related to the phase retardation as well as to the applied modulating voltage.

The ratio between modulated intensity, I_m, and the dc signal, I_c, can be expressed as:

$$I_m/I_c = (2p/\lambda) \times 2d \times \Delta n \times (\sin^2\theta/\cos\theta) \quad (7.6)$$

where q is the angle of incidence, d is the thickness of the polymer film, Dn is the birefringence introduced by the modulation field, and I_c is the output intensity at $p/2$ phase shift. This intensity can be obtained by rotating the 1/2 plate. This birefringence is given by:

$$\Delta n = 1/2(n_3^3 \times \gamma_{33} - n_1^3 \times \gamma_{31})V/d \quad (7.7)$$

where n_1 and n_3 are refractive indices along the directions perpendicular and parallel to the poling electric field, respectively. For an initially isotropic system, the corresponding EO coefficients for light polarized perpendicular and parallel to the surface plane are related to γ_{31} and γf_{33}, $\gamma_{33} = 3\gamma_{31}$. This assumption can be justified because nitrophenol is randomly dispersed in the polymer matrix. Therefore, we have:

$$\lambda_{33} = 3\lambda I_m \cos\left(\frac{\theta}{4\pi}\right) n^3 V_m I_c \sin^2\theta \quad (7.8)$$

where V_m is the modulating voltage. The EO effect of NP/gelatin polymer was observed with a poling field of 400V across a 10-mm thick film. The optical modulation signal from a HeNe laser was displayed on an oscilloscope (see Fig. 7.13). The upper curve is 3.3 MHz RF modulation with a V_{p-p} of 90V. The bottom curve is the corresponding modulated optical signal. Based on (7.8), the EO coefficient of NP/gelatin, g_{33}, of 10 to 40 pm/V was obtained at 632.8 nm with a direct current. The spread in EO coefficient is due to variations in film thicknesses, dye concentrations, and poling fields. The index change introduced by the modulation field is 1–4 x 10^{-4}. This value is comparable to many poled polymers and inorganic $LiNbO_3$. The major reason for the large nonlinear effect is the high concentration of EO moieties, 35% by weight of NP, that can be incorporated into the gelatin matrix without any cluster appearing. This concentration is much higher than that in most guest/host systems (~15%). Effective poling may account for additional EO enhancement. For guest/host systems, the typical poling efficiency is between 10% and 20%. Poling efficiency for short-axis molecules, such as NP, is believed

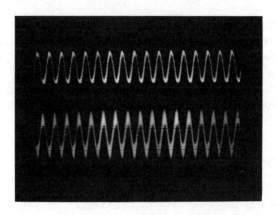

Figure 7.13 The optical signal from a HeNe laser is modulated and displayed on an oscilloscope.

to be higher. This is because the short-axis molecule is subjected to a smaller axial force potential. Thus, the molecules can be rotated and aligned more easily.

The thermal stability of the poled EO response has been a critical issue in the practical application of poled polymers. The high T_g and cross-linked polymers have been used to reduce the diffusion of EO groups and thus increase the thermal stability. The typical EO decay curve of guest/host polymers shows an initial fast decay to 20% within five days, followed by a slower decay. The EO stability of NP/gelatin shows that 40% of the original birefringence persists five days after poling at room temperature. As shown in Figure 7.11, gelatin has two-dimensional sheet-like structures held by intermolecular hydrogen bonds. During the thermal curing process, these chemical bonds in NP/gelatin can be further cross-linked and enhanced to form a rigid and irreversible thermosetting polymer. The observed results indicate that the highly cross-linked hydrogen bonds are responsible for the EO stability. In addition to the intrinsic cross-linking, it is known that gelatin can be hardened and rendered insoluble by cross-linking between chromium ions and amino acid groups (in DCG) into a more stable three-dimensional network. We are currently characterizing the stability enhancement by cross-linking the poled gelatin with dichromate under exposure to laser radiation.

We have successfully constructed and demonstrated an EO modulator in biopolymeric gelatin. Nitrophenol was chosen as the active EO moiety and was shown to have a large EO effect. The planar structure of the gelatin plays an important role in the stability of the EO group in the polymer matrix. Optical birefringence was generated with a poling field at temperatures between 60° and 70°C. The results illustrate that active EO devices can be integrated into holographic gelatin films.

7.6 RARE-EARTH ION-DOPED POLYMER WAVEGUIDE AMPLIFIER

All the demonstrated GRIN polymer waveguide devices reported thus far are summarized in Table 7.1. These devices include planar waveguides, channel waveguides, EO modulators, multiplexed gratings, multiple guiding layers, waveguide lenses and waveguide amplifier. It is clear from Table 7.1 that all the guided wave elements made on $LiNbO_3$ and GaAs substrates can be replaced by polymer-based devices. More important, however, waveguide devices such as highly multiplexed gratings and large area multiple guiding layers can be realized only by using polymer-based technology. Formation of high-efficiency holograms is important for the realization of the proposed architecture (Figure 7.14). Each individual building block shown in Figure 7.14 has been successfully developed, and all the results can be found in the cited references.

Most of the waveguide lasers recently reported are Er^{+++}-doped lasers, primarily due to the 1500-nm lasing band. Depending on the wavelength of stimulated emission, the energy levels of Er^{+++}- and Nd^{+++}-doped glass waveguides labeled with the dominant Russell-Saunders $^{S}L_J$ term are shown in Figure 7.15.

The Nd glass laser is a well-known four-level system ($^4F_{3/2}$, $^4I_{11/2}$). The lasing action occurs between the metastable state $^4I_{13/2}$ to $^4I_{15/2}$ for Er^{+++}-doped glass waveguides. The fluorescence of higher states is almost fully quenched by the process of nonradiative multiphonon relaxation [45], which has a rate of more than 10^5–10^7/sec. The performance of waveguide lasers and amplifiers is governed by the relevant electronic and optical characteristics of active ions, such as cross-sections, spectral shapes of excitation and absorption bands, lifetime at the metastable state, and their immunity to concentration quenching.

The host material has an important influence on all these properties. Note that the observation of Nd^{+++}- and Er^{+++}-doped waveguide amplifiers working at 1.06 and 1.50 mm, respectively, on various waveguide substrates, such as glass (amorphous state) and $LiNbO_3$ (single crystal), implies that the associated metastable states do exist in various host materials.

The polymer introduced is soluble in water. As a result, the chemical compounds containing REIs can be mixed with the host polymer as long as they are also soluble in water. The concentration will be below the level of microscopic clustering, which quenches the active ions.

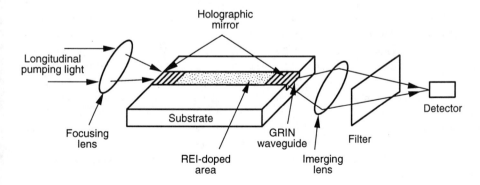

Figure 7.14 REI-doped polymer waveguide laser/amplifier.

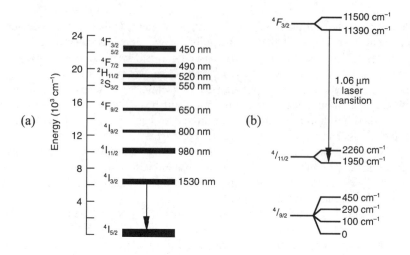

Figure 7.15 Energy levels of (a) Er^{+++}-doped and (b) Nd^{+++}-doped glass waveguides. Metastable states and the corresponding stimulated Emission lines are shown.

Fluorescence lifetimes for $^4I_{3/2}$ for various glasses are 10^{-2}sec (except for borate). There are certain concentration quenchers that basically shorten the lifetime of metastable states. The most serious quenchers are the admixed O-H groups whose concentration must be less than $3 - 5 \times 10^{18}/\text{cm}^3$ [46] for a glass waveguide amplifier. The existence of O-H groups generates a number of intermediate states between $^4I_{13/2}$ and $^4I_{15/2}$ [47]. Such states, primarily from water molecules, significantly reduce the lifetime of the metastable state $^4I_{13/2}$. The general rule of thumb is that the lifetime of the excited state will be temperature dependent if the gap is less than 10 times the effective phonon frequency and completely quenched if less than four times [47]. The water molecules within the photolime polymer can be totally eliminated using the dehydration process. Such a dehydration process provides us with not only an H$_2$O-free GRIN polymer waveguide but also a much higher temperature dynamic range ($-100°$ to $+180°$C), which ensures device survivability under intense pumping situations.

The elimination of the excited state absorption (ESA) of pumping photons (e.g., $^4I_{13/2} > {^4}S_{3/2}$ transition) in an Er^{+++}-doped sample is always a potential problem. In three-level lasers, for there to be a gain in the first place, a significant fraction of the doping population must reside in the upper laser level. It is feasible to reduce the effects of ESA by detuning the pump wavelength away from the center of the ground-state absorption feature [48]. The lasing level terminates at a thermally populated level above the ground state, and the laser therefore behaves as a quasi-four-level system. Consequently, this inherent loss of a waveguide laser was reported [48]. The result demonstrated that lasting efficiency increased with increased lasing wavelength.

The host material (i.e., photolime polymer gel) is an excellent guiding medium due to its wide transmission bandwidth (300 to 2,700 nm). The GRIN characteristics of this material (Figure 7.16) allow the formation of high-quality (loss < 0.1 dB/cm) single-mode devices on an array of substrates, as shown in Table 7.3. The formation of a waveguide laser cavity is realized by recording two reflection holograms at the two end faces (Figure 7.14) of the waveguide. The hologram fabricated is similar to a Lippmann hologram made for laser goggles, which has an optical density (OD) of up to six (10^{-6}). We have demonstrated waveguide transmission holograms by locally sensitizing the emulsions [49]. By

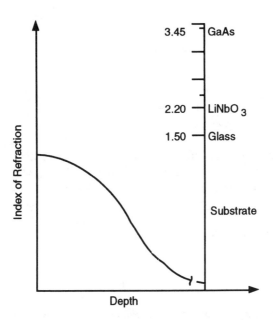

Figure 7.16 Graded index profile of the polymer waveguide. (This characteristic provides us with a universal method of implementing a Waveguide laser on any substrate of interest (Table 7.3)).

Table 7.3
Substrates Upon Which GRIN Polymer Waveguides Have Been Demonstrated

Waveguide	Substrate
Semiconductors	Si, GaAs
Insulators	Glass, LiNbO$_3$, quartz, fused silica, plastic thin film (PC board)
Conductors	Aluminum, chromium, gold, Copper, kovar
Ceramics	Al$_2$O$_3$, BeO, AlN

changing the recording geometry, a reflection waveguide hologram can easily be made for this purpose.

For the three-level laser system, the equation governing the laser action is given by

$$\frac{\partial n_1}{\partial t} = \left(n_2 - \frac{g_2}{g_1}n_1\right)c\phi\sigma + \frac{n_2}{T_{21}} - W_p n_1 \quad (7.9)$$

and

$$\frac{\partial n_2}{\partial t} = -\frac{\partial n_1}{\partial t} \quad (7.10)$$

In deriving (7.10), $n_{tot} = n_1 + n_2$ is assumed. In (7.9) and (7.10), n_1 and n_2 are the respective population densities of the upper and lower levels corresponding to the lasing wavelength, g_1 and g_2 are the associated degeneracy, c is the speed of light in the medium, ϕ is the photon density (photons/cm^3), s is the stimulated emission cross section, t_{21} is the mean lifetime of the metastable state ($^4I_{13/2}$ for an E^{+++}-doped waveguide laser), and W_p is the effective pumping rate. Note that longitudinal pumping provides an extra large interaction length and thus a much higher absorption rate than transverse pumping. The terms on the right side of equation (7.9) express the net stimulated emission, spontaneous emission, and optical pumping. For a laser working at CW operation under constant pumping power (i.e., $dW_p/dt = 0$), the inversion population density

$$n = n_2 - \frac{g_2}{g_1}n_1 \quad (7.11)$$

needs to be achieved and maintained at a constant value. The value of ϕ, photon density, also needs to be maintained. Optimum REI concentration in conjunction with pumping power provides a desired n value (7.11). A stable cavity design keeps the value of ϕ, which is linearly proportional to the intensity of stimulated emission, at a high level.

Similar to polymer waveguide preparation, the Nd^{+++}/photolime gel was spin-coated on top of a substrate. The Nd^{+++} concentration was determined to be 1.03×10^{20}/cm^3. On standing at temperatures below 30°C, solutions containing more than 1% photolime gels become rigid through material cross-linking and exhibit rubber-like mechanical properties. This gelation process holds for both pure gel and doped gel alloy. The absorption spectrum, which determines the optimal pumping wavelengths, was experimentally confirmed using a spectrophotometer. Two major peaks shown in Figure 7.17(a) are in the 750-nm and 790-nm neighborhood. The film thickness used for this measurement was 250 mm. These absorption peaks are very similar to those shown in the Nd:glass ED-2. Due to the amorphous nature of the host material (i.e., photolime gel), the absorption width of each peak is much wider than single crystal Nd:Yag laser. The fluorescent spectrum of Nd^{+++}:photolime gel thin film was also observed by pumping the active medium using a

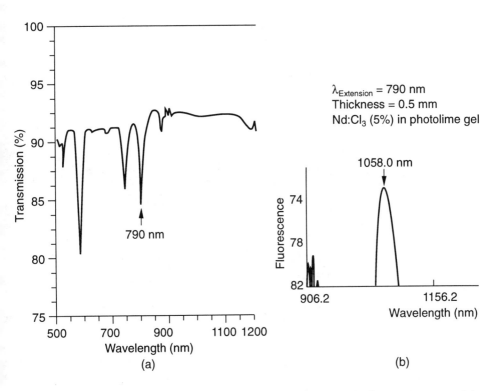

Figure 7.17 (a) The absorption spectrum of the Nd^{+++}-doped polymer film; (b) the fluorescent spectrum of the Nd^{+++}-doped polymer film.

Ti:-Sapphire laser working at 790 nm. The fluorescent spectrum detected by an optical spectrum analyzer is illustrated in Figure 7.17(b) where a fluorescent peak centered at 1.06 mm is indicated. Note that the broadening effect due to the amorphous structure of the photolime polymer thin film is also observed. It is equivalent to the $^4F_{3/2}$ to $^4I_{11/2}$ transition of an Nd:glass laser. A wider line increases the laser threshold because a larger population inversion is required to obtain the threshold value of amplification. However, this broadening effect has an advantage. A broader line offers the possibility of obtaining and of amplifying shorter light pulses. In addition, it permits the storage of larger amounts of energy in the amplifying medium for the same linear amplification coefficient.

The setup for the demonstration of a GRIN polymer waveguide amplifier is shown in Figure 7.18. The interaction length (i.e., the waveguide region for two counter-propagating laser beams) is 2.2 cm. The propagation loss at 750 nm was measured to be 9.3 dB/cm, which is approximately two orders of magnitude higher than pure photolime gel waveguide. The pumping laser beam and the 1.06-mm Nd:Yag laser beam are both coupled into the Nd:photolime gel waveguide using prism coupling method. The input prism coupler for 750-nm and 790-nm laser beams also functions as an output prism coupler

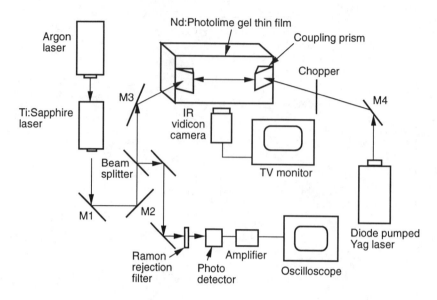

Figure 7.18 Setup for the demonstration of the GRIN polymer waveguide amplifier.

for the amplified 1.06-mm laser beam. An IR vidicon camera is used in the alignment process. The Nd:photolime gel medium is a planar waveguide without horizontal confinement. Therefore, the overlap between the pumping beam and the amplified laser beam (1.06 mm) plays an important role in achieving the maximal value of amplification. The measured gain for the demonstration shown in Figure 7.18 is illustrated in Figure 7.19. An 8.5 dB gain is clearly observed [50].

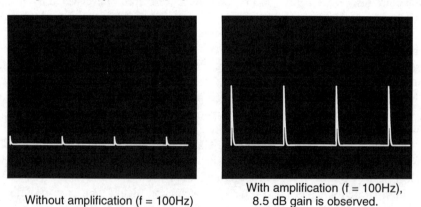

Figure 7.19 Measured results of GRIN polymer waveguide amplifier.

7.7 POLYMER-BASED EO WAVEGUIDE MODULATOR

An important characteristic of modulators and switches is the bandwidth or range of modulation frequencies over which the EO material can be operated. By convention, the bandwidth of a modulator is usually taken as the difference between the upper and lower frequency responses, which is 3 dB less than the maximum modulation depth. Minimizing the switching time is most important when large-scale arrays or switches and modulators are used to route optical signals over desired paths. Similarly, modulation bandwidth is a critical factor when many information channels are to be multiplexed onto the same optical beam. Thus, the usually fast switching speed and large bandwidths of waveguide switches and modulators make them particularly useful in large communications systems such as phased-array antennas. There are two different types of electrode structures that have been realized to date. They are the lumped electrode structure and the traveling-wave. The first regards the modulator as a capacitor and usually has a 50Ω resistance in parallel with the device. The second type treats the electrode pair as a continuation of a transmission line. The 3-dB bandwidth of the second type is usually higher than the first. The bandwidth of the lumped electrode-type is limited by the RC constant, whereas the bandwidth of the traveling-wave electrode-type is constrained by the difference of the effective indices of the optical guided wave and the microwave, and the interaction length.

A GRIN polymer waveguide modulator based on current injection was reported previously [51]. The polymer-based EO device reported herein provides wider modulation bandwidth due to its small walk-off when compared with $LiNbO_3$ and GaAs counter parts. The device structure is shown in Figure 7.20 [52].

A channel waveguide array is made using the method described in Section 7.2.1. The HeNe laser beam coupled into the channel waveguide array is circular polarized. Mode conversion is activated through the electric field induced phase velocity difference between transverse electric (TE) and transverse magnetic (TM) modes. A third section

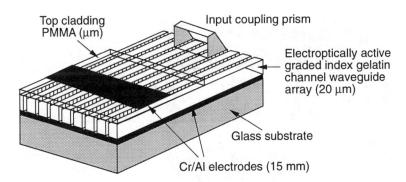

Figure 7.20 Photolime polymer-based electro-optic mode converter [52].

of the waveguide with aluminum cladding is also portrayed in Figure 7.20. The TM wave has a much stronger propagation loss in this section. The modulation effect is represented by:

$$U_\pi = \frac{3d}{L} \frac{\lambda}{2n^3 \gamma_{33}} \qquad (7.12)$$

The combination of these effects results in a modulation depth of 12 dB with an applied voltage of 10V (p-to-p). The result is shown in Figure 7.21. Emitter-coupled logic (ECL) and TTL compatible electrical-to-optical digital signal conversion can also be generated when appropriate logic input is provided.

A photolime polymer-based traveling wave EO Mach-Zehnder interferometer with 40-GHz electrical bandwidth was reported recently [53]. This device structure was based on the $\chi^{(2)}$ nonlinear polymer described in Section 7.4. It is clear from the measured result depicted in Figure 7.22 that the modulation bandwidth can be higher than 40 GHz. The data shown are limited by the availability of the microwave source. Performance improvement relies on a new procedure and material combination through which the performance of each individual integrated optical device is upgraded. Polymer-based guided wave devices provide both architecture (due to GRIN characteristics) and device-level improvements not achievable through conventional GaAs- and $LiNbO_3$-based devices.

The major stumbling block for polymer-based EO devices is their relatively short life time, usually weeks or months, due to the relaxation of the polarization density. X^2 nonlinear polymers are prepared by either the guest/host mixing method or side chain polymerization or cross-linking. If a strong cross-link can be generated after the formation of nonlinear polymer, long-term stability of the polymer matrix will be achievable.

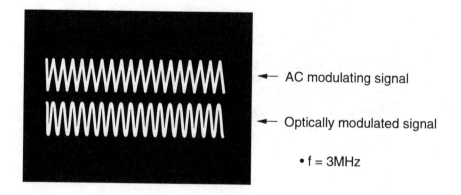

Figure 7.21 Low-frequency modulated signal with 12-dB modulation bandwidth; applied voltage is 10V.

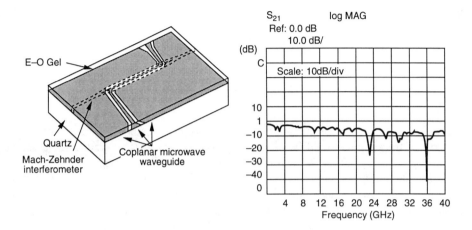

Figure 7.22 A photolime gel polymer-based traveling wave electro-optic Mach-Zehnder interferometer with measured 40-GHz electrical bandwidth.

7.8 FANOUT DENSITY AND OPTICAL INTERCONNECTION

The techniques used to develop the 12-channel demultiplexer shown in Figure 7.10 can also be used to create one-to-N fanout elements for optical interconnection. Such elements will undoubtedly require high efficiencies and low channel crosstalk. Improvements in diffraction efficiency can be expected through careful control of the phase-grating parameters, while enhancements in crosstalk figures will depend on the ability to achieve minimal angular overlap between adjacent fanout directions. The modulation index of the polymer film and its dependence on exposure dosage will ultimately determine the maximum number of phase gratings that can be exposed within a given region of the polymer microstructure waveguide (PMSW) as well as the overall grating efficiency.

Using coupled-mode theory as it applies to a lossless step-index waveguide medium containing slanted phase gratings [54], the diffraction efficiency and the angular and wavelength selectivities can be determined as a function of grating and waveguide parameters, including the interaction length, d; index modulation, Δn; and overlap integral for the TE mode of the waveguide. The diffraction efficiency is given by the expression:

$$\eta = \frac{4\kappa(\mathbf{r}\cdot\mathbf{s})^2}{(C_r/C_s)\,\nu^2 + 4\kappa(\mathbf{r}\cdot\mathbf{s})^2}\sin^2\left\{\frac{1}{2}\left[\frac{\vartheta^2}{C_s^2} + \frac{4\kappa(\mathbf{r}\cdot\mathbf{s})^2}{C_rC_s}\right]^{1/2} d\right\} \tag{7.13}$$

where

$$C_r = \frac{\beta_{mz}}{\beta_m} \qquad C_s = \frac{\alpha_z}{\beta_l} \qquad \vartheta = \frac{\beta_l^2 - \sigma^2}{2\beta_l} \tag{7.14}$$

and

$$\sigma = \beta_m + K \tag{7.15}$$

$$\kappa = \frac{2\pi^2}{\lambda^2} \int_{-\infty}^{+\infty} n_0(\lambda)\Delta n(\lambda) E_m(y) dy \tag{7.16}$$

In this equation, $\hat{r}$ and $\hat{s}$ are the unit polarization vectors, β_m and β_l are the propagation constants of the incident and diffracted guided modes, and κ is the grating wave vector. The propagation constant of the diffracted mode is σ, and a small dephasing constant, ϑ, is introduced, as in (7.14), to characterize the angular and wavelength selectivities of the planar hologram. The Bragg condition is given in (7.15), which can be used to determine the diffraction angle and the constant, σ. Equation (7.16) gives the coupling strength, which depends on the overlap between the incident guided-mode profile $E_m(y)$, the diffracted guided-mode profile $E_l(y)$, and the grating-index profiles, $n_o(y)$ and $\Delta n(y)$. $E_m(y)$ and $E_l(y)$ satisfy the orthogonality relationship:

$$\int_{-\infty}^{+\infty} E_m(y) E_l(y) dy = \frac{\delta_{m,l}}{\beta_m} \qquad m, l = 0, 1, 2, \ldots \tag{7.17}$$

For the calculation, it is assumed that there is complete overlap between the guided-mode and the index perturbation and that the modulation index for each individual grating has a value of $\Delta n \sim 0.01$. We note that the effects of having a finite laser beam width and a graded refractive index profile have not been accounted for in the present analysis. Additional losses that might have been introduced during the polymer thin film sensitization process have similarly been neglected. As expected, the diffraction efficiency of the induced transmission hologram undergoes a periodic transition between a maximum and a minimum value, as the diffraction angle is changed. The diffraction efficiency of each individual waveguide hologram can be improved by changing the exposure dosage during the recording process [55]. The modulation index, as a function of exposure time T_i for the ith hologram, can be determined from the expression:

$$\Delta n_i = \left(\Delta n_{max} - \sum_{j}^{i-1} \Delta n_j \right) (1 - e^{-\gamma E T_i}) \tag{7.18}$$

where γ is a sensitivity constant associated with the DCG material, E is the exposure intensity of the laser beam, Δn is the modulation index value achieved for each exposure, and Δn_{max} (which is the maximum index modulation that can be achieved in DCG holographic material) assumes a value of 0.2. The exposure of the ith hologram is therefore dependent on the exposure parameters of all previously described exposed $(i - 1)$ holograms. Based upon the gelatin waveguide and exposure parameters used in our experiments, we estimate that a maximum of ~300 multiply exposed gratings can be fabricated before the modulation index response of the locally sensitized PMSW begins to saturate.

The angular selectivity of our device structure, calculated using (7.13-7.17), is shown in Figure 2.23. The diffraction angles are chosen such that all curves are normalized to unity for fixed index modulation Δn. To achieve maximum efficiency at other diffraction angles, Δn can be adjusted. The dependence of the angular width and fanout channel density, as a function of grating interaction length and diffraction angle, is shown in Figure 7.24. In this case, the modulation index for each individual waveguide hologram is fixed at a value of 0.01. A mode effective index, $N_{eff} = 1.517$; a waveguide thickness,

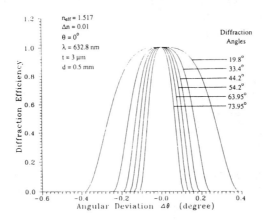

Figure 7.23 Angular selectivity for TE guided wave at different diffraction angles.

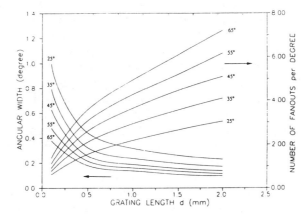

Figure 7.24 Angular bandwidth and fanout density, plotted as a function of grating interaction length, d. Mode effective index $N_{eff} = 1.517$, center wavelength $\lambda = 0.6328 \mu m$, and waveguide thickness $t = 3\mu m$ were used in the calculation.

$t = 3$ mm; and a center wavelength, $\lambda = 632.8$ nm; were also used in the calculation. As seen from Figure 7.24, a decrease in angular bandwidth can be achieved by increasing either the grating interaction length or the corresponding Bragg angle. In comparison to waveguide-based devices, where light propagation occurs within the plane of the waveguide layer over large distances, the smaller interaction lengths (< 60 mm) of bulk holographic one-to-N fanouts and WDM devices actually preclude the achievement of very high channel densities. Based on the previous calculations, a maximum channel density of ~7 fanouts per degree can be expected for a grating interaction length of 2 mm and a Bragg diffraction angle of 65°. By virtue of the polymer material and waveguide geometry, the interaction length can vary from millimeters to centimeters. Hence, the modulation index needed to form a high-efficiency deflection device in PMSWs is smaller than in corresponding bulk holographic devices. For example, an interaction length of 0.5 mm would require an modulation index of only ~4×10^{-4} to achieve 100% diffraction efficiency at an angle of 30°. Because the fanout density is largely dependent on the maximum modulation index of the material, DCG is expected to outperform other photorefractive materials, such as Fe-doped $LiNbO_3$ and SBN, which can achieve maximum modulation indices that are, at best, two to three orders of magnitude smaller.

7.9 RELIABILITY TEST

To further evaluate the reliability of waveguide devices discussed in this chapter, semihermetically sealed samples were made using EPO-TEK 353ND epoxy. The advantages of using this epoxy include its characteristics of curing at lower temperature and surviving at a higher temperature, low cost, and easy handling. The temperature stability results of an unslanted grating hologram under hermetic sealing using Aremco-Bond 631 have been demonstrated. The long drying time and serious out-gas problem have limited the potential of using Aremco-Bond 631 for our hermetic sealing purposes. Thus, our attention has switched to alternative adhesive materials. EPO-TEK 353ND epoxy from Epoxy Technology, Inc., has been found to be an excellent candidate for our purposes.

EPO-TEK 353ND has a flexural strength of 10,600 psi, a compressive strength of 20,200 psi, and a lap shear strength for glass-to-glass bonding of 1,180 psi. It has excellent moisture resistance and temperature stability range. It is important to note that the curing time for this epoxy is very short: 1 min at 150°C or 2 min at 120°C. In case there is some minor out-gas during this short drying period, the box can always be flushed simultaneously during drying using N_2 gas, and then the purging hole can be sealed. It is assumed that this does not generate serious out-gas problems due to the small volume of epoxy. In fact, the waveguide device was not damaged by the direct contact of this epoxy. Damaging chemical reactions between the waveguide hologram and the epoxy or its out-gasing are anticipated. During the box fabrication period, the holograms are continuously baked between 80° to 120°C; this eliminates the potential for moisture penetration or absorption by the hologram. This procedure helps to stabilize the hologram before encapsulation.

Various baking temperatures are used to test the device stability under sealing. For each temperature testing, we use three different PMSW holograms including an unslanted grating hologram used for the substrate fanout optical interconnects [55], a slanted Littrow grating hologram, and a Lippmann grating hologram. The results are general and applicable to all other types of microstructure waveguide devices.

The diffraction efficiency, diffraction angle, transmission characteristic, OD, and center wavelength, if applicable, to each corresponding grating type are tested before and after each temperature treatment. The diffraction efficiency and diffraction angle are measured using an HeNe laser of 632.8-nm wavelength. Reproducible results in OD, transmission bandwidth, diffraction efficiency, and the diffraction angle are found in a Littrow grating of up to 180°C. The shift in OD peak wavelength toward lower wavelength values of ~18 nm and 24 nm occurs after baking at 160° and 180°C, respectively. The slightly larger fluctuation was found to be caused by a relatively large hologram nonuniformity. The shift of OD center wavelength to lower wavelength values is consistent with the results of the Littrow grating due to the possible emulsion shrinkage in the depth direction, which reduces the grating fringe spacing.

The same temperature cycling test has been performed on both transmission and reflection holograms at 120°, 140°, 160°, and 180°C. Table 7.4 shows the results of a temperature treatment at 180°C. The Littrow grating shows a better response than the Lippmann grating. This may be explained by the better hologram preparation condition for the Littrow grating. The stabilization of holograms within a few baking cycles suggests that all holograms should be prebaked before using them for practical applications. Prebaking at temperatures similar to or higher than the application temperature is required.

Table 7.4
Grating Response after Being Baked in an Oven at 180°C for Two Hours

Baking Cycle	Duration (Hours)	Temperature (°C)	Transmission Bandwidth (nm)	Optical Density	Optical Density Peak, λ (Center)	Diffraction Efficiency (%)	Diffraction Angle (°)
Littrow grating hologram:							
Before			145	2.83	642	88.4	11.8
1	1	180	140	2.6	620	88.5	11.86
2	1	180	129	2.57	610	86.5	11.83
Lippman grating hologram:							
Before			140	1.55	(660)		
1	1	180	130	2.84	(630)		
2	1	180	140	3.45	(628)		

Military specifications require a stability between −62° to +125°C. We passed this requirement on the positive temperature side. On the negative temperature side, our experiments were conducted at −66°C, −100°C, and a liquid nitrogen temperature of −196°C. There were no observable changes after a 1-hr treatment at −66°C, −100°C, and −196°C, respectively.

Figure 7.25 shows the transmission and OD plots for a Lippmann hologram before and after treatment at −196°C. Stable results for other holograms have also been observed. At −196°C, considerable care is needed when submerging the holograms into the liquid nitrogen. Submerging the holograms too fast will result in a temperature shock, which will, in turn, either damage the hologram, the hermetically sealed box, or both. The experimental result has proven that the hermetically-sealed hologram can survive at liquid nitrogen temperatures.

All of our polymer-based active and passive device samples have failed after baking at 200°C. An irreversible phase transition occurred at this temperature. Partial damage and complete damage have been found after 10 min and 1 hr bakings, respectively.

The EPO-TEK 353ND epoxy has excellent moisture resistance. As specified, its weight increased by 0.03% after seven days at a 96% humidity level. After submerging the boxed holograms in water for two days at room temperature, no observable damage occurred.

If properly prepared, waveguide devices, including our unslanted grating hologram, can survive temperatures between −196°C to +180°C under hermetic sealing if they are

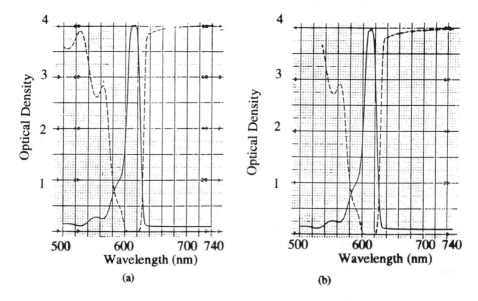

Figure 7.25 Response in transmission (dashed curve) and optical density (solid curve) of a Lippmann hologram (a) before and (b) after one hour temperature treatment at −196°C.

internally filled with dry N_2 gas. Survival at −196°C, the liquid nitrogen temperature, is also anticipated. In addition, the two-day water test showed no changes to the photolime gel-based waveguide devices.

7.10 FURTHER APPLICATIONS

Many far-reaching applications can be realized based on the PMSW technology. Basically, the PMSW we proposed and then developed in this program will find its suitability in all optical and EO systems that involve microstructure waveguides. The capability of locally sensitizing and selectively doping the PMSW with the desired materials provides us with a new way to develop passive and active optoelectronic systems. Some of the more highly plausible applications are described in this section.

Where the speed and the number of fanouts increase, there is a need to employ optical interconnects on the backplane to perform board-to-board interconnection. We have already concluded that a PMSW can be constructed on any smooth surface including insulators, semiconductors, and conductors. As a result, PMSWs are an attractive alternative to realizing the backplane interconnection. The PMSW is deposited on the backplane, which holds the layers of IC boards. If the interconnectivity and parallelity need to be further increased, several integrated PMSW backplanes can be added on the same board. Optical components such as laser diodes, photodetectors, and multiplexed holograms can be integrated onto the same PMSW backplane to transmit, route, and receive the optical signals.

7.10.1 Nonlinear All Optical Switch

The capability of constructing PMSWs on any surface and of selectively doping the PMSW provides some alternatives to build active optical and electro-optical devices. Figure 7.26 shows an optical switch/modulator. In this case, the substrate material needs to be a nonlinear material with a large $\chi^{(3)}$ (e.g., semiconductor-doped glass commercially available from Corning). The PMSW on top of the nonlinear material is locally sensitized with Ammonia Dichromate. Two gratings that function as the rejection filter are accordingly formed. A single-mode guided wave Fabry-Perot etalon is created. Light is the pumping beam creating the nonlinear effect. The substrate index of the PMSW is thus modulated. This effect changes the effective index of the guided wave of PMSW and, therefore, the optical path, OP, which is defined as:

$$OP = L \cdot N_{\mathit{eff}} \qquad (7.19)$$

where L is the physical length of the PMSW. The maximum transmission of a Fabry-Perot etalon occurs when:

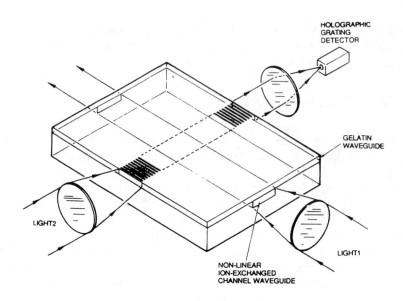

Figure 7.26 Nonlinear all-optical switch based on PMSW on nonlinear substrates.

$$\frac{2 \times OP}{\lambda} = m \tag{7.20}$$

where λ is the wavelength of the signal carrier and m is an integer. Because

$$\lambda \ll OP \tag{7.21}$$

holds for a real device, a small change of the intensity of the pumping power will drastically change the throughput of the signal wave. Another possibility for active device application is to selectively dope the PMSW with nonlinear materials with large $\chi^{(2)}$ or $\chi^{(3)}$, and then an optical device or an EO device can easily be built.

7.10.2 Systolic Arrays for Optical Computing

Vector and matrix multiplication can be implemented by integrating passive and active devices such as waveguides, grating lenses, and acousto-optic Bragg cells on the same substrate. A systolic array is shown in Figure 7.27. All devices can be built on top of the PMSW. Acousto-optic and piezoelectric effects of PMSWs need to be further studied to realize such a system application. By integrating a large array of holographic gratings, a large-dimension matrix and vector multiplication can be produced and performed. Realization of optical communications, signal processing, and computing functions in a PMSW

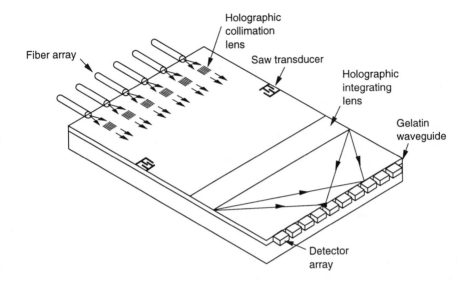

Figure 7.27 A systolic array for optical computing.

is certainly the major application of integrated optics. Previously, such computing and signal processing modules only existed in LiNbO$_3$. The capability of applying PMSW to various substrates and the reproducibility of the experiments provide a promising future for such architectures on different EO materials.

PMSW itself is transparent over a wide range of the optical spectrum. Accordingly, the PMSW functions as a good optical path to route optical waves with various wavelengths (~0.3μm to ~2.7 μm). A high-density wavelength division multiplexing and demultiplexing device based on PMSW technology and multiplexed holograms is feasible. Integration of holographic lenses and highly multiplexed Bragg holograms into PMSWs is shown in both the transmitter and the receiver. Because the refractive index modulation of DCG can be as high as 0.2, a large number of gratings can be multiplexed on the same holographic emulsion. Local area networks, highly parallel computer interconnections, and long distance, wide-band communications are some of the areas in which the PMSW-based WD(D)M will find its use.

7.10.3 Position Sensor

Recently, fiber optic sensors based on wavelength division multiplexing (WDM) techniques have been a major area of interest. By using WDM techniques, rotary and linear position sensing, rotary speed sensing, and pressure and temperature sensing information is provided. WD(D)M devices are used to create a chromatically dispersed strip of light, which can be either an LED or a laser diode array. The dispersed light is then focused onto the code tracks, which contain binary information (i.e., transmission, 0; reflection, 1). Different binary codes correspond to different sensed parameters. The detected optical signal is coupled back into a fiber though the same WDM device. The topology of this sensor system is shown in Figure 7.28. The PMSW WDM we propose here is an outstanding device for this application. For example, the four-channel WDM we demonstrated can be used as a linear or rotary position sensor for 16 different positions. Each individual position is identified by different binary codes. Figure 7.29(a-c) represents (1,0,0,1), (1,1,0,0), and (1,0,1,1), respectively.

7.11 CONCLUSIONS

This chapter has reported the research and development results of the photolime gel-based passive and active guide wave devices, including: high-density linear and curved channel waveguide arrays; EO modulator array; highly multiplexed waveguide holograms for WDM and interconnection; and rare-earth-doped polymer waveguide amplifiers. A single-mode linear channel waveguide array with device packaging density of 1,250 channels/cm was achieved. Further, the first 12-channel wavelength division demultiplexer working at 830, 840, 850, 860, 870, 880, 890, 900, 910, 920, 930, and 940 nm wavelengths was also described in this chapter. A polymer-based EO traveling wave modulator with 40 GHz electrical bandwidth was also described. Finally, a rare-earth ion doped polymer waveguide amplifier working at 1.06 μm with 8.5-dB optical gain was also described.

Laminated guided wave devices demonstrated outstanding temperature stability from −196°C to +180°C. Humidity tests also showed the survivability of the devices. Table 7.1 summarizes the demonstrated features of the polymer-based guided wave passive and

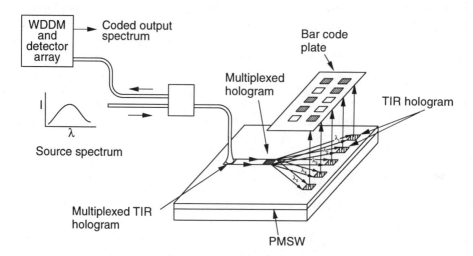

Figure 7.28 Broadband energy being transmitted through a WDW system to a fiber sensor.

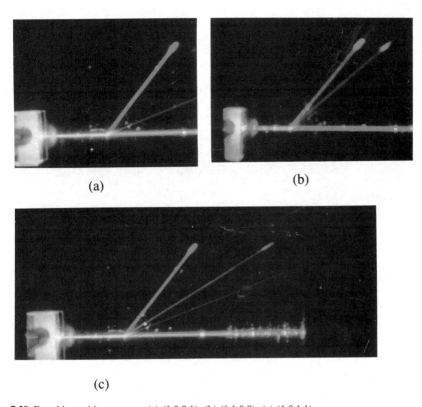

Figure 7.29 Four big position sensors: (a) (1,0,0,1); (b) (1,1,0,0); (c) (1,0,1,1).

active devices with III-V and LiNbO$_3$ as references. All the devices listed in Table 7.1 can be substituted with the reported technologies. Other devices, such as the highly multiplexed waveguide hologram and the large coating area and compression-molded optical channel waveguide, can only be achieved by the polymeric material reported herein. These device characteristics make polymer-based guided wave devices much more suitable for optoelectronic interconnection applications such as optical backplane, wafer-scale optical interconnects, and multichip-module optical interconnects.

This research is partially sponsored by SDIO/IST, Army SDC, Army Research Lab, DARPA, WPAFB, Army Research Office, NSF, DOE, NASA, and the faculty start-up funding of the School of Engineering at the University of Texas, Austin. Assistance from research scientists, engineers, and technicians involved in this research is also acknowledged, including those whose names are found in the references.

REFERENCES

[1] Man, H. T., K. Chiang, D. Haas, C. C. Teng and H. N. Yoon, "Polymeric Materials for High Speed Electrooptic Waveguide Modulators," *SPIE Proc.*, Vol. 1213, 1990, p. 13.

[2] McDonach, A., M. Copeland, "Polymeric Guided Wave Optics," *SPIE Proc.*, Vol. 1177, 1989, p.67.
[3] Cross, G. H., A. Donaldson, R. W. Gymer, S. Mann, N. J. Parsons, D. R. Haas, H. T. Man and H. N. Yoon, "Polymeric Integrated Electrooptic Modulator," *SPIE Proc.*, Vol. 1177, 1989, p. 79.
[4] Haas, D., H. Yoon, H. T. Man, G. Cross, S. Mann and Nicholas Parsons, "Polymeric Electrooptic Waveguide Modulator: Materials and Fabrication," *SPIE Proc.*, Vol. 1147, 1989, p. 222.
[5] Beeson, K. W., K. A. Horn, M. McFarland, A. Nahata, C. Wu and J. T. Yardley, "Polymeric Materials for Guided-Wave Devices," *SPIE Proc.*, Vol.1337, p. 1.
[6] Lytel, R., G. F. Lipscomb, "Nonlinear and Electrooptic Organic Devices," *Optical Nonlinearities in Organic Materials: Fundamentals and Device Applications*, P. N. Prasad and D. R. Ulrich, ed., Plenum Press, New York, 1988, p. 415.
[7] Hacker, N. P. and K. M. Welsh, "Photochemistry and Fluorescence Spectroscopy of Polymeric Materials containing Triphenylsulfonium Salts," *IBM Research Report*, RJ 8278 (75639), Chemistry, 1991.
[8] Sullivan, C. T. and A. Husain, "Guided-Wave Optical Interconnects for VLSI Systems," *SPIE Proc.*, Vol. 881, 1988, p. 172–176.
[9] Sullivan, C. T., B. L. Booth and A. Husain, "Polymeric Waveguide," *IEEE Circuits and Devices*, 1992, p. 27.
[10] Chen, Ray T., "Optical Interconnects: A Solution to very High Speed Integrated Circuits and Systems," *SPIE Proc.*, Vol. 1374, 1990, p. 20.
[11] " Guide wave Planar Optical Interconnections using Highly Multiplexed Polymer Waveguide Holograms," *J. of Lightwave Technology*, Vol. 10, 1992, p. 888.
[12] "Optical Interconnection using Polymer Microstructure Waveguides," *Optical Engineering*, Vol. 30, 1991, p. 622.
[13] Zyss, J., "Nonlinear Organic Materials for Integrated Optics, A Review," *J. Molecular Electron.*, 1985, p. 25.
[14] Booth, B. L., "Low Loss Channel Waveguides in Polymers," *IEEE J. Lightwave Tech.*, Vol. 7, p. 1445.
[15] Thackara, J. I., et al., "Poled Electrooptic Waveguide Formation in Thin Film Organic Media," *Appl. Phys. Lett.* Vol. 52, 1988, p. 1031.
[16] Sullivan, C. T., "Optical Waveguide Circuits for Printed Wire-Board Interconnections," *SPIE Proc.*, Vol. 994, 1988, p. 14.
[17] Franke, H. and J. D. Crow, "Optical Waveguiding in Polyimide," *SPIE Proc.*, Vol. 651, 1986, p. 102.
[18] Herman, W. N., W. A. Rosen, L. H. Sperling, C. J. Murphy and H. Jain, "A High Tg Nonlinear Optical Polymer: Poly(N-MNA Acrylamide), *SPIE Proc.*, Vol. 1560, 1991, p. 225.
[19] Smith, B. A., Herminghaus and J. D. Swalen, "Electrooptic Coefficients in Electric Field Poled Polymer Waveguides," *SPIE Proc.*, Vol. 1560, 1991, p. 400.
[20] Yankelevich, D., Andre Knoesen, C. A. Eldering and S. T. Kowel, "Reflection-mode Polymeric Interference Modulator," *SPIE Proc.*, Vol. 1560, 1991, p. 406.
[21] Chen, Ray T., "Graded Index Cross-link Induced Linear and Curved Channel Waveguide Array for High Density Optical Interconnects," *Appl. Phys. Lett.*, November, 1992.
[22] Chen, Ray T., "Polymer-based Waveguide Devices for Optical Interconnects," *Final Report to Army Strategic Defense Command and Strategic Defense Initiative Office*, Contract No. DASG60-90-C-0047, 1992.
[23] Chen, Ray T., "Polymer-based Guided Wave Devices and Their Applications," *Invited Talk at the Symposium on Polymeric Materials for Electrooptic Applications*, organized by The Rank Prize Funds, London, Dec. 7–10, 1992.
[24] Chen, Ray T., W. Phillips, D. Pelka and T. Jannson, "Integration of Polymer Waveguide with GaAs, LiNbO$_3$, Glass and Aluminum to Provide massive Fanout Optical Interconnect for Computing," Chen, Ray T., 1989, Postdeadline Paper, IEEE/OSA Topical Meeting on Optical Computing, PD-1, Post conference edition, Vol.9, 1991, 425.
[25] Chen, Ray T., M. Wang and T. Jannson, "Polymer Microstructure Waveguide on BeO and Al$_2$O$_3$ Substrates for Optical Interconnection," *Appl. Phys. Lett.*, Vol. 56, 1990, p. 709.

[26] Chen, Ray T., M. R. Wang and T. Jannson, "Intraplane Guided Wave Massive Fanout Optical Interconnects," *Appl. Phys. Lett.*, Vol. 57, 1990, p. 2071.
[27] Chen, Ray T., W. Phillips, D. Pelka and T. Jannson, "Gelatin Waveguides in Conjunction with Integrated Holographic Optical Elements in GaAs, LiNbO$_3$, Glass and Aluminum Substrates for Optical Signal Processing," *Opt. Lett.*, Vol. 14, 1989, p. 892.
[28] Chen, Ray T., "Gelatin Waveguide for Optical Interconnection," *SPIE Proc.*, Vol. 1151, 1989, p. 60.
[29] Chen, Ray T., "Polymer Microstructure Waveguides on Various Substrates for Optical Interconnection and Communication," *SPIE Proc.*, Vol. 1213, 1991, p. 100.
[30] Chen, Ray T., "Cross-link Induced Linear and Curved Polymer Channel Waveguides for High Density Optical Interconnects, "*SPIE Proc.*, Vol. 1774, 1992, p. 10.
[31] Chen, Ray T., H. Lu and T. Jannson, "Highly Multiplexed Graded Index Polymer Waveguide Hologram for Near-Infrared Eight-Channel Wavelength Division Demultiplexing," *Appl. Phys. Lett.*, Vol., 59, 1991, p. 1144.
[32] Hocker, B. and W. K. Burns, "Mode Dispersion in Diffused Channel Waveguides by Effective Index Method," *Appl. Opt.* Vol. 16, 1977, p. 113.
[33] Marcatili, E. A. J., "Dielectric Waveguide and Directional Coupler for Integrated Optics," *Bell Syst. Tech. J.*, Vol. 48, No. 7, 1969, p. 2071.
[34] Chen, Ray T., "Microprism Array for Large-scale, Wide-band Optical Interconnection for Optoelectronic systems," *Final Report to ISDO/IST and Army Research Office*, Contract No. DAAL03-91-0030, 1992.
[35] Chen, Ray T. and C. S. Tsai, "Thermally Annealed Single-mode Proton-Exchanged Channel Waveguide Cutoff Modulator," *Opt. Lett.*, Vol. 11, 1986, p. 546.
[36] Marcatili, E. A. J., "Bends in Optical Dielectric Guides," *Bell Syst. Tech. J.*, Vol. 48, No.7, 1969, p. 2162.
[37] Marcatili, E. A. J. and S. E. Miller, "Improved Relations Describing Directional Control in Electromagnetic Waveguidance," *Bell Syst. Tech. J.*, Vol. 48, No. 7, 1969 p. 2161.
[38] Digital Bus Handbook, edited by J. De Giacomo (McGraw-Hill, New York (1990) and Steve Heath, "VMEbus User's Handbook, CRC Press, Inc., Florida(1989).
[39] Chen, Ray T., H. Lu, D. Robinson, Z. Sun, T. Jannson, D. Plant and H. Fetterman, "60GHz Board to Board Optical Interconnection Using Polymer Optical Buses in Conjunction with Microprism Coupler," *Appl. Phys. Lett.* Vol. 60, 1992, p.. 536.
[40] Chen, Ray T., H. Lu and T. Jannson, "Eight-Channel Wavelength Division Demultiplexer Using Multiplexed GRIN Polymer Waveguide Hologram," *Post-deadline paper for OSA/IEEE Topical Meeting on GRIN Optical Systems*, 1991, PD2-1.
[41] Chen, Ray T. and H. Lu, "Polymer-based 12-Channel Single-mode Wavelength Division Multiplexer on GaAs Substrate, "*SPIE Proc.*, 1992, pp.1794-44.
[42] Lubensky, T. C. and P. A. Pincus, "Supermolecules," *Phys. Today*, Oct., 1984, p. 44.
[43] Williams, D. J., Ed. "Nonlinear Optical Properties of Organic and Polymeric Materials", *ACS Symposium Series 233*, Washington, DC, 1983.
[44] Ho, Z. Z., R. T. Chen and R.Shih, "Electrooptic Effect in Gelatin-based Nonlinear Polymer," *SPIE Proc.*, 1774–29(1992). and Z. Z. Ho, R. T. Chen and R. Shih, " Electro-optic Phenomena in Gelatin-based Poled Polymer," *Applied Physics Lett.*, Vol. 61, 4(1992).
[45] Layne, C. B. Layne and M. J. Weber, "Multiphonon Relaxation of Rare-Earth Ions in Beryllium-Fluoride Glass," *Phys. Rev. B*, Vol. 16, 1977. pp. 3259–3261.
[46] Gapontsev, V. P., S. M. Matitsin, A. A. Isineev, and V. B. Kravchenko, "Erbium Glass Lasers and Their Applications," *Opt. and Laser Technol.*, Aug. 1982, pp. 189–196.
[47] Miniscalo, W. J., "Erbium-Doped Glasses for Fiber Amplifiers at 1500nm," *IEEE J. Lightwave Tech.*, Vol. 9, 1991, p. 234.
[48] Armitage, J. R., "Three-Level Fiber Laser Amplifier: A Theoretical Model," *Appl. Opt.*, Vol. 27, 1988, pp. 4831–4836.

[49] Wang, M. R., Ray T. Chen, G. Sonek and T. Jannson, "Wavelength Division Demultiplexing Device on Polymer Microstructure Waveguide," *Opt. Lett.*, Vol. 15, 1992, p. 363.
[50] Chen, Ray T., Z. Z. Ho and D.Robinson, "Graded Index Polymer Waveguide Amplifier," Postdeadline Paper, *SPIE Proc.*, 1992, pp. 1774-39.
[51] Chen, Ray T., L. Sadovnic, T. Jannson and J. Jannson, "Single-mode Polymer Waveguide Modulator," *Appl. Physics Lett.*, Vol. 57, 1991, p. 2071.
[52] Shih, R., Ray T. Chen, and Z. Z. Ho, "Traveling Wave Polymer Waveguide Modulator using Coplanar Electrode Structure," *SPIE Proc.*, 1992, 1774-11.
[53] Shih, R., Ray T. Chen and Z. Z. Ho, "Traveling-wave Electrooptic Polymer Waveguide Modulator using Coplanar Electrodes,", *SPIE Proc.*, 1992, pp. 1794-53.
[54] Chen, Ray T., H. Lu and T. Jannson, "Highly Multiplexed Graded Index Polymer waveguide Hologram for Near Infrared Eight-channel Wavelength Division Demultiplexing," *Appl. Phys. Lett.*, Vol. 59, 1992, p. 1144.
[55] Chen, Ray T., M. R. Wang and T. Jannson, "Intra-plane Guided Wave Massive Fanout Optical Interconnects," *Appl. Phys. Lett.*, Vol. 57, 1990, p. 2071.

Chapter 8
Optical Interconnection for Use in Hostile Environments

Christopher S. Tocci and Stanley Reich

8.1 INTRODUCTION

It has often has been said that the invention of the laser was a technology looking for an application. Fiber optics on the other hand found instant application in the communications arena, initially in the area of computer networking and then expanding to consumer applications such as broadcast and interactive mode cable TV. This instant applications bonanza rapidly spawned a vast array of devices and hardware that challenged the designers' imagination. As the variety of optical interconnect components increased and became readily available, the range of interconnect applications spread well beyond that of computer networking, with its predominantly single-terminus, usually "butt-to-butt" interface, to parallel interconnects formed by waveguides, integrated couplers, wavelength division multiplexers (WDM), and an array of active components.

In a hostile environment, the fiber optic interconnect found fairly immediate acceptance due to its advantages in the areas of crosstalk, electromagnetic pulse (EMP), and data security as compared to conventional electronic connection methods. The first acceptance roadblocks to fiber technology were primarily the connectorization issues. Field reliability and backward compatibility were the two fundamental issues facing interconnection conversion within existing programs. Another salient area for the defense industry was the radiation hardness (rad-hard) of the new interconnecting medium. Pure silica fiber gave the best performance under testing involving neutron bombardment without generating color-centers or extensive core fluorescence (i.e., continued fiber core "lighting" after the irradiation source is removed). The color-center generation was found to be caused by contaminating OH⁻ radicals within the fiber core due to water contamination. The fluorescence was also due to OH⁻ and other contaminant mechanisms associated with the fabrication of the fiber.

Today, militarized fiber and cabling have been made sufficiently rugged so that routine use in battlefield and space-based applications is now routine. The acceptance of mature conventional fiber optic connector technology, such as the SMA, ST, or FCPC connector types, has assisted in the system solutions associated with the hostile (e.g., military) application arena when reliability is considered in comparison to electronic methods in a broad range of applications.

In the following subsections, a brief statement is made about where optical interconnection technology has found use (both military and commercial). Sections 8.3 and 8.4 will give two specific applications in which optical interconnection is well suited to overcoming hostile environmental situations. The first case study develops a concept that couples energy onto and out of an environmentally severe cryogenic region. The second case study details a packaging issue associated with very high bandwidth I/O MCMs and how the I/O count was reduced. The second case design has the added burden in that many of these high-bandwidth MCMs are located remotely (hundreds of meters, in some cases) from their respective processing nodes.

8.2 OPTICAL INTERCONNECTION APPLICATIONS IN HOSTILE ENVIRONMENTS—A BRIEF OVERVIEW

8.2.1 Backplane Interconnect Applications

Generally, interconnects in backplane applications provide more than a physical connection between remote points. In many applications, the signal path is a mix between bi-directional transfer between arbitrary nodes and broadcast-only communications. Other applications require selective transmission between nodes. Bi-directional and selective connections between nodes can, for example, be created by a combination of devices such as reflective splitters, WDMs, and time division multiplexers (TDM). The TDM, in some fashion, enables signals to leave and return to any node from any other node.

As indicated, the optical backplane can take on a variety of forms depending on the architecture selected; the multiplexing requirements of the bus; and whether all the nodes must talk to each other, or the network is broadcast only from a single source. These conditions determine the degree of complexity in forming the network. In general, limitations arise because of the impact on signal-to-noise (S/N) or bit error rate (BER) caused by a combination of data rate (bandwidth), insertion loss, and dynamic range. In applications where data can be received from adjacent and distant nodes or from a companion transmitter in an application requiring a bi-directional transceiver with greatly differing power levels, the signal dynamic range can be the performance-limiting factor. This condition is encountered in fiber optic FDDI LAN applications as well as in optical backplane architectures having a single, or wavelength multiplexed, reflective Star coupler configuration where every node can communicate full duplex with all other nodes.

Optical backplanes can be formed with a number of architectures, one being a single wavelength (nominal) multimode system with a transmissive Star coupler connecting any one of $N(i)$ inputs to all $N(o)$ outputs. The simplest form of this architecture is the case of a single input, which is the broadcast application. The point-to-point insertion loss in this application is made up of the cable, connector, and Star coupler losses. For the digital case, the transmitter power, less the insertion loss, need only exceed the receiver sensitivity by an amount necessary to meet the BER requirements. Assuming a reasonably stable source and interconnects, which is usually the condition for commercial connectors such as the SMA, STC, and FCPC types, dynamic range is a secondary issue.

In applications where the requirement is to communicate bi-directional (full duplex) data over an N-bit bus, the architecture used in the optical backplane becomes more complex and is dependent on a balance between the system parameters, particularly the trade-off between dynamic range, sensitivity, and bandwidth. The general system limitations are primary reasons for the constant quest to develop improved all optical cross-connect networks. In order for all nodes to communicate bidirectional (full duplex) data over an N-bit optical bus formed in a backplane, an array of N reflective-type star couplers, with a split ratio equal to the number of nodes and a combination of transceivers in conjunction with a 3-db coupler, is required. This holds for multi- and single-mode configurations, as well as the case in which there is physically no backplane, and the Star coupler is integral with an MCM or similar device.

8.2.2 Smart Skins And Structures

An application such as smartskins and structures, which includes a fiber optic distribution system, is a good example of how interconnections extend beyond just single and multitermini connectors. The following will discuss some typical application examples.

Fiber optic interconnects associated with smart skins and structures are unique in that in some applications, the optical fibers or waveguides are embedded within the structural elements of the aircraft along with a variety of application-dependent sensors and electronic modules. For ease of removal, however, some electronic modules are mounted on the internal surfaces of the structure. For composite-type structures, the embedding process usually takes place during the manufacturing, which includes an autoclave cycle that develops a high temperature and pressure. Obviously, only those sensors, modules, and fibers that can survive the autoclave environment are candidates for embedded applications. Additionally, aircraft structures are subject to shock and vibration brought on by the flight dynamics, which can adversely affect the strength and performance of the interconnect as well as that of the fiber itself.

Within the domain of smart skins and structures, fiber optic interconnects are very application-dependent because a variety of parameters, such as location on the aircraft, the type of construction used at a particular location, fiber-type, and performance, are important drivers. In some applications, it may be necessary to locate a particular sensor element and fiber in a removable composite section of an aircraft skin, thus requiring a fiber optic interconnect be associated with the skin. The connector in this application can either be edge-mounted flush with the surface of the structure, recessed in a well within the structure, or pigtailed through the rear (unexposed) surface of the skin. In this location, it could be connected or disconnected via an access hole before removing the skin panel. Connectors associated with this class of application are typically specialty items developed to meet a specific requirement.

Regardless of which connection technique is used, the transition region between the skin panel and the fiber or connector must be designed so that the fiber is protected as it passes through the region from the stiff, inflexible section of the structure to the flexible fiber pigtail or connector. All these conditions lead to the need for connectors, termini, or pigtails that are integral with the structure, flexible at the transition point, able to withstand harsh environments, and able to provide a low-optical loss. The requirements become even more severe when specialty fibers such as polarization maintaining fiber (PMF) are used.

The ideal sensors for embedding applications are the intrinsic types that are part of and integral with the fiber. Their smaller cross sections cause less interference with the media being sensed. Additionally, their integral construction makes them more rugged than extrinsic-type sensors, which are larger and fabricated separately from the distribution fiber.

8.2.3 Additional Candidates for Embedding

More sophisticated connection requirements may occur in a smart skins application such as a corporate-fed phased-array radar or communications antenna. Interconnect devices such as single-mode polarization maintaining WDMs, Star couplers, microlens systems, gratings, fiber optic, transmitters and receivers, and fibers might be required in a corporate feed that distributes optical signals, then down-converts them to RF radiation. It is likely such a system might initially be fielded in a metal aircraft, where it would be attached to the internal surface of the structure. In some future aircraft, however, it would be embedded directly into a composite structure. In all likelihood, the composite structure would be combined with other materials to create the required structural rigidity and dielectric constant that would match the phased-array physical and electrical characteristics. With the possible exception of the extremely low operating temperature of the focal plane environment described elsewhere, the autoclave curing cycle, encountered during fabrication, is probably the most severe environment to which the interconnects (and the other integrated optical components) will be subjected. In many applications, the limiting case for reliability will be a composite structure design with an embedded fiber system that will survive the manufacturing cycle.

8.3 CASE EXAMPLE 1: EXTRACTION OF CRYOGENICALLY GENERATED SIGNALS

One of the most difficult areas for electronic operation and interconnection is cryogenic cooled environments for infrared sensors. In particular, sensors fabricated with Mercury Cadmium Telluride (MCT) operating at 77K have created specific types of design problems for system engineers trying to implement standard electronic processing architectures into the cryogenic front end.

Figure 8.1 shows the present cryogenic signal-handling approach. This approach has a number of shortcomings for the front-end of the sensor, in terms of electrical performance, real estate consumed, and cost [1]. The primary problems associated with conventional bipolar and CMOS signal-handling in cryogenic region are [2,6]:

1. Electronically generated heat, which raises the MCT sensor's noise floor. (This also increases cryostatic loading.)
2. Compromised performance of CMOS circuitry due to cryogenically induced gain and dc instabilities.
3. Electromagnetic interference between the low-level MCT and supporting electronic control signals.
4. Channel-to-channel crosstalk due to individual channel leakage and multiplexing control signals.

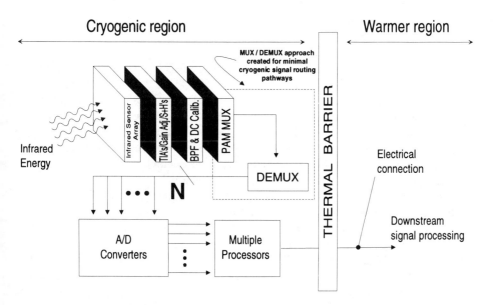

Figure 8.1 Current cryogenic signal coupling approach.

Problems 3 and 4 significantly contribute to a lowering of the effective signal-to-noise ratio for any given sensor channel beyond the infrared transduction noise.

Another common problem with conventional multiplexing of MCT signals is the increased switching speeds required with a large number of channels. These increased switching speeds increase crosstalk amongst channels and require increasingly smaller sampling times to allow for transient settling at the end of the cascaded multiplexer chain. This puts design pressure on the following A/D conversion and signal processing for real-time systems.

Since the early 1980s, industry has studied CMOS technology operating at cryogenic temperatures to process focal plane array data before the signal-to-noise ratio could be corrupted [3,6]. The study of CMOS operation at liquid nitrogen temperatures (77K) has shown that there were "freezeout" effects, or hot-electron device degradation [4,7], in depletion and enhancement-mode CMOS devices. This effect becomes apparent when the temperature is lowered to a point where the free electrons required for conduction do not have enough thermal energy to keep free. Consequently, the electrons fall back into the semiconductor lattice and become lodged, hence useless for modulation or any control of the transistor (both bipolar and CMOS structures). When the drive voltage is increased, some of these frozen electrons break free from the lattice but in a very abrupt fashion, which causes large dc offset and ac gain instabilities. Another problem associated with cryogenic operation is increased interconnection resistance among devices that are fabricated at the submicron (<0.5 micron geometry) scale [5]. This phenomenon creates signal bandwidth limitations. Finally, the physical interconnection between the cryogenic and noncryogenic regions imposes a thermal leakage path when electrical connections are anchored to a backplane. This puts extra heat into the sensor and requires greater cryostat pumping requirements.

One solution to these problems is to offload as much of the conventional electronics as possible from the cryogenic region into a warmer region as shown in Figure 8.2. This approach will have some electronics on the cryogenic side of the system, but it will eliminate the biggest offender, the A/D converter with its associated switching induced crosstalk noise. In the case of very large infrared sensors, this could mean hundreds, even thousands, of converters removed from the noise sensitive cryogenic region. It would be especially advantageous if this method could retain the acquired signal's dynamic range and SNR while reducing sensor heating. This would greatly lighten the cryostatic loading while allowing a majority of the electronics to operate in a much less hostile environment, thereby increasing reliability and performance while reducing cost. These figures of merit become particularly acute in light of a new breed of ultrasensitive infrared sensor composed of Copper doped Germanium whose optimal operating temperature is around 4K to 10K.

Comparing Figures 8.1 and 8.2, notice that both the A/D converters, the bandpass filters/dc calibration, and pulse amplitude multiplexer have been removed to the warmer region of Figure 8.2. Basically, the cryogenic region has been restructured to only acquire, modulate, and transfer the infrared sensor array information. All basic and advanced signal processing functions have been put into the warmer region where the aforementioned system advantages can be realized.

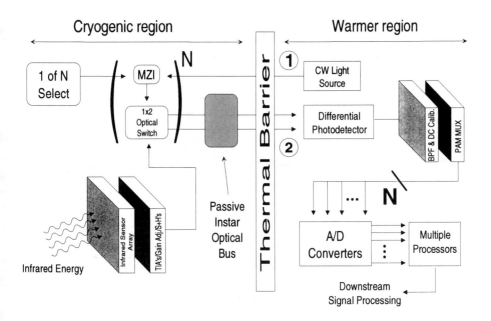

Figure 8.2 Alternate cryogenic signal coupling approach.

The heart of the approach in Figure 8.2 is the Mach-Zehnder Interferometer (MZI) and 1-×-2 optical switch combination acting as both multiplexer and electro-optical transducer. The passive instar optical bus will be explained later, but its function is to sum the multiple electro-optical transduced signals from the entire infrared sensor array into a single differential optical output.

It is not within the scope of this discussion to derive a rigorous electro-optical analysis associated with either the MZI or 1-by-2 optical switch's operation. Rather, we will assume the normalized transfer functions shown in Figure 8.3(a,b) as givens for design purposes [9,10].

Figure 8.3(a,b) shows the individual structures of the MZI and 1-×-2 optical switch. These integrated electro-optical devices basically act as intensity modulators whose operational optical output power, P_{out}, is a function of the control drive voltage shown.

The control drive voltage required to extinguish the output of the MZI or reverse the optical output in the 1-by-2s differential output is called the V_π, or extinction voltage [9,10]. The equation shown in Figure 8.3(a) relates the different parameters important to the operation of the MZI as:

$$\frac{P_{OUT}}{P_{IN}} = \cos\left(\frac{\pi V}{2V_\pi}\right)^2 \tag{8.1}$$

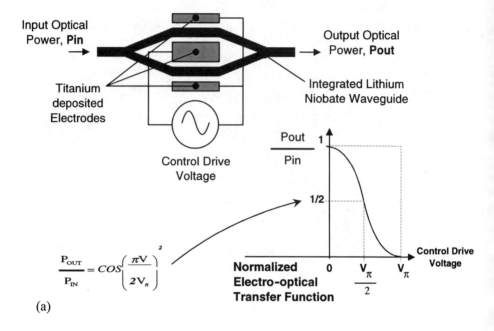

(a)

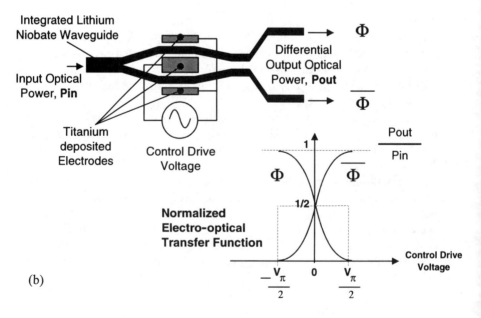

(b)

Figure 8.3 (a) Mach-Zehnder interferometer; (b) one-by-two optical switch.

where V denotes the control drive voltage (volts) and V_π denotes the extinction voltage (volts). This particular voltage derives its name from the 180° phase shift it produces between the two parallel legs in both integrated optical structures. When this phase differential exists, the output optical waveguide will not sustain the newly created mode at the summing junction of the MZI, and power is leaked into the device substrate and lost from the remaining output waveguide. In the case of the 1-by-2 optical switch, the 180° phase differential causes a shift of optical power from one output leg to the other. Hence, the MZI loses the optical power into the substrate, and the 1-by-2 switches the power between the two output waveguides.

Given this brief understanding of these integrated optical devices, Figure 8.4(a) shows the MZI's equivalent topology used as a single-pole-single-throw (SPST) [11] switch to control power into an equivalent amplifier (1-by-2) whose gain is either 1 or 0, depending on the digital enabling signal. The 1-by-2 switch can be modeled as a single-ended input amplifier with a differential output. Thus, optical intensity modulation emerges from the differential optical output as a differential optical signal whose variance is directly proportional to the applied sensor voltage and is directly gated (enabled/disabled) via a digital signal applied to the MZI.

Because each of these electro-optical structures controls and converts one infrared sensor signal, there would be many such structures to handle an infrared sensor array, as shown in Figure 8.4(b). Obviously, available optical power, unlike electrical power, is

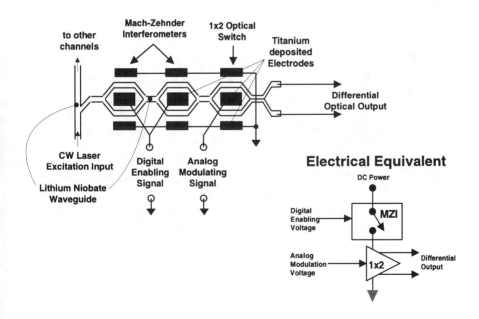

Figure 8.4(a) Integrated titanium-doped lithium niobate MZI/one-by-two optical switch combination.

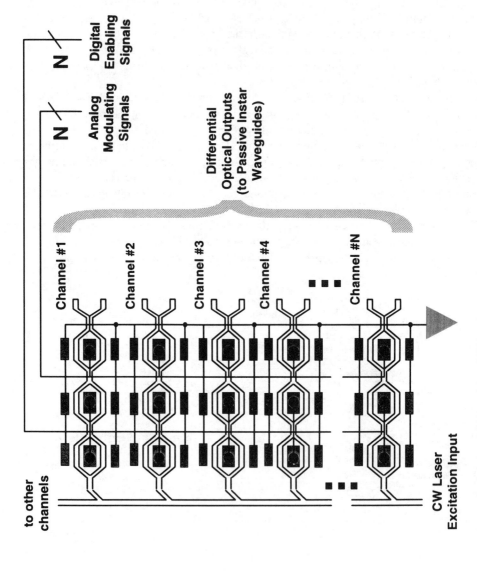

Figure 8.4(b) Integrated titanium-doped lithium niobate MZI/one-by-two optical switch planar array combination.

reduced each time a pathway is created. This indicates that there is an upper limit of channels that can be accommodated from a single "CW laser excitation input." Above this channel number, the reduction of available optical power causes a reduced "differential optical output" signal-level relative to the noise generated by the following and front-end electronics. This results in a reduced signal-to-noise (SNR), and consequent loss is usable dynamic range for A/D conversion. In Figure 8.4(b), the collected group of analog modulating signals are represented with a single line representing N parallel lines. Similar reasoning exists for the digital enable signals shown.

Figure 8.4(c) shows how the electro-optically modulated differential sensor signals are collected and routed out of any particular layer. Because only one channel is passing light at any time (per layer), then simply merging the optical waveguides acts like a wired-OR output for analog information. From Figure 8.4(c), if we let the upper and lower legs of the MZI/1-by-2 optical switch be the $+ \Phi$) and $- (\overline{\Phi})$ output, respectively, this is topologically identical to the electrical differential amplifier in many integrated operational amplifiers and many emitter-coupled-logic (ECL) output drivers using current steering logic.

Figure 8.4(d) takes Figure 8.4(c) into the third-dimension by showing how the individual layers are sandwiched to produce a two-dimensional array of electro-optical transducers for cryogenic signal conversion. Typically, the sandwiching in Figure 8.4(d) is accomplished with an alumina substrate with intersubstrate adhesive comprised of cryogenically compatible epoxies. Another approach does not mechanically bond the individual layers together but keeps a small separation by "gapping" at the four corners of each layer. This approach has some immediate advantages. First, thermomechanical expansion differences do not cause any generalized stress field over the alumina (or other) substrate. Consequently, stress relief points at the corners could be built using certain elastomers that retain sufficient elasticity at cryogenic temperatures. Second, replacement and testing of individual layers can be easily performed with the gapping approach. However, this gain is only useful for infant mortalities at production. Once fielded, especially in space-based applications, this approach would do nothing for system reliability.

Figure 8.4(e) depicts how the individual layers are fed optical power via the weighted taps, which can be fabricated from thermally fused multimode fibers. The use of multimode fiber is important from the standpoint of mode-matching efficiency. If one were to use single-mode fiber or waveguides, then precise geometries must be used to ensure efficient energy transference. If the modes (dependent on operating wavelength) are not matched due to physical changes caused by thermal or mechanical stresses, then uniform power delivery (to all layers) and distribution (to each layer) would become unreliable.

Looking at the power-splitting dynamics on the optical feeders, Figure 8.4(f) shows $1/N$ of the total optical power onto each layer involved in the cryogenic array.

Assuming a unity power input, then the tapping power fraction (TPF) of the fused fiber at the ith node would be:

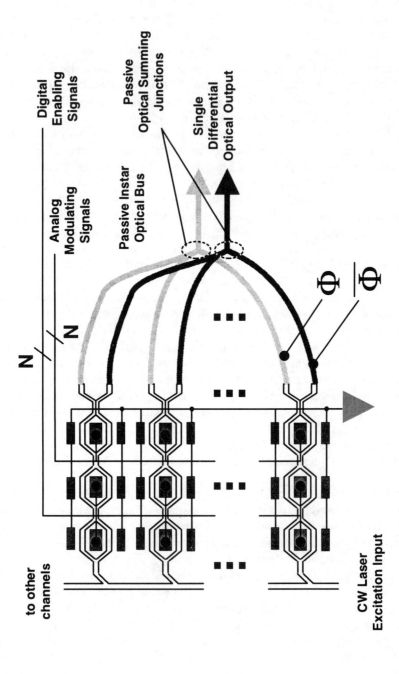

Figure 8.4(c) Waveguide structure for interconnecting and summing the individual differential MZI/one-by-two optical switch combinatorial outputs on individual layers.

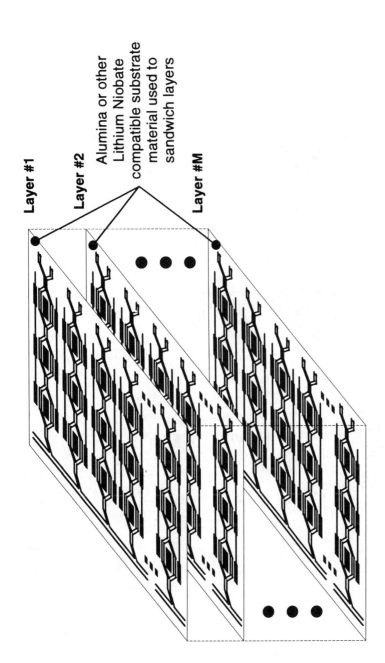

Figure 8.4(d) Exploded view of the layering of integrated MZI/one-by-two optical switch planar array for a two-dimensional interconnecting surface.

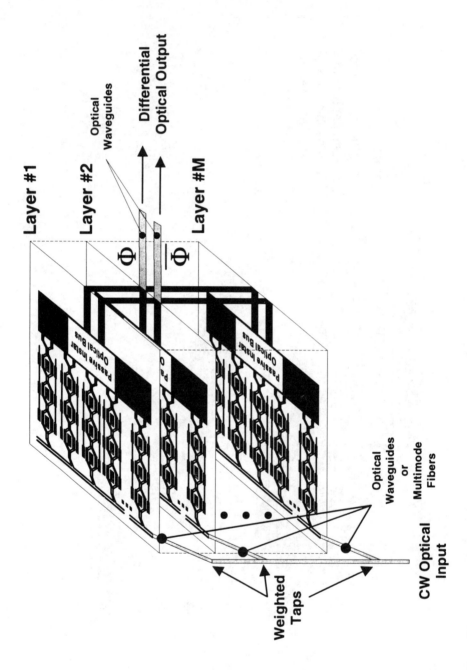

Figure 8.4(e) Layered planar arrays with passive instar optical bus and weighted optical waveguide feeder structures.

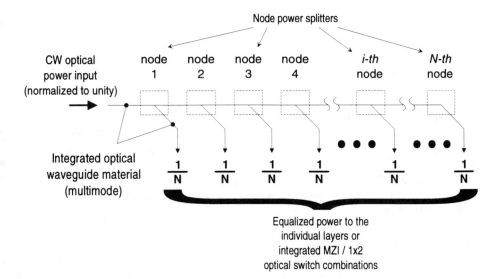

Figure 8.4(f) Approach for equalizing distributed optical power within the cryogenic coupler by using variably weighted power splitters.

$$\text{TPF}_i = \frac{1}{N - i + 1} \tag{8.2}$$

This assumes that there are N total nodes on any particular layer. This is evident when you are at the Nth node ($i = N$); then $TPF = 1$. Of particular importance here is the differential tapping power between nodes. This gives an indication of how precise the optical fiber fusion process has to be to ensure fairly uniform power distribution between and on individual layers.

Starting with (8.3), then performing the differential:

$$\frac{\partial}{\partial i}(\text{TPF}_i) = \frac{\partial}{\partial i}\left(\frac{1}{N - i + 1}\right) = \frac{1}{(N - i + 1)^2} \tag{8.3}$$

This clearly indicates the increasing difficulty associated with large node numbered layers. If we were to determine the TPF sensitivity versus node position, i, it would be derived from the standard definition of functional sensitivity [8]:

$$S_B^A \equiv \frac{B}{A}\frac{\partial A}{\partial B} \tag{8.4}$$

Then, performing the mathematics and using (8.2) and (8.3), the sensitivity of TPF (8.2), S_i^{TPF}, to node position, i, is:

$$S_i^{TPF} = \frac{i}{(N - i + 1)} \qquad (8.5)$$

In Figure 8.5, which shows a plot of TPF sensitivity as a function of i, we can see the increasing difficulty associated with tapping appropriate power levels with large N-node layer structures. The differential power tapping factor between contiguous nodes becomes small (tight geometric control), but small variations do not cause substantial power variations at low-node positions, i. However, inadequate tapping control at the earlier nodes can create severe constraints of remaining power for latter nodes in the layer or between layers.

On the individual layer basis, the same tapping of optical power can be performed as between layers. Figure 8.6 shows how the individual layers have their individual MZI and 1-by-2 electro-optical components associated with each channel addressed and abstracted into the passive instar waveguide structure. This structure is simply a fused fiber combiner or integrated waveguide material working in the multimode region. Looking back at Figure 8.4(e), we can see this passive instar optical bus structure on each layer. Further, each layer communicates off the layer by combining its respective differential

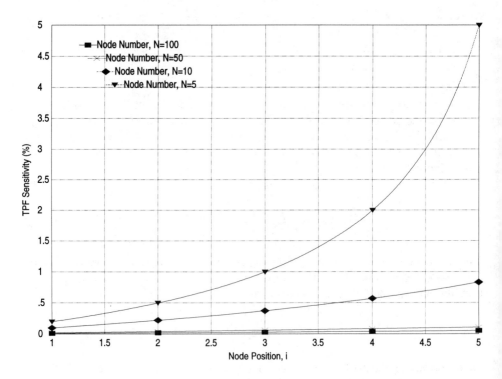

Figure 8.5 TPF sensitivity versus node position.

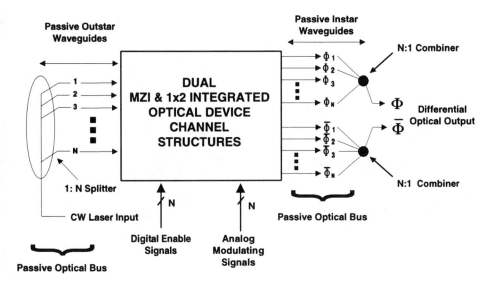

Figure 8.6 Passive optical bus structure.

output (Φ, $\overline{\Phi}$). Consequently, this individual layer instaring and outstaring is repeated for M-layers (Figure 8.4(d,e)), resulting in a sequential readout process, which, in effect, creates an $N \times M$ 2-D image.

At this point, if we refer back to Figure 8.2, there is one optical feed crossing from the warmer side to the cryogenic side, while there are two optical feeds crossing back to the warmer side. Obviously, this structure requires light to function at all and that light is more economically and reliably generated from the noncryogenic region associated with this design. The two returns are the differential optical signals that have been time-multiplexed and differentially passed back to the warmer region for further processing.

Figure 8.7(a) shows the inner structure involved with the noncontacting optical interconnection in the thermal barrier. Notice the microlens position relative to the light's propagating direction. The reason for this orientation is that optical alignment is better achieved when the lens is used as a light collector rather than a light concentrator and alignment tool. To show this statement, assume that the light-sending side of the integrated optical waveguide in Figure 8.7(b) is governed by geometric optics. This means that we are interested in the optical energy that is about ≤5° off-optical axis. This is a reasonable assumption because the numerical aperture (NA) of these waveguides is on the order of 0.05 to 0.1, hence corresponding to a diverging light angle of:

$$\text{optical emittance angle } (\theta) = \text{SIN}^{-1}(\text{NA}) \tag{8.6a}$$

or

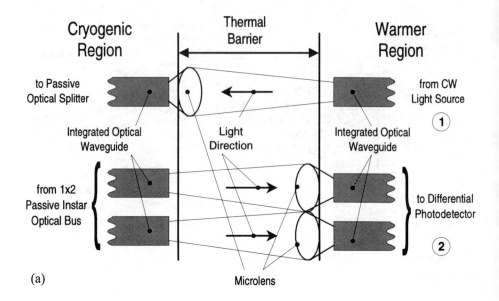

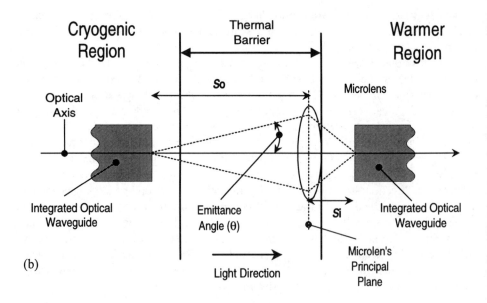

Figure 8.7 (a) Details of optical interconnection in the thermal barrier; (b) geometric optical analysis in the thermal barrier.

$$\theta = \mathrm{SIN}^{-1}(.05-.1) \approx 2.87°-5.74° \qquad (8.6\mathrm{b})$$

Then we can use the simple or Gaussian lens formula:

$$\frac{1}{f} = \frac{1}{S_o} + \frac{1}{S_i} \qquad (8.7)$$

and defining the amplification or gain of the imaging lens as:

$$M_T \equiv -\frac{S_i}{S_o} \qquad (8.8)$$

where $f \equiv$ focal length of the lens (meters); $S_o \equiv$ object to principal plane of lens distance (meters); $S_i \equiv$ image to principal plane of lens distance (meters); and $M_T \equiv$ transverse magnification (dimensionless). Because M_T also indicates the relative reaction of object-to-image movement, it is possible to define a relative measure of optical misalignment sensitivity in two forms: namely, depth and lateral misalignment sensitivities. The negative sign indicates image inversion is not important for this analysis.

Depth misalignment occurs when the microlens at the receiving optical waveguide is placed at the nonoptimal imaging position. When the microlens is too close to the receiver optical waveguide, the optical energy has not converged or focused to its minimal beam diameter. This could cause some of the received energy to miss the receiver waveguide and be wasted. When the microlens is too far away from the receiver waveguide, the optical energy is diverging off the optical axis centerline. This may cause much of the received energy to miss the waveguide and, again, be lost.

Lateral misalignment occurs when the microlens, transmitting, and receiving optical waveguides are no longer on the centerline of the optical axis in Figure 8.7(b). When this occurs, the imaging energy is offset from the center of the receiving optical waveguide. A direct sensitivity measure of lateral misalignment is given in (8.8). If M_T is less than unity, then this type of misalignment is desensitized. Conversely, if M_T is greater than unity, then the reverse is true. Hence, if M_T is numerically equal to 0.5, then two units of lateral displacement at the transmitter waveguide corresponds to 0.5-by-2 units or 1 unit of displacement at the receiver waveguide.

Depth misalignment sensitivity is a little more involved because we must use (8.7) and (8.8) to determine the analytical expression required to characterize this type of misalignment's effect on received optical energy imaging. First, we manipulate (8.7) into the following forms:

$$S_i = \frac{fS_o}{S_o - f} \qquad S_o = \frac{fS_i}{S_i - f} \qquad (8.9)$$

Next, a standard definition of sensitivity, δ, of a composite function, $f(x)$, on any of its constituent parameters, x, is shown as [8]:

$$\delta_x^{f(x)} \equiv \frac{x}{f(x)} \frac{\partial f(x)}{\partial x} \qquad (8.10)$$

Letting $f(x) = S_i$, $x = S_o$, and using (8.7) and (8.8) yields:

$$\delta_{S_o}^{S_i} = M_T \qquad (8.11)$$

Consequently, (8.11) indicates that even though lateral misalignment can be desensitized when the transverse gain is greater than unity, the depth misalignment sensitivity increases by the same gain factor. However, in the actual construction of an optical interconnection in the thermal barrier, the depth misalignment is much less sensitive than the lateral. This is due to the rigid structure of dewars, in which the thermal barrier is only 1- or 2-cm across while lateral misalignments can be cumulative over the length of the thermal bottle. Therefore, it is advantageous to have the microlens on the receiving side of the thermal barrier for optimal optical energy coupling.

8.3.1 Problems With the Proposed Integrated Optical Cryogenic Coupler

One of the fundamental limitations to this approach is leaked light when the dual MZI structure is suppose to be in an off state. Light that is allowed to pass into the 1×2 optical switch structure is passed onto the differential optical output. This light, along with other parallel channel light leakage, is summed into the passive instar waveguide, shown in Figure 8.5. When enough leaked light is dumped into this passive bus structure, the individual channel that is delivering the sensor signal is swamped out. This leaked light is considered optical bus noise.

8.3.2 One Solution to the Leaked Channel Light

Because of this leaked light issue, two *MZI* structures are cascaded to add a greater degree of optical extinction during the off state. In fact, both the *MZI* and 1-by-2 switch structures have a figure of merit called the extinction ratio which is defined as:

$$\text{extinction ratio} \equiv 10 \log\left(\frac{\text{output power in on state}}{\text{output power in off state}}\right) \text{ (db)} \qquad (8.12)$$

Obviously, a higher extinction ratio means a lower amount of leaked light into the waveguide, hence, a lower optical noise floor in the passive instar waveguide structure in Figure 8.5.

The launching of optical intensity modulation from a large number of individual *MZI*/1-×-2 switches requires an integrated waveguide structure that performs two primary functions: routing and summing. Figure 8.4(c) shows this approach where the individual signal (Φ) and signal complement ($\overline{\Phi}$) are separately routed and combined. The combination junction is referred to as an $N{:}1$ combiner and passive optical summing junction in Figures 8.6 and 8.4(c), respectively. The N refers to the number of channels that are present in the infrared sensor and is shown individual digital and analog electrical control signals in Figures 8.4(b,c), and 8.6.

Figure 8.4(d) shows that layering of the planar array can be built to accommodate a two-dimensional signal-handling in a compact, sandwiched structure. Obviously, light energy would have to be routed into the individual layers via an orthogonal waveguide taper with appropriately weighted energy taps. This, as in the individual layers, would cause the amount of optical power delivered to each layer to be about equal. In consideration of Figure 8.4(c), each layer would have its own passive instar optical bus, which, in turn, would join to form another passive instar optical bus for the individual layers. Figure 8.4(e) continues this mentality by incorporating the passive instar optical bus structures, individual layer optical feeders, and final differential optical output associated with this two-dimensional approach to cryogenic signal coupling.

As mentioned previously, the passive optical bus collects each of the differential optical outputs associated with each channel. Figure 8.6 shows an overview of the cryogenic coupler's integrated optical structure with required fan-in and fan-out waveguides. The passive instar waveguides are multimode-type waveguides. This is done so that there is minimal optical backscattering due to mode mismatch at the combiners. The optical backscattering will produce coupling losses and bus noise from excessive leaky light in the unused channels.

8.3.3 Inefficient Handling of Unused Channel Optical Power Solution: Recycle Light

One particular problem with the stated design of the cryogenic integrated optical interconnection is the loss of optical energy into the MZI substrate. Remember that the 1-by-2 optical switch only performed rerouting between two optically guided pathways while the MZI extinguished energy by pouring it into the substrate. This is reasonable for small arrays. However, with larger arrays (e.g., > 64-by-64 sensors or 1-by-100 linear arrays), the loss becomes intolerable for three distinct reasons:

1. Inefficient use of limited power, especially in severe remote operations (e.g., space-based, and deep-submersible applications).
2. Heating of the cryostatic region due to photonic substrate absorption.
3. Reduction of individual channel SNRs due to environmental illumination by dumped light.

Figure 8.8 shows one approach to addressing these weaknesses. Notice that the cascaded MZIs have been replaced with cascaded 1-×-2 integrated optical switches. This architecture now recycles the unwanted light back into the system for use at downstream channels. Consequently, when the shown channel is selected, both 1-×-2 switches conduct

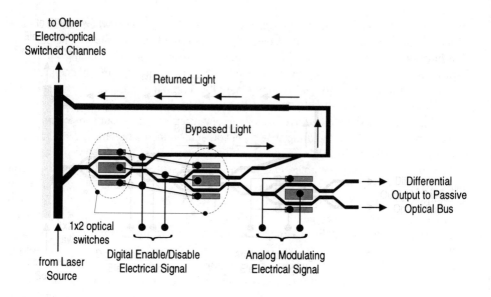

Figure 8.8 Alternative to MZI optical dumping used by previous scheme.

the light to their lower legs and onto the third 1-×-2 for analog modulation. During the off state, the tapped light is rerouted back into the main optical waveguide feed. Obviously, the waveguide structure is not drawn to geometric scale because such sharp bend would cause severe energy loss out of the returned light path shown. However, even if waveguide layout were not given strong consideration, at least most of the unwanted channel light would not enter the substrate to degrade neighboring channel SNR due to background illumination. This approach of recycling light greatly ameliorates the three previously mentioned problem areas. Two-dimensional arrays were experimentally doubled in size, and linear arrays were nearly quadrupled.

8.4 CASE EXAMPLE 2: MCM HIGH-BANDWIDTH INTERCONNECTIONS

Another hostile interconnection area is in the area of high-bandwidth MCMs subjected to military environments (i.e., high-temperature, humidity, radiation, and vibration). Figure 8.9(a) shows the basic idea of a passive directional optical splitter that would allow most of the transmitted light to pass without flooding the local or neighboring receivers on the network. With reference to Figure 8.9(a), the bidirectional optical bus carries the network information, while individual nodes tap about 10% of the passing energy. Each node needs most of this tapped energy to enter its receiver and little into its transmitter subsections. By selectively splitting the power and geometrically shaping the split, it is fairly easy to distribute the transmitted and received light according to the diagram. Using the numbers in Figure 8.8(a), about 9.9% of the bidirectional optical bus's power is carried into each

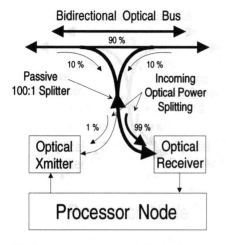

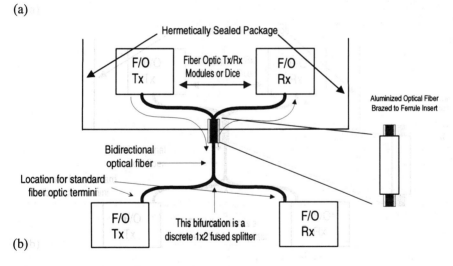

Figure 8.9 (a) Integrated passive optical node power splitter/equalizer; (b) utilization of integrated waveguide technology to reduce fiber I/O count in hybrid packaging.

node's receiver subsection. Conversely, only .1% of that bus's power is wasted in the transmitter subsection. This constitutes a 99-to-1 node receiver-to-transmitter power tap-off ratio. Other ratios can be constructed for smaller communication networks. The upper limit to this approach is the nearest neighbor's sensitivity. This means that if there is too much power on the bi-directional optical bus, then even 10% (or some other designed percentage) is too much light for the nearest node's receiver to handle without going into severe saturation. As energy is tapped further from the transmitting node, then less power is left that could cause a node's receiver subsection to saturate. Needless to say, this approach is generally only useful for digital signals. In using this approach, it is important to consider how to package and MCM to allow an optical fiber to enter an otherwise sterile hybrid environment to maintain signal integrity and component reliability.

Figure 8.8(b) shows how a shipboard optical fiber SONET ring was designed and constructed. Fundamentally, the greatest problem was keeping the hermeticity of the MCM packages while having a fiber penetrate the housing. This was accomplished by using aluminized optical fiber and then brazing the fiber to an aluminum ferrule shown in Figure 8.8(b). Concerns for bending were considered because a sharply bent fiber emerging from one the MCMs could easily crack or disrupt the optical power flow through the 62.5 micron fiber.

Another problem was inside the MCM package with the bend radius and the fiber's mechanical hitting of surface-mounted hybrid electronic components. This problem was avoided by using a special RTV adhesive within the MCM package and allowing fiber looping because attenuation in 62.5 micron core is low (2 dB/Km) at 1,300 nm. Also, physically positioning the F/O Tx and F/O Rx subsections close to the fiber bifurcation as it entered the MCM package helps in certain hybrids that were amenable to that chip-and-wire layout.

For a more detailed analysis of the cryogenic optical coupler, see the Appendix.

REFERENCES

[1] Longsworth, R. C. and Steyert, W. A., "Technology for Liquid-Nitrogen-Cooled Computers," *IEEE Trans. on Elec. Dev.*, Vol. 34, No. 1, Jan. 1987.
[2] Gaensslen, F. H. and Jaeger, R. C., "Foreward" in *IEEE Trans. on Elec. Dev.*, Vol. 34, No. 1, Jan. 1987, pp. 1–3.
[3] Bibyk, S. B., Wang, H. and Borton, P., "Analyzing Hot-Carrier Effects on Cold CMOS Devices," *IEEE Trans. on Elec. Dev.*, Vol. 34, No. 1, Jan. 1987, pp. 83–88.
[4] Watt, J. T. and Plummer, J. D., "The Effect of Interconnection Resistance on the Performance Enhancement of Liquid-Nitrogen-Cooled CMOS Circuits," *IEEE Trans. on Elec. Dev.*, Vol. 36, No. 8, Aug. 1989, pp. 1510–1520.
[5] Chen, Bor-Uei, "Integrated Optical Logic Devices," *Integrated Optical Circuits and Components*, ed. L. D. Hutcheson Marcel Dekker, Inc., 1987, pp. 289–316.
[6] Hanamura, S., et. al., "Operation of Bulk CMOS Devices at Very Low Temperatures," *Proc. 1983 Int. Symp. VLSI Technology*, Vol. 46, 1983.
[7] Pimbley, J. M. and Gildenblat, G., "Effect of Hot Electron Stress on Low Frequency MOSFET Noise," *IEEE Elec. Dev. Lett.*, EDL-5, 1984, pp. 345–347.
[8] Dorf, Richard C., *Modern Control Systems*, Addison-Wesley Publishing Company, 1967, pp. 62–65.
[9] Korotky, S. K. and Alferness, R. C.,"Ti:LiNbO$_3$ Integrated Optic Technology: Considerations, Capabilities, Fundamentals and Design," edited by Hutcheson, L.D., *Integrated Optical Circuits and Components*, Marcel Dekker, Inc., 1987, pp. 169–228.
[10] Crystal Technology Application Notes, "1 × 2 and 2 × 2 OGW Switches," Newton, MA, 1989.
[11] Crystal Technology, Siemans Corp., "1 × 2 Switch Application Note," Newton, MA, 1990.

Chapter 9
An Approach for Implementing Reconfigurable Optical Interconnection Networks for Massively Parallel Computing

William R. Michalson and Eric G. Schneider

Recent improvements in the state of optical interconnection technology allow creation of interconnection networks that are larger, faster, and more power-efficient than ever before. These developments allow optical interconnection in applications that were previously unimaginable. One such application is the field of optical computing, where millions of signals must be directed among arbitrary sources and destinations. In this application, not only is the requirement for massive interconnect present, but there is often the additional requirement that the available interconnect be rapidly reconfigurable, meaning that an appropriate interconnection scheme must be both dense and adaptable.

This chapter compares several different network topologies that have been proposed for use in integrated optical interconnection networks in an effort to identify an optimal network configuration for a massively parallel computer system. The focus is, in many respects, similar to the focus of previous researchers in that networks are compared on the basis of the available amount of connectivity, network complexity, and simplicity of routing algorithms [1-4]. In performing these comparisons, however, we found that only a limited amount of research has been done with guided-wave networks [1], and that this work tended to consider only small interconnection networks (less than a few dozen communications channels).

In contrast, a realistic optical computer system will require thousands or millions of available communications channels to supply nontrivial amounts of computational capacity. Thus, much larger interconnection networks are required. This has an important effect relative to prior network studies because any network parameters that have negligible effects in small networks become critical whenever the network size is increased dramatically.

Thus, this chapter reviews several network topologies that were the topics of prior studies and extends this previous research into the realm of networks that can provide the interconnection needed for optical computation systems. By extending interconnection techniques into large topologies, it is found that the optimal networks determined in previous investigations were biased by only considering small networks. As will be shown, the tradeoffs for large networks may be significantly different.

The next section provides a brief introduction to the architecture of the optical computer's architecture that provides the motivation for this study and illustrates the reliance of the architecture on the availability of a large number of reconfigurable communications channels [5]. Following this introduction, an analysis of several network topologies is presented and several properties of these networks are detailed. The last section illustrates the application of a selected network to the construction of a three-dimensional integrated optical computer. In this section, the manner in which network characteristics, such as switch size, impact computational capacity is illustrated.

9.1 A 3-D INTEGRATED OPTICAL COMPUTER ARCHITECTURE

For an optical computer architecture to be practical it must be scalable, reprogrammable, easily manufacturable, and reliable. Based on the architecture of Sawchuk [6], Michalson and Tocci [7] attempted to attain these desired attributes. This architecture was later extended into a full-scale three-dimensional integrated optical computer that can achieve a complexity comparable to a modern microprocessor [8].

The basic Michalson and Tocci architecture, which is illustrated in Figure 9.1, consists of an I/O plane, an integrated optical interconnection network (which is referred to as a virtual three-dimensional hologram), and an array of logic elements. The I/O plane is an array of LEDs or microlasers and an array of photodetectors. Signals pass into the architecture through a semireflective mirror and then enter the switching network. This network consists of a set of stacked semiconductor substrates, each substrate containing an integrated optical switching network. Each substrate facilitates communications in the two dimensions that lie in the plane of the substrate. If each substrate corresponds to a row of the input array (as illustrated in the figure), then each substrate can perform a permutation along the rows of the input array. A single stack of n substrates, then, allows signals to be permuted along each row of an $n \times n$ input array. An additional stack of substrates, rotated 90° with respect to the first, allows a permutation along columns, thus enabling an input signal to be redirected in three dimensions. The redirected input signals emerge from the virtual hologram and impinge on an array of reflective self-electro-optic effect devices (S-SEEDs), which are used to perform Boolean logic operations on the input signals.

Because the S-SEED devices are reflective, the logical result of combining input signals will be reflected back into the virtual hologram. As before, these signals can be redirected to specific locations on the output photodetector array by permuting along

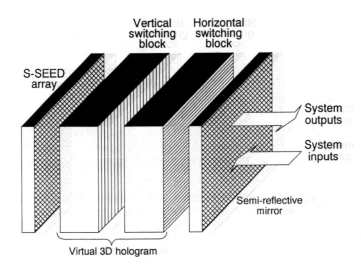

Figure 9.1 The 3-D integrated guided-wave optical computer of Michalson and Tocci.

columns and then rows. The semireflective mirror on the output side of the array provides a means for returning output signals back into the array, providing a feedback mechanism that can be used as a mechanism for implementing sequential logic.

As a computational entity, the system illustrated in Figure 9.1 is similar to a programmable logic array in that it provides a set of available logic operations that are accessed by establishing interconnections between the inputs, outputs, and gates. Unlike a typical programmable logic array, however, the architecture also provides the flexibility of modifying the interconnections between gates and of modifying the logic function performed by each individual gate (based on the way individual SEEDs are initialized). By allowing the virtual hologram to be reconfigurable by providing optical switches on each substrate, the system becomes reprogrammable, allowing different functions to execute at different points in time.

Although the architecture of Michalson and Tocci had not been fully developed and simulated, by targeting the attributes that a real computer must possess, it raised a number of issues critical to the design of realistic integrated optical computers and, in particular, with the necessary interconnection networks. For example, while the two-stack virtual hologram provided three-dimensional interconnectivity, the actual connectivity was limited because regardless of the network topology used, the proposed network would still not provide the flexibility to connect any channel on any layer to any other channel. Efficient use of the available logic devices in the SEED array demands such connectivity so that devices do not become unavailable due to an inability to route signals through the network.

Thus, the architecture of Figure 9.1 was refined by Schneider [8] to remove bottlenecks that might exist in the interconnection network. The new architecture, illustrated in Figure 9.2, addresses several of the limitations of the Michalson and Tocci design and the limitations of other, previously proposed, optical computing systems [6,9–12]. In addition, the architecture has been simulated in order to verify correct operation of the logic array and the interconnection network.

The fundamental problem with the architecture of Figure 9.1 is that with only two permutations, the network is only rearrangably nonblocking. That is, any single input may be routed to any output as long as no previously routed connection is using a necessary interconnection. If such a connection exists, it must be removed and rerouted (if possible), or it becomes disconnected. As the required number of interconnections increases, configuring the switching network to implement the desired communications topology becomes an increasingly difficult problem because a large percentage of the available communication paths are consumed. This problem is compounded by the fact that achieving high SEED utilization requires a large number of signals, possibly all in the worst case, to be connected simultaneously. Introducing a third substrate stack to the architecture solves this problem by allowing the virtual hologram to be nonblocking in the wide sense. This

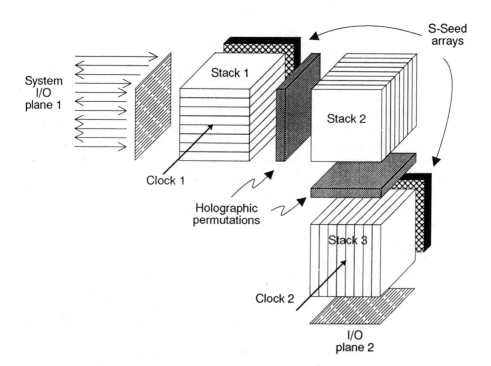

Figure 9.2 Optical computer schematic.

means that as long as a prescribed routing algorithm is followed, all of the necessary signal paths will be routable.

In the architecture of Figure 9.2, each of the interconnection stacks connect two $n \times n$ arrays of signals in a programmable manner, such that flexible access to each element in two n by n SEED arrays and two $n \times n$ I/O planes is provided. The orientation of the components allows nonblocking in the wide-sense connection between all elements of the SEED arrays. Stack 1 determines the x-coordinate of the SEED in array 1; stack 2 determines the z-coordinate of the SEED in array 2; and stack 3 determines the z-coordinate of the SEED in array 1 and the x-coordinate of the SEED in array 2. Connections between I/O plane 1 and S-SEED array 2 are rearrangably nonblocking because stack 1 simply passes the signals straight through, while stack 3 determines the x-coordinate of the S-SEED in array 2, and stack 2 determines the z-coordinate of the S-SEED in array 2. Connections between I/O plane 2 and S-SEED array 1 are similarly rearrangably nonblocking.

Figure 9.3 shows a functional block diagram for the system architecture depicted in Figure 9.2. In order to operate, the program to be executed must first be compiled. Because the logic array is essentially fixed and the interconnect variable, the compiler must generate the configuration information for the switching network. This requires decomposing a user program, specified in a high-order language, into three parts: a set of required logic functions, a list of the connections that must be established to interconnect the functions, and a schedule that determines when these connections must be established. A prototype for a compiler that uses genetic algorithms to optimize these competing factors is currently under development [13]. The schedule derived by the compiler is then passed to the control sequencer, and the required interconnections are converted into a set of switch settings for the three-dimensional network. The counter/sequencer then generates the necessary timing strobes so that the desired set of switch settings is applied to the switches in the optical computer, the correct set of inputs is applied to the optical computer, and the clock beams are applied to the optical computer.

In order to realize the optical computer described in this section, it is necessary to select a switching topology for use in the communications network. To preserve maximum

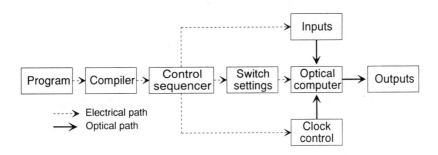

Figure 9.3 System block diagram.

flexibility, this network must be capable of nonblocking interconnect and must be capable of high channel density because an $n \times n$–SEED array requires a minimum of n^2 connections to uniquely address each SEED. To achieve a level of performance (gate-bandwidth product) roughly equivalent to a modern microprocessor would translate to a value of $n \approx 1,000$. This large value of n requires an integrated optical network that is capable of supporting independent interconnections.

This level of connectivity is extraordinarily high compared to most networked systems, electrical or optical. Thus, it is necessary to carefully consider all available options when selecting a topology for implementing this network. The following section discusses many of the available options.

9.2 LARGE GUIDED-WAVE NETWORKS

The basic component for any guided-wave interconnection network is the switching element. Figure 9.4 illustrates the basic operation of a four-port bi-directional switch. A control signal, either optical or electrical, determines whether ports A and B are connected in a pass-through mode (Fig. 9.4(a)) (which connects port A to port C and port B to port D) or in a crossover mode (which connects ports A to D and ports B to C). This four-port switch provides the fundamental switching element for most network topologies.

Presently, lithium niobate (LiNbO$_3$) is the most widespread technology used to construct switches like the one shown in Figure 9.4 [14–19]. A common implementation uses LiNbO$_3$ waveguides and exploits the Mach-Zehnder interferometer principle to perform switching between two adjacent waveguides as shown in Figure 9.5. In this type of switch, two light beams are channeled through two independent LiNbO$_3$ waveguides. These waveguides are arranged such that they have sections that are in close proximity to each other for a distance known as the interaction length. When an electric field is created around the waveguides, their refractive index in the vicinity of the electric field is changed, causing a phase shift in the signal carried by the waveguide. As a consequence

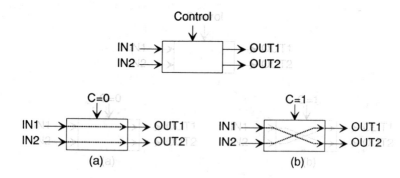

Figure 9.4 Bidirectional switch in (a) pass-through and (b) crossover modes.

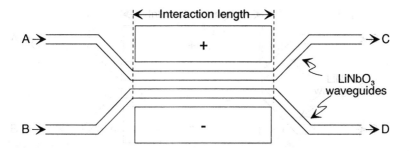

Figure 9.5 An LiNbO$_3$ switch.

of this phase shift and the proximity of the waveguides, the light beams in each waveguide can interfere—constructively or destructively depending on their phase shift—with one another. Thus, a switch with the two desired modes can be constructed in which the control signal determines the presence or absence of the electric field.

Although an LiNbO$_3$ switch performs the operations illustrated in Figure 9.4, these switches have the disadvantage that the length of this interaction region is rather long, typically on the order of millimeters [14]. This extremely long length can severely restrict switch density, as will be seen in Section 3.3.5.

Some emerging alternatives to the Mach-Zehnder approach are the design of switches using electro-optic polymers [20–22] or multiple quantum-well (MQW) heterostructures [23,24]. Unlike switches made using LiNbO$_3$ technology, multiple quantum-well switches utilize orthogonally intersecting waveguides and have relatively small interaction regions. The MQW switch is comprised of two orthogonally intersecting quantum-well rib waveguides on a silicon substrate, as shown in Figure 9.6 [23]. Switching occurs within a nonlinear modulation region at the junction of the two waveguides. This modulation region is approximately the same size as the width of each of the two waveguides, approximately

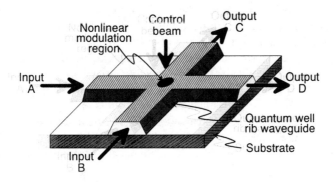

Figure 9.6 An MQW switch.

5- to 10-mm [23,24]. The relatively small size of this interaction region allows many more switches to be implemented per unit area relative to the LiNbO$_3$ approach. An additional advantage of quantum-well switches is the ability to engineer desirable switch properties by modifying the material properties of the waveguide junction. Currently, there are two basic types of MQW switches under development. One is activated optically by a beam focused at the waveguide junction [23], and another is controllable using an electric field [24].

The choice of switch technology employed is dependent on the specific requirements of the network topology that the switch is used in. Some of the more significant performance parameters by which a switch is evaluated include switching speed and power, the amount of crosstalk between signal beams, maintenance of a high signal-to-noise ratio, low coupling losses, ease and repeatability of fabrication, and compact physical size. Relative to these characteristics, MQW switches promise to fare much better than their LiNbO$_3$ counterparts [25], making them excellent candidates for optical computing applications. Indeed, it will be shown that while all of the network topologies reviewed may be used with either switching technology, the demands for interconnect density made by integrated optical computing systems dictate the use of switches with the dimensions of MQW devices.

9.3 NETWORK ANALYSIS

Given a technology for implementing an optical switch, it is necessary to determine a switching topology that can address the issues raised in Section 3.3.1. Much research has been done to characterize both electrical and fiber optical switching networks. In general, interconnection networks are characterized by their connectivity, complexity, path length, power loss, and routing algorithm. Connectivity provides a measure of the network's flexibility in terms of its ability to provide a desired signal path through the network. In general, there are four different classes of network interconnection topologies. The first class of networks is referred to as strictly nonblocking. In a strictly nonblocking network, every input has a dedicated path through the network to each output. Thus, in this class of network, any input can be routed to any unused output regardless of the way other input signals are routed. The second class of networks are those that are wide-sense nonblocking. Like a strictly nonblocking network, a network that is wide-sense nonblocking allows any input to be routed to any unused output. The difference between the two is that in a wide-sense nonblocking network, while each input has a path to each output, these paths are not dedicated. However, if a specific routing algorithm is followed, all of the desired interconnections can be established. A third class of networks are those that are rearrangably nonblocking. That is, any input can be routed to any unused output, but one or more existing connections may have to be rerouted to establish the path. The last class of networks are referred to as blocking networks. In a blocking network, there exists at least one input that cannot be routed to a given unused output unless some existing connection is removed. This means that not all of the desired connections may exist at the same time. In effect, some portion of a path through the network is shared between two different available signal paths. Because a path through the network may only be used by a single signal at a time, the shared portion of the network must be multiplexed between the two signals.

Network complexity is proportional to the total number of switching elements required to implement the network. Path length refers to the total number of switching elements a signal passes through in order to traverse the network. The availability of a routing algorithm for the network is also important because it provides a mechanism for automatically determining the required switch configuration to implement a desired routing of signals through the network.

In addition to the previously described factors, large guided-wave networks must also be evaluated with respect to several other parameters. In particular, it is important to evaluate a large guided-wave network on the basis of the number of waveguide crossings that a signal encounters when passing from one side of a network to the other. These crossings become significant in large networks because each crossing provides an opportunity for crosstalk between signal paths. If a particular network has a large number of crossings, signal integrity will degrade sooner than in a network with fewer crossings. A related concern is the angle at which the waveguides cross. In many of the topologies reviewed, waveguides must cross in order to achieve sufficient interconnection density, and thus, some signal degradation will result. For other topologies, the network can be configured such that waveguides cross at right angles, thus minimizing the amount of signal degradation resulting from crosstalk. In general for a network with identical connectivity, those topologies having nonorthogonal crossings or which require substantial redirection of the signal to achieve orthogonal crossings (by using integrated prisms, for example) would be inferior to topologies that inherently promote right-angle crossings.

In addition to classifying networks according to their signal routing capabilities, networks may also be classified according to the number of stages in the network and by the permutation that can be made by each stage. For example, the network configuration illustrated in Figure 9.7 is referred to as a shuffle-exchange network, which consists of a shuffle section that permutes the input signals before their entry to the switches and an exchange section that can operate in the pass-through or crossover modes. Networks are created by cascading several such stages together in such a way that connectivity is improved over that which is possible using just a single stage. In this paper, the size of a network is referenced to the number of inputs and outputs that the network provides. This size is referred to as the network's dimension, n. Thus, Figure 9.7 provides an example of a network with $n = 8$, meaning that there are 8 possible inputs and outputs.

In order to determine the optimal network architecture for large ($n > 1,000$) integrated reconfigurable guided-wave networks, all the previously described performance and complexity measures will be applied to several common network topologies. In particular, crossbar, n-stage, buddy-type multistage interconnection networks, duobanyan, active splitter/active combiner, passive splitter/active combiner, and benes topologies will be

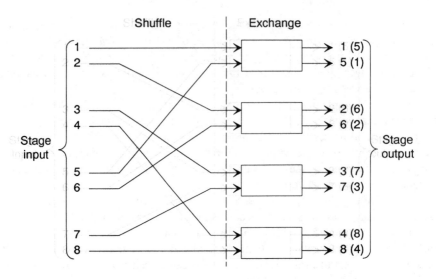

Figure 9.7 Example of a single-stage shuffle-exchange network.

compared with respect to the previously described factors. It is interesting to note that most previous studies of optical networks were mostly limited to the interconnections needed for fiber optic communication [18]. Relative to the needs of optical computation, telecommunications applications require fewer interconnections and have density (interconnections per unit area) requirements that are much less severe. Thus, much of the prior work was biased towards networks in which $n \ll 1{,}000$. When n is increased to the levels required for optical computation, a tradespace is revealed that is both complicated and previously not well explored. While much of this work verifies prior findings, a major result of exploring large networks is that in the region where $n > 1{,}000$, some networks that appeared optimal for small n are no longer optimal. In other cases, it is found that networks that appear equivalent for small n are not equivalent as n increases. The following sections describe each of the network topologies already presented and quantify some of the characteristics that become important when these networks are applied to large, integrated, reprogrammable, guided-wave networks for optical computing.

9.3.1 Crossbar Networks

A crossbar interconnection, illustrated in Figure 9.8, is a topology that is wide-sense nonblocking. It uses the four-port switch primitive described earlier, but requires a large number, n^2, of these switches.

In terms of the relationship to computer systems, the use of crossbar switches is minimal. The reason for this is twofold. First, the n^2 nature of the interconnect means that a large number of wires are needed to implement such a switch using electronics

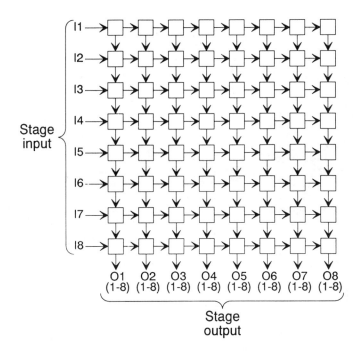

Figure 9.8(a) Crossbar network: topology.

(10^6 for $n = 1,000$, as compared to the more common hypercubic structure which would require $\approx 10^5$). Second, although a crossbar has the highly desirable property of being strictly nonblocking, in the electrical domain, each of the n^2 switches dissipates power. Because only one switch must actually be active, this characteristic makes the crossbar extremely inefficient in terms of power dissipation.

Some of these problems are resolved in free-space optical systems where it is possible to achieve high interconnection densities. The usual approach involves fanning out an input signal vector through a cylindrical lens and using a spatial light modulator (SLM) to allow specific signals to pass or be blocked. A cylindrical lens on the opposite side of the spatial light modulator reconstructs the output vector [26]. There are two important disadvantages with this arrangement, however. First, there is a large power loss associated with fanning out the original signal. At best, only $(1/n)$th of the original signal is available at each output of the SLM, in addition to power losses on the order of $(1/n^2)$ between the source lens and SLM and between the SLM and the output lens. Second, the free-space approach requires using three dimensions in space to implement a switching network that is inherently two-dimensional.

When implemented with guided-wave optics, these problems are reduced. Because waveguides on a single substrate may cross, the interconnect complexity relative to electronic systems is reduced—although the same number of interconnections is required.

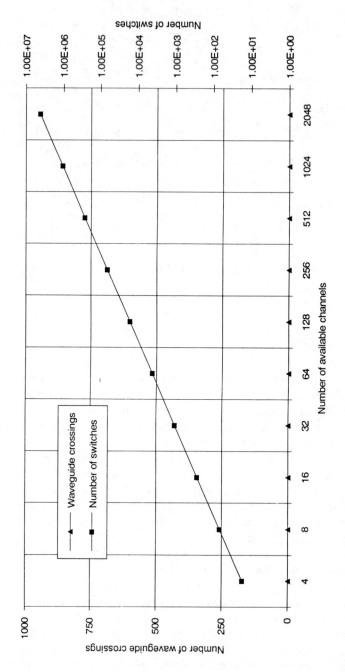

Figure 9.8(b) Crossbar network: analysis.

Compared to free-space optical systems, an integrated optical crossbar can be implemented as a planar device, eliminating any requirement for a third dimension. A major advantage of the crossbar is that no waveguides cross outside of the switches themselves. In a crossbar network, the signal path length through the topology ranges from 1 (for the path between I8 and O1) to $2n - 1$ switches (for the path from I1 to O8), with an average path length of n. Although the path length can vary a great deal, any particular signal passes straight through all but one of these switches. However, because the signal power loss is significantly greater for a signal being actively redirected by a switch than it is for a signal passing straight through, the effective path length in the case of an integrated optical network is essentially one switch traversal. This means that the power efficiency remains high because only a single switch need be active on each path, and optical losses are minimal.

Another important characteristic of crossbar networks is their extremely simple routing algorithm, which, when followed, guarantees nonblocking interconnection. In addition, the crossbar network has the unique property that during normal use, when interconnecting two sets of n signals, only two of the four sets of n waveguides (left and bottom of Figure 9.8) are used. Thus, it is possible to gain an additional degree of flexibility by exploiting the waveguide connections along the top and right of the array. These waveguide connections can be used to pass signals from top to right concurrently with the signals being passed left to bottom. However, it should be noted that when the crossbar topology is used in this manner, it is no longer fully nonblocking. Table 9.1 provides a summary of the characteristics of a crossbar network.

9.3.2 n-Stage Network

The n-Stage is a rearrangably nonblocking architecture (see Figure 9.9). As its name implies, this network has n stages, resulting in a worst-case path length of n switches. As shown in the figure, the n-stage network consists of two different types of stages. Half of these stages contain an even number, $n/2$, of switches, while the other half of the stages contain an odd number, $n/2 - 1$, of switches. In total, there are n stages of switches in the network, resulting is a total of $n(n - 1)/2$ switches.

Like the crossbar, this network has no waveguide crossings and can be implemented using integrated optical waveguides in a single plane. In comparison to the crossbar

Table 9.1
Crossbar Network Parameters

Parameter	Value
Connectivity	Nonblocking in the wide sense
Total number of switches	n^2
Signal path length	1 to $2n - 1$
Total number of waveguide crossings	0

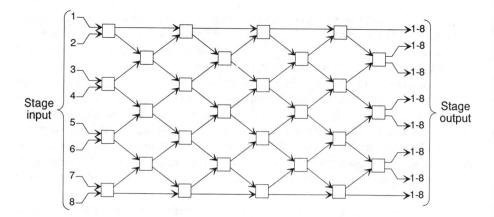

Figure 9.9(a) 8 × 8 *n*-stage network: topology.

network, the *N*-stage topology has far fewer switches than the crossbar, $O(n)$ versus $O(n^2)$. Thus, the *N*-stage network represents an attempt to economize on switches. While this reduction in switch count provides a significant reduction in network complexity, particularly at large values of *n*, this advantage is somewhat offset by the fact that the *n*-stage network trades fewer switches for less connectivity and has a slightly more complicated routing algorithm compared to the trivial algorithm of the crossbar[1]. Table 9.2 provides a summary of the characteristics of the *n*-stage network.

9.3.3 Multistage Interconnection Networks

Another important class of interconnection topologies is those that contain $O(n \log_2(n))$ switches. Like the *n*-Stage network, these networks attempt to economize on the number of switches required but, at the same time, attempt to retain as much of the interconnectivity and simplicity of routing of the crossbar network as possible. These networks are referred to as multistage interconnection networks (MIN) if they have a maximum path length of $\log_2(n)$ switches and contain $\log_2(n)$ stages with $n/2$ switches in each stage [3].

An interesting property of multistage interconnection networks is that all networks which conform to the previous rules are topologically equivalent; that is, they will provide the same level of interconnectivity for a given network size regardless of the actual configuration of the connections. There are many variations of multistage interconnection networks that have been studied in the literature. Most notable are the baseline, omega, and banyan networks, which are illustrated in Figure 9.10.

All multistage interconnection networks are blocking architectures, like the *n*-stage network, and thus are limited in the amount of connectivity they can support. The biggest advantage is the substantial reduction in the path length, $O(\log_2(n))$, as opposed to $O(2n - 1)$ for the crossbar and $O(n)$ for the *n*-stage topologies. This property has made MINs

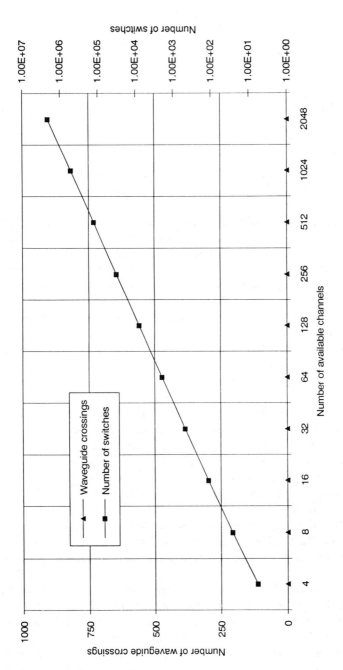

Figure 9.9(b) 8 × 8 *n*-stage network: analysis.

Table 9.2
n-Stage Network Parameters

Parameter	Value
Connectivity	Rearrangably nonblocking
Total number of switches	$n(n-1)/2$
Signal path length	n
Total number of waveguide crossings	0

extremely popular in electrical switching systems in telecommunications and computing applications because network complexity for a given level of connectivity is very low relative to other topologies.

From the perspective of integrated optical communications networks, MINs can be implemented on a single plane, but unlike their crossbar and *n*-stage counterparts, they cannot be implemented without waveguides crossing. Determining an expression for the number of waveguide crossings in an omega network of arbitrary n (assuming n is a power of two) is relatively simple because the same permutation occurs in each stage. Counting the number of crossings for the first, second, ... $(n/2-1)$th, and $n/2$th waveguides results in $0, 1, \ldots, n/2 - 2$, and $n/2 - 1$ crossings, respectively. Adding the first and last terms in this series, the second and second to last, the third and third to last, and so on, for all $(n/4)$ pairs of terms, clearly results in a total of $(n/4)(n/2 - 1)$ crossings in each stage of the network. Because the omega network has a total of $\log_2(n) - 1$ stages, the total number of waveguide crossings equals $(n/4)(n/2 - 1)(\log_2(n) - 1)$.

For the baseline network, calculating the number of waveguide crossings in general is more involved due to the manner in which the network splits in half with every stage traversed. Beginning with the left-most stage, the number of crossings is simply $C_n = (n/4)(n/2 - 1))$, as in the omega network. However, the second stage from the left must be treated as two separate blocks of size $(n/2)$. The third layer would be treated as four separate blocks of size $(n/4)$, and so on. Thus, the total number of crossings for stage i is $2^{i-1} C_{(n/2^{i-1})}$. The total crossings for the entire baseline network is then the sum for each stage over the $\log_2(n)$ stages.

The number of waveguide crossings in the banyan network is calculated almost identically to that of the baseline, except that the number of crossings per block in a given stage is different. The crossings in a given stage can be classified into two groups depending on whether the crossings are caused by two diagonal waveguides or one diagonal and one horizontal waveguide. For the right-most stage there are $(n/4)$ diagonal waveguides sloping down to the right and $(n/4)$ sloping down to the left. These diagonal waveguides result in $(n/4)^2$ crossings. There are a total of $(n/2)$ diagonal waveguides sloping to either direction, each of which crosses $(n/4 - 1)$ horizontal waveguides. Therefore, the total crossings for any individual block is $C_n = (n/4)^2 + (n/2)(n/4 - 1)$, and the total number of crossings for a given stage i is again $2^{i-1} C_{(n/2^{i-1})}$. The total number of waveguide crossings

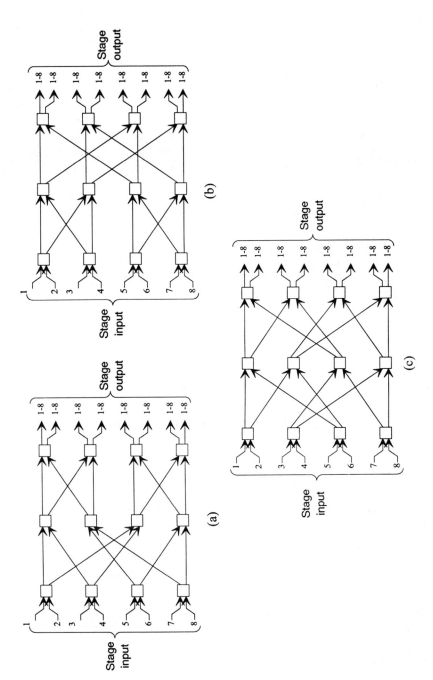

Figure 9.10 8 × 8 MINs: (a) baseline network; (b) banyan network; (c) omega network.

for the entire banyan network is simply the sum for each of the $\log_2(n)$ stages. The number of crossings and the total number of switches in each of the three networks is shown in Figure 9.11 and is summarized in Table 9.3.

Recall that unlike electrical etches, optical waveguides can cross, but there is a penalty in the form of power loss and crosstalk. It is interesting to note that for large multistage networks, although they are topologically equivalent, they are not at all equivalent in terms of the number of waveguide channels that must cross. For example, in the eight-by-eight networks presented in Figure 9.10, the baseline network has 8 crossings, the banyan has 10, and the omega network has 12. This phenomena is even more dramatically illustrated in Figure 9.11 where it is seen that for $n = 2,048$ there is an order of magnitude difference in the number of crossings between the omega and baseline networks.

In guided-wave networks, not only is the number of waveguide crossings important, but so are the angles at which they cross. Inspection of Figure 9.10 shows that as the network size is increased, either waveguide crossings will occur at increasingly oblique angles, resulting in very high levels of crosstalk, or fabrication will be complicated by a need to try to route signals in such a way that orthogonal crossings can be maintained. It is interesting to note that of all the MINs reviewed, the baseline topology incurs the smallest number of waveguide crossings for any value of n.

9.3.4 Duobanyan Network

The duobanyan, illustrated in Figure 9.12, is another example of a popular blocking network architecture. This network is based on the banyan topology but uses only one port on the outer side of each four-port switch in the first and last stages. By reducing the number of I/O signals, the duobanyan requires approximately twice the number of

Table 9.3
Multistage Interconnection Network Parameters

Parameter	Value
Connectivity	Blocking
Total number of switches	$n\log_2(n)/2$
Signal path length	$\log_2(n)$
Baseline network	$C_{tot} = \sum_{i=1}^{\log_2(n)-1} 2^{i-1} C_{n/2}^{i-1}$, where $C_n = \frac{n}{4}\left(\frac{n}{2} - 1\right)$
Banyan network	$C_{tot} = \sum_{i=1}^{\log_2(n)-1} 2^{i-1} C_{n/2}^{i-1}$, where $C_n = \left(\frac{n}{4}\right)^2 + \frac{n}{2}\left(\frac{n}{4} - 1\right)$
Total number of waveguide crossings	
Omega network	$C_{tot} = (\log_2(n) - 1)C_n$, where $C_n = \frac{n}{4}\left(\frac{n}{2} - 1\right)$

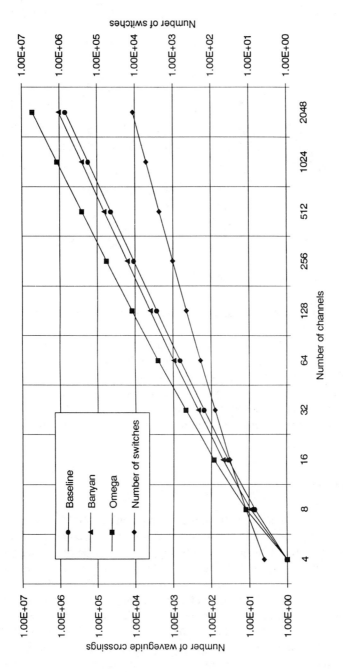

Figure 9.11 Number of waveguide crossings and switches for baseline, banyan, and omega MINs.

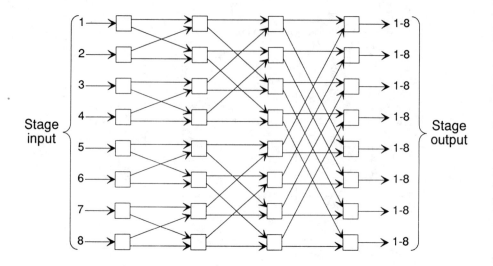

Figure 9.12(a) 8 × 8 duobanyan network: topology.

switches, $n \log_2(n) + n$, as the banyan and has a somewhat longer path length containing $\log_2(n) + 1$ switches.

Although the complexity of the duobanyan is increased relative to the banyan network, some advantage is gained because the additional switches provide additional routing flexibility. However, the increase in complexity is substantial. An $n \times n$ duobanyan has the same number of waveguide crossings as a banyan network twice its size, meaning that any problems associated with crosstalk and power loss will be exacerbated in the duobanyan network. Table 9.4 provides a summary of the characteristics for the duobanyan network.

9.3.5 Benes Network

The benes network, shown in Figure 9.13, is also based on the MIN but has the advantage of being a rearrangably nonblocking architecture. It is formed by combining $\log_2(n)$ stages of a multistage interconnection network with $\log_2(n) - 1$ stages of a reversed MIN. In total, the benes network contains $n(2 \log_2(n) - 1)/2$ switches and has a path length of $2 \log_2(n) - 1$.

The specific number of waveguide crossings depends on the MIN used as a basis for the benes network but, in general, is the sum of the number of crossings from the MIN and reversed MIN, or twice that of the MIN from which it formed. The benes network can also be implemented in a single plane, and the number of crossings is smaller than a MIN of equivalent size. Being rearrangably nonblocking is an advantage, but the

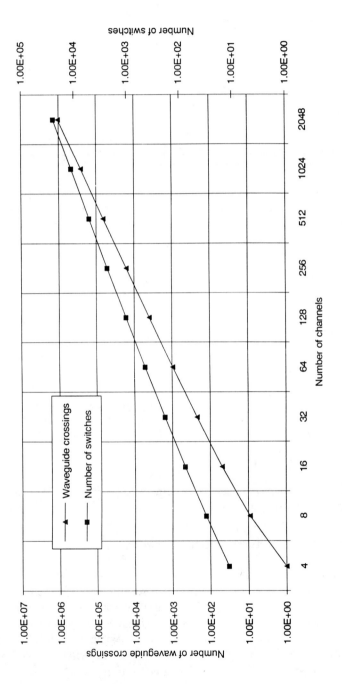

Figure 9.12(b) 8 × 8 duobanyan network: analysis.

Table 9.4
Duobanyan Network Parameters

Parameter	Value
Connectivity	Blocking
Total number of switches	$n\log_2(n) + n$
Signal path length	$\log_2(n) + 1$
Total number of waveguide crossings	$C_{tot} = \sum_{i=1}^{\log_2(n)-1} 2^{i-1} C_{n/2^i}^{i-1}$, where $C_n = \left(\frac{n}{2}\right)^2 + \frac{n}{2}\left(\frac{n}{4} - 1\right)$

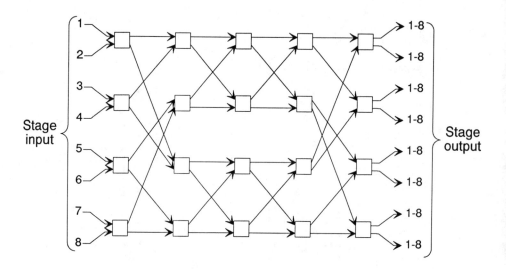

Figure 9.13(a) 8 × 8 benes network: topology.

routing algorithm for the benes network is extremely complex [20]. The benes network characteristics are summarized in Table 5.

9.3.6 Active and Passive Splitter/Combiner Networks

The networks studied in the previous sections all used a four-port switch as the basis for redirecting signals. Another class of networks often seen in integrated optical systems is a combination of signal splitters and combiners that, in general, may be active or passive. These splitter and combiner networks can be thought of as using a three-port switch, which either splits a signal between two waveguides or combines a signal from either of two waveguides into a single waveguide. Using this type of switch, a signal is routed from port A to either port B or C, or signals appearing on ports B or C are merged and

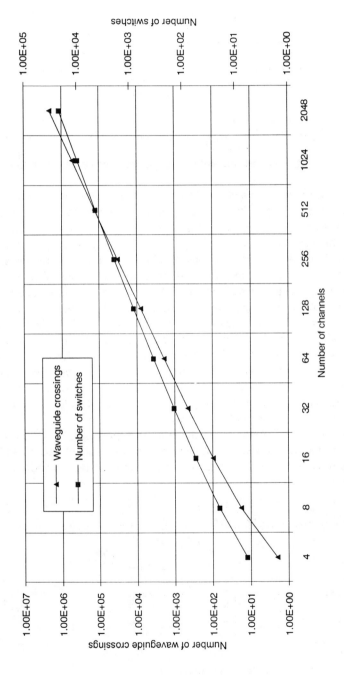

Figure 9.13(b) 8 × 8 benes network: analysis.

Table 9.5
Benes Network Parameters

Parameter	Value
Connectivity: Nonblocking	Rearrangably
Total number of switches	$n(2 \log_2(n) - 1)/2$
Signal path length	$2 \log_2(n) - 1$
Total number of waveguide crossings	$C_{tot} = 2C_{tot}$ of the MINs that form benes

routed to port A. Several network topologies can be constructed based on this switching primitive.

The active splitter/active combiner (ASAC) network, detailed in Figure 9.14, is a strictly nonblocking architecture. It is termed a multilayer architecture because an n input by n output network is typically implemented using $2n$ switching planes. By implementing the network in this manner, the need for any waveguide crossings may be eliminated.

Each plane in the ASAC network consists of a binary tree of depth containing $n-1$ switches and is configured to either fan a single input across one of n outputs or to coalesce one of n inputs into a single-output signal. When a fanout, or splitter, precedes a combiner stage and the stages are aligned as shown in Figure 9.14, it becomes possible to connect on n inputs to any of n output channels. Thus, the total ASAC network requires two stages of switches for each of n inputs, making the total number of switches equal to $2n(n - 1)$. Each signal traverses $2\log_2(n)$ switches and also suffers some power loss when crossing between waveguides on different substrates. The benefits of the ASAC multilayer implementation are an absence of waveguide crossings and an extremely simple routing algorithm because each input has exactly one dedicated path to each output. The characteristics of the ASAC network are provided in Table 9.6.

Similar to the ASAC topology is the passive splitter active combiner (PSAC) network, where signals are passively split between two waveguide branches and actively combined by a three-port switch. For signals propagating in the reverse direction, the

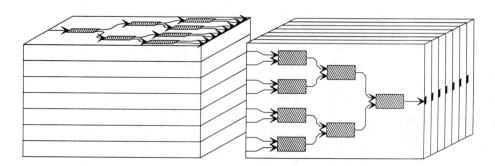

Figure 9.14(a) 8 × 1 active splitter/active combiner: topology.

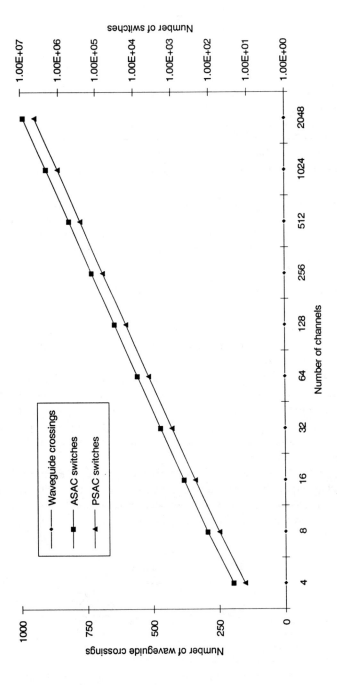

Figure 9.14(b) 8 × 1 active splitter/active combiner: analysis.

Table 9.6
ASAC Network Parameters

Parameter	Value
Connectivity	Strictly nonblocking
Total number of switches	$2n(n-1)$
Signal path length	$2 \log_2(n)$
Total number of waveguide crossings	0

network is better named an active splitter passive combiner, where three-port switches actively split signals, and waveguides passively combine them. This network requires the same number of planes as the ASAC but only half the total number of switches, which also halves the signal's path length. An additional feature of this topology is that the passive splitting (or combining) provides the network with a limited ability to implement a broadcast capability. The PSAC characteristics are provided in Table 9.7.

9.4 COMPARISON OF NETWORKS

Previous studies generally concluded that networks that provide nonblocking connectivity with as few switching elements per channel as possible were the ideal topology. While these results were valuable for the systems investigated, they are not necessarily applicable to the networks required for general purpose optical computers. The previous section reviewed several different network configurations and developed expressions that characterize not only the complexity of the networks but also allow comparison of networks on the basis of waveguide crossings, which will act to reduce signal integrity in a large network. The result of this review is the identification of a network trade space which has not previously been considered.

This trade space is illustrated in Figures 9.15 and 9.16. In Figure 9.15, the total number of switches required to implement a network of a given size is plotted for each class of network studied. As expected, this graph confirms that the nonblocking networks, like the ASAC and crossbar, have very high switch counts relative to the number of

Table 9.7
PSAC Network Parameters

Parameter	Value
Connectivity: nonblocking	Strictly
Total number of switches	$n(n-1)$
Signal path length	$\log_2(n)$
Total number of waveguide crossings	0

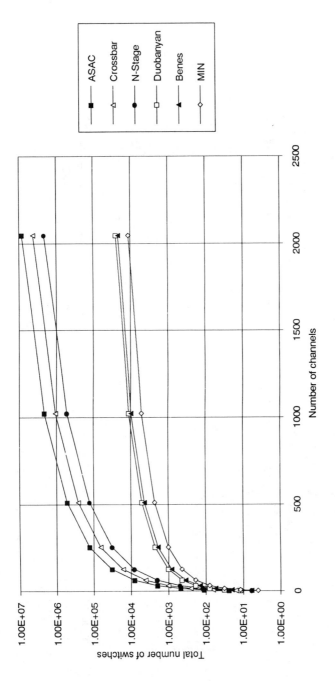

Figure 9.15 Total number of switches versus total number of channels for 6 $n \times n$ networks.

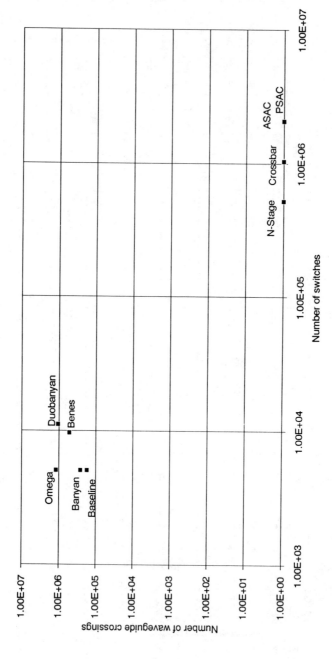

Figure 9.16 Total number of waveguide crossings versus total number of switches (for $N = 1{,}024$).

channels provided. Indeed it is graphs like Figure 9.15 that are typically used to discount nonblocking networks from consideration in integrated systems.

Figure 9.16 illustrates the same tradespace when the effects of waveguide crossings are considered for a network of size $n = 1,024$. In this figure, we see that despite the differences in configuration, the entire spectrum of networks investigated falls into essentially two classes: those networks with a rather huge number of waveguide crossings and those with virtually none.

The duobanyan and benes are representative of this first group of networks. Both require only on the $O[n \log_2(n)]$ switches for an n-input-×-n-output network. However, they are penalized with $O(n^2)$ waveguide crossings. For most of these waveguide crossings, the angle between guides is less than 90°, increasing the loss and crosstalk beyond that of the ideal right-angle crossing. In addition, as n increases, either the spacing between switching elements increases, or the angle between waveguides decreases even further.

The crossbar, N-stage, and splitter-combiner topologies belong to the second group of networks and are often shunned due to their requirement for $O(n^2)$ switching elements. However, these networks have no signal crossings outside of the switches themselves, resulting in low crosstalk and thus preserving signal integrity. The single-layer crossbar and n-stage networks are also most likely to achieve a high packing density for the switches due to the lack of waveguide crossings.

The vast tradespace between the two classes of networks illustrated by Figure 9.16 is somewhat worrisome because this region represents potential network topologies that have advantages that are uniquely linked to the characteristics of integrated optical systems but which remain undiscovered. Also, contrary to popular belief, given the currently available network topologies, Figure 9.16 strongly indicates that the large penalty in signal-to-noise ratio incurred by a topology with a high number of waveguide crossings may offset any gains attained by limiting the number of switching elements.

This means that the often-shunned crossbar network may be well suited for application to large, programmable, integrated, guided-wave interconnection networks on the basis of preservation of signal integrity. In addition, the crossbar network is rearrangably nonblocking, has a trivial routing algorithm, and has the unique feature that it can route signals between more than just two of its edges. Thus, based on these properties, the crossbar network will be discussed in the remaining section.

9.5 A 3-D INTEGRATED OPTICAL CROSSBAR NETWORK

The fundamental 2-D integrated guided-wave crossbar network is formed from an array of n-×-n switches as illustrated in Figure 9.17. This network connects the n signals on the left side of the crossbar to the n signals on the bottom of the crossbar or the n signals on the top of the crossbar to the n signals on the right side of the crossbar in a nonblocking-in-the-wide-sense manner and can be integrated onto a single die or substrate.

Because it is desired to make the network fully programmable, a scheme for reconfiguring the network must be found. A straightforward but costly method would be to provide

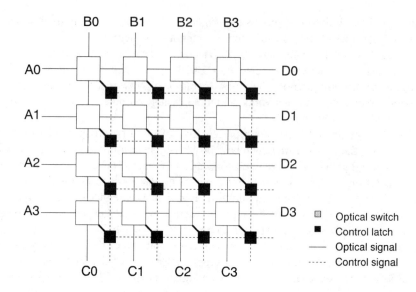

Figure 9.17 A crossbar switching array with control.

one control signal for each of the n^2 switches. A more economical approach, as shown in Figure 9.17, is to provide latching control elements at each switch and to arrange the latches so that they can be accessed uniquely by addressing a particular row and column of the array. Using this approach, each switch can be uniquely accessed using only $2n$ signals.

This network can then be expanded to create the three-dimensional network building blocks first illustrated in Figure 9.2 by stacking m such substrates on top of one another as shown in Figure 9.18. This creates a cube-like structure capable of connecting two n-×-m arrays of signals. However, the connections are still limited to signals in the same plane. That is, connections can only be established between inputs and outputs on the same substrate. To achieve connectivity between planes, a second stack of substrates, each containing an integrated crossbar network, is required. This second stack of substrates is aligned such that it is rotated by 90° with respect to the first. The result is a rearrangably nonblocking interconnection for n-×-m signals. With these two building blocks, any of the n signals from each of the m horizontal substrates in the first stack can be connected to one of the n signals from each of the m vertical substrates from the second stack for square signal arrays where $n = m$.

Unfortunately, because the wafer thickness is much greater than the minimum possible spacing between waveguides on the same wafer, the waveguide density using this approach will essentially be determined by the wafer thickness. To improve signal density, a fixed holographic interconnection is introduced between the network building blocks as shown in Figure 9.19. This hologram provides a mapping between each horizontal

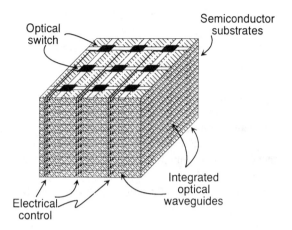

Figure 9.18 Integrated 3-D network building block.

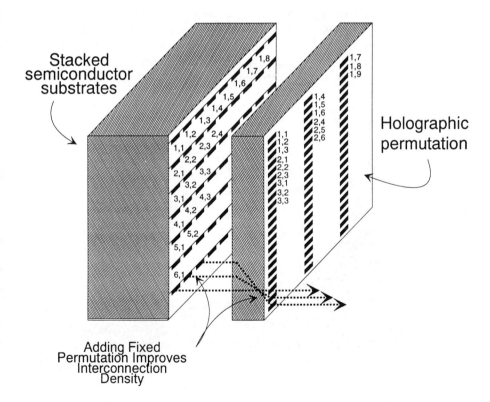

Figure 9.19 Close up of a wafer stack and fixed interconnection hologram (for $n/m = 3$).

group of signals from one cube to the corresponding vertical group of (*n/m*) signals on the other cube. This maximizes waveguide density by mapping *n* signals of one stack to *n* waveguides of the other stack instead of merely the m waveguides opposite the n signals of the first stack.

The holographic permutation illustrated Figure 9.19 would be appropriate if the substrate thickness, *T*, was approximately three times the waveguide spacing, *G*. The resolution of the hologram, 0.5mm per light beam [27], ensures that it will not be a limiting factor because this is much smaller than both the substrate thickness and the waveguide spacing. A major advantage of this proposed guided-wave system over free-space optical implementations is its small physical size. Given a substrate of width, *w* = 1.27cm, and a thickness, *T*, of 2.5 to 15.0 mils (0.064 to 0.381 cm), stacks of tens to hundreds of layers are possible. Figure 9.20 illustrates the relationship between substrate thickness, waveguide spacing, and the number of interconnection channels available for a stack of dimension $1.27cm^3$. Interconnection density can be further improved (by a factor of 2 times) by fabricating switches on both sides of each substrate without modifying substrate thickness.

At present, it does not appear that the signal density will be limited by the S-SEED spacing because this dimension is already on the order of 20 mm [28]. While substrates could be further reduced in thickness, this alternative becomes undesirable due to the difficulties of handling substrates that are much thinner than about 2 mils. Thus, the major parameter controlling signal density is the obtainable waveguide spacing.

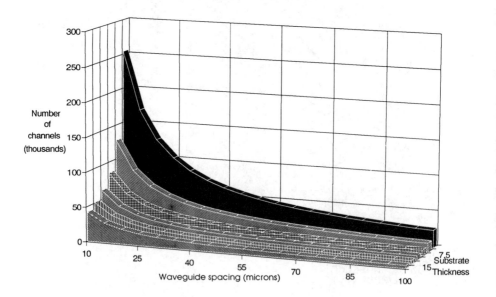

Figure 9.20 Total number of channels versus die thickness versus waveguide spacing.

In a directional coupler-based crossbar network, the waveguide spacing is determined primarily by the size of the switches, the bending radius of the waveguides, and any requirements for interaction lengths. As has been shown, the waveguide spacing in a crossbar network is primarily a function of the switch size because all the waveguides can run in parallel in a two-dimensional mesh. Thus, Figure 9.20 can be interpreted as a function that relates switch size to interconnection density. Under this interpretation, it becomes obvious that switch sizes on the order of 20 mm or less are needed to maintain reasonable interconnection densities.

This means that switching technologies requiring long interaction regions, such as lithium niobate, are inappropriate for use in high-density integrated networks because their requirement for interaction regions of several millimeters precludes achieving high switching density. In contrast, the interaction region needed in an MQW switch is approximately 5 mm, making them ideal alternatives for high-density interconnection systems. Using MQW switches, it is reasonable to achieve a waveguide spacing, G, of 10 mm, which corresponds to approximately $n = 1,200$ waveguides per layer. Together with a substrate thickness, T, of 5 mils (127 mm) or less, ~ 100 layers can be stacked to achieve a channel count of $\geqslant 120,000$. Such a network would allow a few hundred thousand S-SEEDs per cubic inch to be connected in a wide-sense nonblocking manner.

9.6 SYSTEM OPERATION

This section focuses on the operation of the optical computer and network that was presented in Section 9.1. Recall that the optical computer is composed of two I/O planes and two S-SEED arrays interconnected by a three-dimensional integrated guided-wave crossbar network. This network or virtual three-dimensional hologram realizes arbitrary connectivity between the logic arrays, allowing execution of complex functions.

Depending on the complexity and nature of the program to be executed, the system may use a single set of inputs and the corresponding switch settings, or it may cycle through a large number of sets of inputs and switch settings. During each cycle, the clock beams generate signals that are either routed, internal to the optical computer, to the other S-SEED array or are routed to the system output through the three-dimensional interconnect. Because the interconnection patterns determine the logic functions of the S-SEEDs in each array, a dynamic or variable interconnection allows execution of different functions using the same S-SEEDs.

The system accepts data from two I/O planes and stores the data into the S-SEED arrays. By programming the switches in each of the three stacks and applying a clock beam, the S-SEED array can transfer its data to an I/O plane or to the other S-SEED array. In this way, the logic functions performed by the S-SEEDs are interconnected, resulting in computing structure similar to conventional programmable array logic. The main difference is that the resultant structure is redefinable on a clock-cycle-by-clock-cycle basis. Basic system operation follows these steps (refer to Figure 9.2):

1. Configure switches in stacks two and three so that signals from I/O-plane one can be routed to S-SEED-array two.
2. Inputs are applied inputs at I/O-plane one. This puts each S-SEED of array two into one of two stable states. Note: if no inputs are applied, the values of S-SEED-array two remain uninitialized if this is first pass or remain unchanged from step 8, if not.
3. Reconfigure switches in all three stacks so that values stored in S-SEED-array two can be routed to either the desired element in S-SEED-array one or the desired I/O channel in I/O-plane one.
4. Apply clock two. This causes light to be either reflected or absorbed by the SEEDs in array two. Reflected light is routed through the stacks to S-SEED-array one or I/O-plane one, depending on how the switches were set in step 3.
5. Configure switches in stacks one and two so that signals from I/O-plane two can be routed to S-SEED-array one.
6. Apply inputs at I/O-plane-two. If no inputs are applied, the values of S-SEED-array-one remain unchanged from step 4.
7. Reconfigure switches in all three stacks so that the values stored in S-SEED-array-one can be routed to either the desired element in S-SEED-array two or the desired I/O in I/O-plane two.
8. Apply clock one. This causes light to be either reflected or absorbed by the SEEDs in array one. Reflected light is routed through the stacks to S-SEED-array two or I/O-plane two, depending on how the switches were set in step 7.
9. Go to step 1.

To reconfigure the switches in the substrate stacks to connect a given SEED in one array to another SEED in the other array, a routing algorithm was developed. Because the path lengths of signals through the crossbar based interconnection cubes can vary greatly, it is desirable for the routing algorithm to help minimize these differences. The following simple algorithm reduces differences in path length by selecting switch settings, for one signal at a time, so that each signal's path length comes as close to equaling $3n$, the average path length through the network, as possible based on the given idle switches. For simplicity, it is assumed that waveguide spacing, G, and substrate thickness, T, are equal. This implies that the number of waveguides per layer, n, equals the number of layers per stack, m. Thus, the fixed interconnect holograms are not needed. In this case, all coordinates take on integer values between 1 and n.

The coordinates of the source and destination of a light beam are x- and z-values in one of the two S-SEED arrays. For this example, a fanout of two is assumed, meaning that there are two light beams incident on each SEED window. Each switch on every substrate is defined by an x-, y-, and z-coordinate and is in the crossover mode by default, except for those switches determined by application of the routing algorithm to be in the active signal bending mode. The routing algorithm is outlined below:

1. Given a source and destination SEED, determine coordinates x_1, z_1, x_2 and z_2.
2. Calculate the total path length through the three cubes, $(x_1 + y - 1) + (x_2 + z_1 - 1) + (y + z_2 - 1)$, and set it equal to the average path length, $3n$.

3. Solve for y, rounding off to nearest integer.
4. Assign y coordinate to each signal such that total path length through all three stacks is approximately equal to average path length, with the following restriction: no two signals on the same substrate in stacks one or two may share the same y coordinate (i.e., signals with the same z_1- or x_2-coordinates must have unique y coordinates). If so, another value for y must be selected.

The restriction in step 4 is due to the fundamental property that each port of a switching element may only handle one signal at a time. Step 4 of the previous algorithm may have to be repeated several times when the utilization of network connections is high before a y-value that resolves this contention is found. Each new y-value causes the path length to deviate further from the average, but some nonblocking value is guaranteed to exist. Fortunately, this routing algorithm need only be executed at compile time. During run time, the switch setting block of this system only activates the control signals for the switches to be set in a given cycle. Each of the three stacks is nonblocking in the wide sense; any input to a given stack can be routed to any unused output of the stack, provided the trivial algorithm mentioned previously is followed. Thus, a system of three such stacks connected in series must also be nonblocking because an output of one stack is simply the input to the next.

9.7 CONCLUSION

A compact programmable three-dimensional interconnection scheme, suitable for use in an optoelectronic computing system has been described. Several possibilities for implementing the required network were investigated. It was generally found that previous research was only applicable to the construction of relatively small interconnection networks. When these networks are compared with respect to their signal integrity, it is found that not all networks are created equal, especially when the number of required interconnections approaches many thousand.

As a result, it was determined that, by exploiting the recently emerging technology of wafer stacking in conjunction with the latest generation of multiple quantum-well switches, extremely high-density switching networks could be constructed using the often-shunned crossbar interconnection topology. Indeed, is has been shown that the crossbar switch actually offers many distinct advantages as a network topology for optical computing systems.

Another significant finding in this work is that the current bottleneck to high integrated optical network density is the dimension of the switching element used to implement the network. Indeed, it was shown that only by further developing the highly compact multiple quantum-well switch further will it be possible to achieve the interconnection densities required in optical computers.

The proposed network consists of three stacks of substrates, each containing a planar array of crossbar switches. For a given substrate thickness, T, waveguide spacing, G, and substrate width, W, the proposed scheme provides W^2/TG signal channels between two arrays of S-SEEDs and requires $3W^3/TG^2$ switches. A simple routing algorithm for determining the necessary switch settings for the network was also presented.

REFERENCES

[1] Erickson, J. R., "Signal-to-noise ratio simulations for small guided-wave optical networks," *Proc. SPIE Digital Optical Computing*, Vol. 752, 1987, pp. 222–229.
[2] Lin Wu, Chuan and Tse-Yun Feng, "On a Class of Multistage Interconnection Networks," *IEEE Transactions on Computers*, Vol. C-29, No. 8, 1980, pp. 694–702.
[3] Agrawal, Dharma P., "Graph Theoretical Analysis and Design of Multistage Interconnection Networks," *IEEE Transactions on Computers*, Vol. C-32, No. 7, 1983, pp. 637–648.
[4] Kruskal, Clyde P. and Marc Snir, "A Unified Theory of Interconnection Network Structure," *Theoretical Computer Science*, North-Holland Vol. 48, 1986, pp. 75–94.
[5] Schnieder, E., W. R. Michalson, "Integrated Guided-Wave Crossbar Interconnection Of SEED Arrays," *Optoelectronic Interconnects*, Ray T. Chen, ed., SPIE 1849, 1993, pp. 209–220.
[6] Sawchuk, Alexander A. and B. Keith Jenkins, "Dynamic Optical Interconnection for Parallel Processors," *Proc. SPIE Optical Computing* 625, 1986, pp. 143–153.
[7] Michalson, William R. and Christopher S. Tocci, "Optimal Power Efficiency Computing," unpublished, 1990.
[8] Schneider, Eric G., "Design, Simulation, and Analysis of a 3-D Integrated Optical Computer", MS Thesis, Worcester Polytechnic Institute, December, 1992.
[9] Brenner, Karl-Heinz, Alan Huang, and Norbert Streibl, "Digital optical computing with symbolic substitution," *Applied Optics*, Vol. 25, No. 18, 1986, pp. 3054–3060.
[10] Tanida, J. and Y. Ichioka, "Optical logic array processor using shadowgrams," *J. Opt. Soc. Am.*, Vol. 73, No. 6, 1983, pp. 800–809.
[11] Murdocca, Miles J., Alan Huang, Jurgen Jahns, and Norbert Streibl, "Optical design of programmable logic arrays," *Appl. Opt.*, Vol. 27, No. 9, 1988, pp. 1651–1660.
[12] Brenner, Alan F., Harry F. Jordan, and Vincent P. Heuring, "Digital optical computing with optically switched directional couplers," *Optical Eng.* Vol. 30, No. 12, 1991, pp. 1936–1941.
[13] Michalson, W. R., S. Plötner, G. Stevens, "A Logic Compiler With Genetic Algorithm Based Optimization", Technical Report, Department of Electrical and Computer Engineering, Worcester Polytechnic Institute, Worcester, MA, 1993.
[14] Armenise, Mario N. and Beniamino Castagnolo, "Design of Photonic Switches for Optimizing Performances of Interconnection Networks," *Proc. SPIE Integrated Optics and Optoelectronics II*, Vol. 1374, 1990, pp. 186–197.
[15] Thylén, Lars, "Integrated Optics in LiNbO$_3$: Recent Developments in Devices for Telecommunications," *Journal of Lightwave Technology*, Vol. 6, No. 6, 1988, pp. 847–861.
[16] Voges, Edgar and Andreas Neyer, "Integrated-Optic Devices on LiNbO$_3$ for Optical Communication," *Journal of Lightwave Technology*, Vol. LT-5, No. 9, 1987, pp. 1229–1238.
[17] Bogert, G. A., "4x4 Ti:LiNbO$_3$ Switch Array with Full Broadcast Capability," *Topical Meeting on Photonic Switching*, Technical Digest Series (Optical Society of America) Vol. 13, 1987, pp. 68–70.
[18] Alferness, R. C., L. L. Buhl, S. K. Korotky, and R. S. Tucker, "High-Speed—β-Reversal Directional Coupler Switch," *Topical Meeting on Photonic Switching*, Technical Digest Series (Optical Society of America) Vol. 13, 1987, pp. 77–79.
[19] Neyer, A., "Electro-Optic X-Switch Using Single-Mode Ti:LiNbO$_3$ Channel Waveguides," *Electron. Lett.*, Vol. 19, No. 14, 1983, pp. 553–554.
[20] Beesoon, Karl W., et al., "Photochemical Formation of Polymeric Optical Waveguides and Devices for Optical Interconnection Applications," *Proc. SPIE Integrated Optics and Optoelectronics II*, Vol. 1374, 1990, pp. 176–185.

[21] Lytel, R., et al., "Electro-optic Polymers for Optical Interconnects," *Proc. SPIE Digital Optical Computing II*, Vol. 1215, 1990, pp. 252–262.

[22] Chen, Ray T., "Optical Interconnects: A Solution to Very High Speed Integrated Circuits and Systems," *Proc. SPIE Integrated Optics and Optoelectronics II*, Vol. 1374, 1990, pp. 162–175.

[23] Brinkman, M. J. and G. J. Sonek, "Low power response of all-optical crossbar networks in quantum well heterostructures," *Applied Optics*, Vol. 31, No. 3, 1992, pp. 338–349.

[24] Langer, D. W., M. Chen, H. Lee, and M. Chmielowski, "Quantum well structures for integrated optoelectronics," *Proc. SPIE Digital Optical Computing II*, Vol. 1215, 1990, pp. 243–251.

[25] Eichen, E., et al., "Optical Amplifiers for Photonic Switching," *IEEE LEOS Annual Meeting Proceedings*, 1990, pp. 250–254.

[26] Arrathoon, R., "Fiber-Optic Programmable Logic Arrays," *Optical Computing: Digital and Symbolic*, ed. R. Arrathoon, Marcel Dekker Inc., 1989.

[27] Feldman, Michael R., and Clark C. Guest, "Interconnect density capabilities of computer generated holograms for optical interconnection of very large scale integrated circuits," *Applied Optics*, Vol. 28, No. 15, 1989, pp. 3134–3137.

[28] Chirovsky, Leo and Joseph M. Freund, "S-SEED Arrays for Parallel Processing of Digital Optical Information," Proc. Electro/92 Semiconductor Device Technology II, 1992, pp. 116–121.

Appendix
Theoretical Analysis for Cryogenic Coupler's Usable Dynamic Range (Per Channel)

Due to the nonlinear transduction function associated with the analog modulating 1-×-2 switch, there is an interrelational limit of dynamic range and usable binary conversion (digitization) from analog information obtained by this type of IOD. Figure A.1 shows the graphical characteristics of the 1-×-2 IOD's differential output [1]; however, we will use (A.1) since it is the trigonometric equivalent function to the two complimentary equations shown and easier to use in this analysis. In fact, (A.1) is the MZI's exact intensity transfer function, because the device has an offset voltage (that is, $P_O/P_{IN} = 1/2$ when $V = 0V$) associated with its operation. Therefore, with reference to Figure (A.1), the power transfer functions for $\overline{\Phi}$ and Φ, respectively, are:

$$\frac{P_o}{P_{in}} = \cos^2\left(\frac{\pi(2V_{ac} - V_\pi)}{4V_\pi}\right) \qquad \frac{P_o}{P_{in}} = \cos^2\left(\frac{\pi(2V_{ac} + V_\pi)}{4V_\pi}\right) \qquad (A.1)$$

where V_{ac} = applied analog voltage (volts); V_π = extinction voltage (volts); and P_o/P_{in} = normalized power transfer function (dimensionless). To maximize linearity, we operate at the inflection point, that is, $V = V_{ac} + V_{DC}|_{DC=0} = V_{ac}$, then AC operation at the operating point becomes for either Φ, $\overline{\Phi}$:

$$\left(\frac{P_o}{P_{in}}\right) = \cos^2\left(\frac{\pi(2V_{ac} \pm V_{ac})}{4V_{ac}}\right) \qquad (A.2)$$

From the idealized linear transfer shown in Figure (A.1), can be given as $P_o/P_{in} = \pm V_{ac}/V_\pi + 1/2$, hence characterizing the error between the actual and idealized can be stated as

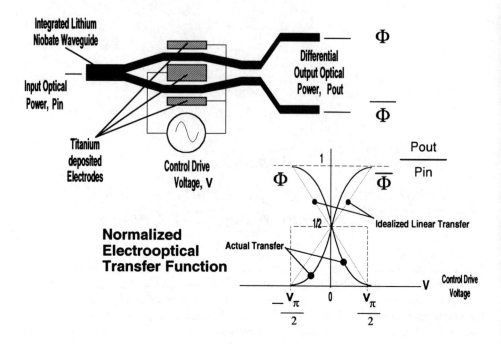

Figure A.1 The 1-×-2 IOD's differential output.

$$\text{error} \equiv \left(\frac{P_O}{P_{IN}}\right)_{linear} - \left(\frac{P_O}{P_{IN}}\right)_{actual}$$

or, after substitution,

$$\text{error} = \left(\mp\frac{V_{ac}}{V_\pi} + \frac{1}{2}\right) - \cos^2\left[\frac{\pi(2V_{ac} \pm V_\pi)}{4V_\pi}\right] \quad (A.3)$$

Since we are interested in the maximum usable bits of analog signal information, then we first state (A.4), which relates the maximum binary conversion given analog error as noise.

$$\text{maximum usable bits} = -\frac{\log_{10}(\text{error})}{\log_{10}(2)} \text{ bits} \quad (A.4)$$

again substituting

$$\text{maximum usable bits} = -\frac{\log_{10}\left\{\left(\mp\frac{V_{ac}}{V_\pi}+\frac{1}{2}\right)-\cos^2\left[\frac{\pi(2V_{ac}\pm V_\pi)}{4V_\pi}\right]\right\}}{\log_{10}(2)} \text{ bits} \quad (A.5)$$

At this point, define the variable, Γ, as the ratio of V_{ac}/V_π, then substituting this variable into (A.5) while choosing one particular sense (that is, either Φ or $\overline{\Phi}$) yields

$$\text{maximum usable bits} = -\frac{\log_{10}\left[\left(\Gamma+\frac{1}{2}\right)-\cos^2\left(\frac{\pi(2\Gamma-1)}{4}\right)\right]}{\log_{10}(2)} \text{ bits}$$

or

$$\text{maximum usable bits} = -3.3219 \times \log_{10}\left\{\left(\Gamma+\frac{1}{2}\right)-\cos^2\left[\frac{\pi(2\Gamma-1)}{4}\right]\right\} \text{ bits} \quad (A.6)$$

If we plot (A.6) as a function of Γ, we will see how the analog control or input voltage shown in Figure (A.1) affects the usable dynamic range and bit resolution. See Figure A.2. (A.6) give the maximum usable number of bits by A/D conversion of the transduced analog signal. This concept of usable bits directly dictates the maximum dynamic range, since *each* usable bit translates to about 6 dB of dynamic range. This concept is inherent from (A.4), since the numerator represents the inverse of the dynamic range capability or dynamic range (dB) = $-20 \cdot \log_{10}$(error) Plotting this gives us Figure A.3.

With reference to Figure A.2, the peak occurs at $\Gamma = .5$ since the error between the linear (idealized) and actual goes to zero. Remember, the $\cos^2(x)$ function has three places where the error *as defined* (see (A.3)) goes to zero: the operating point, $V_{ac} = 0$, and the two extremes, $V_{ac} = \pm V_\pi/2$. (See Figure (A.1).) The log plot in Figure A.2 prohibits showing a similar infinite-bit resolution peaking at $V_{ac} = 0$. At these points, as well as at nonzero error points in the transfer, other noise mechanisms will limit the number of usable bits or dynamic range.

Utilizing what has just been described, suppose, for example, that we desire less than 1% electro-optical transduction error (peak); then, from (A.3),

$$\left(\Gamma+\frac{1}{2}\right)-\cos^2\left(\frac{\pi(2\Gamma-1)}{4}\right) < .01 \quad (A.7)$$

This equation has no analytical closed-form solution, but we can graphically arrive at a value of Γ by plotting (A.7) as a function of Γ. As see from Figure A.4, $\Gamma < .01760$, or $|V_{ac}| < .01760 \, V_\pi$ (volts, peak). Referring to the Crystal Technology model MZ313P 1-×-2 IOD data sheet gives the following specifications at an operating wavelength of

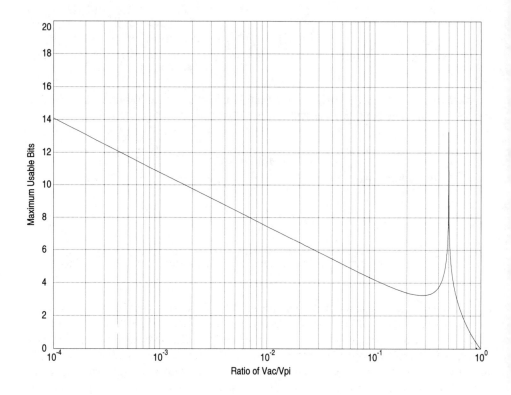

Figure A.2 Maximum usable bits versus the ratio of V_{ac}/V_π.

1,300 nm as [1]: $V_\pi = 8.0$V (dc); hence $|V_{ac}| < .1408$V (peak) or $|V_{ac}| < .2816$V (peak-to-peak)

Table A.1 shows the maximum peak-to-peak (p-p) analog modulation voltage for a given number of usable bits of resolution and associated error based upon $V_\pi = 8$ volts. The p-p drive voltage was determined graphically from Figure A.5 which a numerical solution to (A.7). The table constitutes the upper limit of analog drive of the 1-×-2 switch, based on our definition of error for a given number of bits of resolution and a V_π voltage. Also, this and the remaining analysis does not assume any additive A/D conversion error. The *lower* limit of the signal handling will be the noise floor associated with the photodetector, TIA, and laser. The noise mechanisms associated with this photodetector structure are shown in Figure A.6 and given on a per root Hertz basis as [2]:

$$I_T^2 = 2qI_{sig} + \frac{4KT}{R} + (I_A^\#)^2 \quad (A.5)$$

where I_{sig} denotes the incoming converted optical signal current; K is Boltzmann's constant $(1.38 \times 10^{-23}$ J/K; T is the absolute temperature (Kelvin); R is the equivalent photodiode

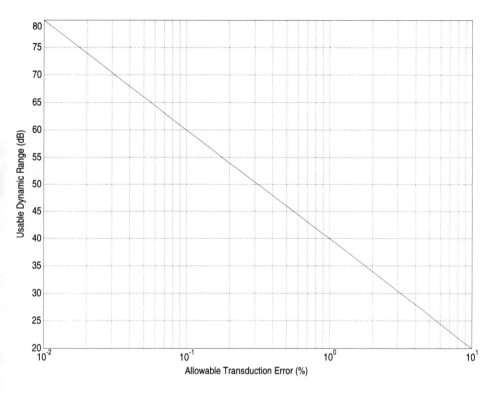

Figure A.3 Usable dynamic range (dB) versus allowable transduction error (%).

parallel resistance; $I_A^\#$ is the equivalent TIA noise current referred to input (RTI); q is the electronic charge (1.6×10^{-19} coulomb); and I_T is the total noise current (RTI). The first two terms of (A.5) are detector diode noise mechanism, namely, $2qI_{sig} \Rightarrow$ signal-induced noise (shot noise) and $4KT/R \Rightarrow$ diode resistive thermal noise (Johnson noise), and the last term is noise from the TIA RTI.

The laser noise is characterized by a measure called relative intensity noise (RIN). For some of the better 1.3-micron excitation sources operating in the CW mode, this value is about −150 dbc/Hz [2]. Given that the responsivity, $\Re$, of 1.3-micron detector technology is about .8 amp/W, and an incident optical power, Φ_{in}, of 1 mW (0 dbm), this translates to

$$\langle I_{RIN}^2 \rangle = (I_{sig})_{RIN}^2 = (\Phi_{in} \times \Re)_{RIN}^2 = 6.4 \times 10^{-22} \text{ amp}^2/\text{Hz}$$

Combining with (A.5), the total receiver noise referred to input (RTI) gives

$$I_T^2 = 2qI_{sig} + 4KT/R + (I_A^\#)^2 + \langle I_{RIN}^2 \rangle \qquad (A.6)$$

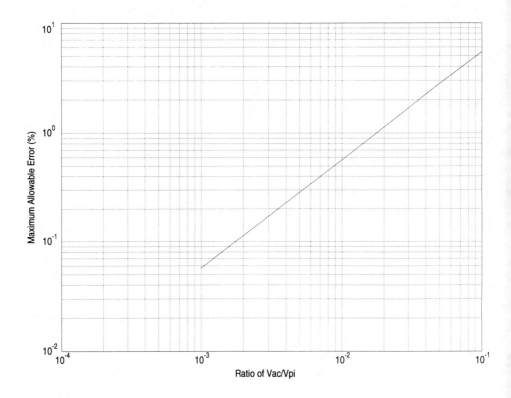

Figure A.4 Maximum allowable error (%) versus the ratio of V_{ac}/V_π.

Table A.1
Maximum Peak-to-Peak Analog Modulation Voltage

Usable Number of Bits	Maximum 1-×-2 IOD Distortion (%)	P-to-Point Drive ($V_\pi = 8V$)
2	25.0	5.358V
4	6.25	.949V
6	1.56	.22V
8	.39	.056V
10	.0977	.0137V
12	.0244	3.4mV
14	.0061	.85mV
16	.0015	.22mV

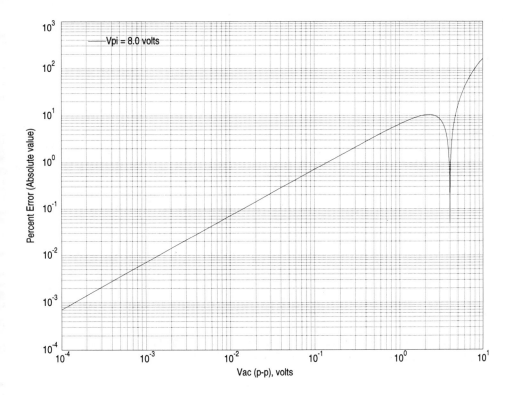

Figure A.5 Percent error (absolute value) versus V_{ac} (point-to-point), volts.

Substituting some typical values (per Hertz basis):

$$2qI_{sig} \Rightarrow 3.2 \times 10^{-19} \, I_{sig} \text{ amp}^2$$

$$(4KT)/R \Rightarrow 3.86 \times 10^{-25} \text{ amp}^2 \quad (T = 70°\,K, R \approx 10K)$$

$$(I_A^{\#})^2 \Rightarrow 1 \times 10^{-18} \text{ amp}^2$$

$$\langle I_{RIN}^2 \rangle \Rightarrow 6.40 \times 10^{-22} \text{ amp}^2$$

Hence,

$$I_T = \sqrt{(3.2 \times 10^{-19})I_{sig} + 10^{-18}} \text{ amp}/\sqrt{\text{Hz}} \tag{A.7}$$

where

$$I_{sig} = \Re\Phi_{in} \tag{A.8}$$

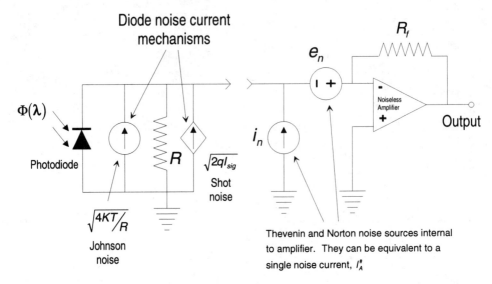

Figure A.6 Noise mechanisms associated with photodiode structure.

and $\mathfrak{R} \equiv$ responsivity of photodiode (amp/watt), and $\Phi_{in} \equiv$ input optical power (watts). Then, since $\mathfrak{R} = .8$ amp/W and generally $\Phi_{in} \ll 1\text{W}$,

$$I_T \approx I_A^{\#} = 10^{-9} \text{ amp}/\sqrt{\text{Hz}} \tag{A.8}$$

Therefore, the predominant electrical noise source in this approach is the photodetector's TIA associated with the cryogenic signal coupling IOD structure.

The approximate front-end dynamic range can now be determined with the aid of Table A.1 for a given bit resolution. This can be accomplished if we define the error function in (A.3) as simply another "noise" component to be added to the above-described electrical noise. Consequently, if we know that the predominant electrical noise component is given by (A.8), we can divide this expression by the responsivity, $\mathfrak{R}$, which gives the equivalent input-noise optical power, P_{noise}^{elect}, as

$$P_{noise}^{elect} = \frac{I_T}{\mathfrak{R}} = \frac{10^{-9} \text{ amp}/\sqrt{\text{Hz}}}{.8 \text{ amp/watt}} \times f(\sqrt{\text{Hz}}) = 1.25 \times 10^{-9} \text{ watt} \times f\left(\sqrt{\text{Hz}}\right)$$

This expression can be considered a weighting of electrically generated noise or error with a bandwidth dependence as

$$X_{error}^{elect} = 1.25 \times 10^{-9} \times f\left(\sqrt{\text{Hz}}\right)$$

If we now consider the harmonic distortion error ("noise"), defined previously as

$$X_{error}^{harmonic} = \left(\Gamma + \frac{1}{2}\right) - \cos^2\left[\frac{\pi(2\Gamma - 1)}{4}\right]$$

The total front-end system noise or error, X_{error}^{total}, can be defined as the harmonically generated noise, $X_{error}^{harmonic}$ added to the electrical base noise, X_{noise}^{elect} or

$$X_{error}^{total} \equiv \left\{\left(\Gamma + \frac{1}{2}\right) - \cos^2\left[\frac{\pi(2\Gamma - 1)}{4}\right]\right\} + 1.25 \times 10^{-9} \times f\left(\sqrt{Hz}\right) \quad (A.9)$$

From (A.9), the total equivalent system noise is dependent on the analog drive voltage level and operating electrical front-end bandwidth, $\sqrt{Hz}$.

Now, defining the dynamic range by performing the 20 log ratio of the intended to the error or "noise" associated with the electro-optical transduction and additive electrical-noise mechanism (all logarithms will be assumed to be base 10) as

$$\text{dynamic range} \equiv 20 \log \left(\frac{1}{X_{error}^{total}}\right) \quad (A.10)$$

As see from (A.9), the relatively small value of electronic noise, the loss in dynamic range, hence usable bits, is insignificant when compared to the harmonic error term.

Another dynamic loss area arises from waveguide losses due to power division. As discussed in the optical passive outstar, there could be N equally power taps feeding any given array layer. Incorporating this factor would affect (A.10), as follows

$$\text{dynamic range (per channel)} \equiv 20 \log \left(\frac{1}{X_{error}^{total}}\right) - 20 \log (N) \quad (A.11)$$

Finally, if there are N channels, then there are $N - 1$ "off" or disabled channels on the analog optical bus which are contributing some "noise" or unwanted analog signal throughput due to incomplete blockage of CW laser excitation at the $N - 1$ MZIs. If each MZI is capable of attenuating the analog throughput by G, then (A.7) has another uncorrelated noise term that is not bandwidth dependent. Thus (A.7) may be rewritten as

$$I_T = [3.2 \times 10^{-19} I_{sig} + 10^{-18} f(Hz) + [G(N - 1)]^2]^{\frac{1}{2}} \quad (A.12)$$

where G is the fractional loss of incoming outstar optical flux ($G \ll 1$); N is the number of channels involved in the outstar/instar optical flux; and $f(Hz)$ is the required signal bandwidth.

Consolidating (A.12) with our dynamic range definition gives

$$\text{dynamic range (per channel)} \equiv 20\log\left(\frac{1}{X_{error}^{total}}\right) - 20\log(N) \\ - 10\log[10^{-18} \times f(\text{Hz}) + G^2(N-1)^2] \quad \text{(A.13)}$$

Clearly, (A.13) shows that the dynamic range is strongly dependent upon the MZI's ability to keep unwanted laser excitation out of the optical bus. This ability is known as the extinction ratio (ER), which is defined as [3]:

$$\text{ER(dB)} = 10\log\left[\frac{\text{MZI output ``on'' flux}}{\text{MZI output ``off'' flux}}\right] \quad \text{(A.14)}$$

From off-the-shelf MZI and 1-×-2 IOD's, an extinction ratio of 30 dB is achievable with vendor screening [4].

In conclusion, (A.13) indicates that, if one wants a certain dynamic range at a given bit resolution and with a certain fanout, then one only needs to adjust the CW laser excitation to attain the solution. Another problem that sensor front-end designers are concerned with is saturation effects on the front-end due to a constant and/or strong input. The transimpedance amplifier design prevents any charge build-up because it will convert a constant current into a constant voltage. There are also very simple scaling or compression schemes integrated within the transimpedance amplifier to effect gain changes under very strong input signal conditions. One of the more common designs uses an operational transconductance amplifier (OTA), which can accommodate up to nine decades of input signal current and guarantee the nonlinearity to be less than .1% in any one decade. These devices are manufactured by RCA, National Semiconductor, Motorola, and other vendors.

REFERENCES

[1] Crystal Technology, Siemans Corp., Application Notes, "1-×-2 Switch Application Note," Newton, MA, 1990.
[2] Waldman, Gary and Wootton, John, *Electro-Optical Systems Performance Modeling*, Artech House, 1993, pp. 39–42.
[3] Weissman, Y. *Optical Network Theory*, Artech House, 1992, pp. 176–181.
[4] Crystal Technology, Siemans Corp., Application Notes, "1-×-2 and 2-×-2 OGW Switches," 1989.

List of Acronyms

$/P	cost per performance
APD	avalanche photodiodes
ASAC	active splitter/active combiner
BER	bit error rate
BIU	bus interface unit
BPM	beam propagation method
BR	bit rate
C4	controlled collapse chip connection
CCPD	charge coupled photo detector
CGH	computer generated holograms
CMOS	complementary metal oxide semiconductor
DCG	dichromate gelatin
E-O	electro-optical
ECL	emitter-coupled logic
ELED	edge emitting LED
EMP	electromagnetic pulse
EO	electro-optic
ERFC	error function
ESA	excited state absorption
FDMA	frequency division multiple access
FET	field-effect transistor
FTTH	fiber to the home
FWHM	full width at half-maximum
GRIN	graded index
GRTH	gradient-thickness
HB	horizontal Bridgeman
HBT	heterojunction bipolar transistors
HiPPI	high-performance parallel interface

HOE	holographic optical element
I/O	input/output
IC	integrated circuit
ISDN	integrated services digital network
ITO	indium tin oxide
IWKB	inverse Wentzel-Kramers Brillouin
LAN	local area network
LD	laser diodes
LEC	liquid-encapsulated Czochralsky
LED	light emitting diodes
LPE	Liquid phase epitaxy
MAC	multifiber array connectors
MBE	molecular beam epitaxy
MCM	multichip module
MCT	mercury cadmium telluride
MESFET	metal-semiconductor field-effect transistor
MIN	multistage interconnection networks
MLB	multilayer board
MM	multimode
MNA	methyl nitroaniline
MOCVD	metal organic chemical vapor deposition
MOS	metal-oxide semiconductor
MOSIS	metal-oxide semiconductor implementation service
MQW	multiquantum well
MSBVW	magnetostatic backward volume waves
MSFVW	magnetostatic forward volume waves
MSI	medium scale integration
MSSW	magnetostatic surface waves
MSW	magnetostatic waves
MUX	multiplexer
MZI	Mach-Zehnder Interferometer
NA	numerical aperture
NP	nitrophenol
NRZ	nonreturn-to-zero
OD	optical density
OEIC	optoelectronic integrated circuit
ONU	optical network units
OR	logical "OR" gate
PCB	printed circuit board
PD	photodetector
PE	processing elements
PIH	pin-in-hole

PIN-FET	PIN-field effect transistor
PMF	polarization maintaining fiber
PMMA	polymethyl methacrylate
PMSW	polymer microstructure waveguide
POF	plastic optical fiber
PON	passive optical network
PSAC	passive splitter active combiner network
QCSE	quantum-confined Stark effect
RC	resistor-capacitor
ROC	radii of curvature
S-SEED	symmetric self-electro-optic device
S/N	signal-to-noise
SAW	surface acoustic wave
SCC	single chip carriers
SCI	scalable coherent interface
SCSI	small computer system interface
SEED	self-electro-optic-effect device
SI	step index
SLM	spatial light modulator
SM	single mode
SPICE	special-purpose in-circuit circuit emulator
SPST	single-pole-single-throw
TAB	tape automated bonding
TDM	time division multiplexers
TDMA	time division multiple access
TE	transverse electric
TIR	total internal reflection
TM	transverse magnetic
TPF	tapping power fraction
TPON	telephony over passive optical network
TS	thermal set
UL	Underwriter Laboratory
VLSI	very large scale integrated
VLSIO	very large scale integrated optoelectronic
WDDM	wavelength division demultiplexer
WDM	wavelength division multiplexing
WDMA	wavelength division multiple access

About the Authors

Christopher S. Tocci is currently senior staff physicist at the Baird Corporation, Bedford, Massachusetts. He is involved with linear and nonlinear system modeling for atomic spectroscopy and consultant to several mathematical software companies in their attempt to merge mathematical engines into a unified simulator for use in system engineering. Dr. Tocci was also one of the cofounders of Applied Research Consortium, Inc., in Charlton, Massachusetts. His past technical experience has been in medical engineering, optical communications, optical device technology, and biophysics. He has been involved with industrial and military designs and analysis of optics and electronics as it applies to infrared sensors, two-dimensional signal processing, communication, and interconnection for massively parallel computing architectures. He has had senior positions at Raytheon, MIT Lincoln Labs, Augat Fiberoptics; and consulted for Ciba-Geigy Diagnostics on optical metrology used in blood analysis. Dr. Tocci has had several patents in cryogenic optical device technology and nearly 30 refereed publications in both trade and professional journals. He received his Ph.D. from Clarkson University in engineering science in 1985. Dr. Tocci is a member of OSA, SPIE, Who's Who in the East in Science & Technology, AAAS, Eta Kappa Nu, Boston Computer Society, and NRA.

H. John Caulfield is the university eminent scholar at Alabama A&M University. Before that he was the founder and first director of the Center for Applied Optics at the University of Alabama in Huntsville and has done research for industries large (Texas Instruments, Raytheon, and Sperry) and small (Block Engineering and Aerodyne Research). He has authored five books, a dozen book chapters, and hundreds of technical papers. A fellow and multiple honoree of both SPIE and OSA, Dr. Caulfield has also edited one journal (*Optical Engineering*) and has served on the editorial boards of 15 others. His 1984 cover story on holography for *National Geographic* was read by 25 million people. *Business Week* calls him "one of America's 10 top scientists" and *Byte* calls him "one of the most influential people in the world in minicomputers." He is also assistant sheep farmer for Far Our Farm in Tennessee.

Ray T. Chen is currently a tenure-track faculty member at the Microelectronics Research Center of the Department of Electrical and Computer Engineering, University of Texas, Austin. He joined the university from the Department of Electrooptic Engineering of the Physical Optics Corporation where he served as the department director from 1989 to 1992. Within the past four years, he has been the author and the principal investigator for over 20 awarded research proposals sponsored by many subdivisions of DOD, NSF, DOE, NASA, and other private industries. His research topics cover 2-D and 3-D optical interconnections, polymer-based integrated optics, polymer waveguide amplifier, graded-index polymer waveguide lens, active optical backplane, traveling wave electro-optic polymer waveguide modulator, GaAs all-optical crossbar switch, holographic lithography, and holographic optical elements. He has served as the chairman and a program committee member for over 10 domestic and international conferences organized by SPIE, IEEE, and PSC. He is also the invited lecturer for the short course of optical interconnects for the international technical meetings organized by SPIE. Dr. Chen has over 80 publications in open literature and has delivered numerous invited talks in the professional societies. Dr. Chen is a member of IEEE, SPIE, OSA, and PSC.

Michael R. Feldman received his M.S. and Ph.D. in electrical engineering (applied physics) from the University of California, San Diego, in 1986 and 1989, respectively, and the B.S.E. degree in electrical engineering from Duke University in 1984. His research accomplishments include the development of new computer generated hologram (CGH) encoding methods (including iterative discrete on-axis encoding), the experimental demonstration of an optically interconnected 1-K CMOS RAM chip, and the development of the Nested Crossbar connection network for optically interconnected processor arrays. Since 1989, Dr. Feldman has been a faculty member at the University of North Carolina at Charlotte where he has been leading an effort in optical interconnects for multichip modules using CGH technology. He is the author of over 30 technical papers, including a dozen papers in refereed technical journals. Dr. Feldman is a member of the Advanced Technology Packaging Group at the Microelectronic Center of North Carolina (MCNC) as well as OSA, SPIE, IEEE, Tau Beta Pi, and Eta Kappa Nu.

Lynn D. Hutcheson for the past five years has been Director, Optical Systems at Raynet Corporation, Menlo Park, California. Before taking the position at Raynet, he was chief scientist at APA Optics, responsible for development of optical interconnects, optoelectronic, and integrated optical devices. He has held managerial and research positions at CyberOptics Corporation, Honeywell, and Westinghouse. Dr. Hutcheson is the editor of the book, *Integrated Optical Circuits and Components: Design and Applications*, published by Marcel Dekker, Inc. He has helped organize and chair numerous articles on various aspects of optics technology. He received his M.S. and Ph.D. degrees in optical sciences from the University of Arizona and B.S. and M.S. degrees in electrical engineering from the University of Colorado.

Tomasz Jannson is chief scientist and cofounder of Physical Optics Corporation (POC). He received his Ph.D. in optoelectronics from Warsaw Technical University, Warsaw, Poland. POC was recently recognized by Inc. Magazine as the fastest growing privately held optoelectronics company in the United States. In his work at POC, Dr. Jannson invented and introduced holoplanar interconnects, which are based on planar Bragg multiplex holograms. Dr. Jannson has also developed a broad variety of commercial and military products in holography, optical information processing, fiber optics, and integrated optics. These products include holographic edge filters, holographic diffusers, holographic Fabry-Perot etalons, fiber optic multimodal WDM LANs, high-speed plasmon modulators, optoelectronic multichip modules, and many others. Dr. Jannson has over 100 publications and 30 patents. He is a member of OSA, SPIE, and IEEE.

William R. Michalson is an assistant professor in the Electrical and Computer Engineering Department at Worcester Polytechnic Institute, Worcester, Massachusetts. His past technical experience involved the design of supercomputer-class embedded computer architectures for performing real-time signal and data processing on space-based platforms. His current research involves computer architectures for highly parallel optical computation, optimizing the performance of scientific applications on massively parallel computers, and computer modeling and algorithm development for GPS-based navigation systems. He has authored numerous papers in the areas of optical computing, parallel numerical methods, and computer architecture. He received his Ph.D. from Worcester Polytechnic Institute in 1988. Dr. Michalson is a member of ACM, IEEE, OSA, SPIE, and ION.

Rick Morrison is currently a member of the technical staff in the Photonic Switching Department at AT&T Bell Laboratories in Naperville, Illinois. He is also a part-time faculty member at the Physics Department at DePaul University. He earned his M.A. and Ph.D. in physics from Yale University in 1979 and 1985, respectively and attained his B.S. in physics from Iowa State University. Since 1987, Dr. Morrison has worked actively with a group at Bell Laboratories that has demonstrated free-space digital optical systems of increasing complexity and functionality. His primary interests include diffractive optics and high-power semiconductor lasers for application in high-bandwidth networks, photonic switching systems, and optical computing. He has demonstrated new designs for diffraction optic devices that generate spot array illumination. He has also developed diagnostic systems necessary to characterize the operation of these new photonic systems.

Ronald Nordin is currently a technical staff member at AT&T Bell Laboratories in Naperville, Illinois. Dr. Nordin works in the Advanced Technology Laboratory's Advanced

Switching Technology Group where he is involved with low-cost optical interconnection technology for high data rate telecommunication systems. Dr. Nordin received his B.S.E.E. in 1977 from Purdue University and his M.S.E.E. and Ph. D. from Northwestern University in 1979 and 1984, respectively. Dr. Nordin's Ph.D. thesis, "Materials for Possible Application as Photo-DeIntercalation Devices," pioneered integrated solid-state solar batteries. He has published over 17 technical papers, has one patent, designed more than 12 analog integrated circuits, 6 digital integrated circuits, and several supporting ancillary circuits and systems for Bell Laboratories. Dr. Nordin is a senior member of the IEEE. and vice chairman of the local Electron Device Society. Since 1984, Dr. Nordin has been teaching solid-state physics, analog/digital integrated circuit design, electromagnetic field theory, network theory, and electronic circuits courses part-time at local colleges.

Stanley Reich received a B.E.E. and M.E.E. from the Polytechnic Institute of Brooklyn. He is an engineering specialist in the Grumman Corporate Exploratory Development Laboratory and is responsible for the design and development of photonic- integrated- and fiber optic-based systems for advanced avionic and space systems. His current activities include the development of high-speed analog and digital integrated and fiber optic communication links for low-temperature environments. Mr. Reich has more than 35 years of electronics and electro-optics experience, including the ongoing development of embedded fiber optic sensors for smart skins and structures; the design and analysis of photonic and integrated optical systems for control, time delay and beam steering of phased array radar and communications antennas employing T/R modules; and the investigation and development of optical backplanes and device interconnects. Before joining Grumman, Mr. Reich was a principal in a startup fiber optic networking company, developed optical scanning devices and systems for copying, inspection laser projection TV, and point of sales applications. Additionally, he was involved in the design and development of solid-state and stabilized HeNe lasers and systems.

Eric G. Schneider received his M.S.E.E. from the Worcester Polytechnic Institute in 1993, where he studied the architecture of optical computing systems. He has authored papers in the areas of optical interconnection and optical computing architecture. Mr. Schneider is currently a design engineer with Quickware Engineering and Design, Incorporated, where he is involved in the design and development of computer systems.

Robert Shih is a currently a cofounder at Alpha Photonics in El Monte, California. His previous position was as a senior research scientist at Physical Optics Corporation (POC). He received his Ph.D. in electrical engineering from the University of California, Los Angeles, where he was the first to demonstrate microwave and millimeter wave phase conjugation by degenerate four-wave mixing in an artificial Kerr medium. His pioneering effort in the characterization of nonlinear millimeter materials has effectively opened up the field of nonlinear optics in the millimeter wave region. At POC, he has developed high-speed polymer waveguide modulators using traveling wave electrodes. Dr. Shih also developed various GaAs-based waveguide modulators and switches. Currently, he is involved in the development of polymer based massive fan-out optical channel waveguides for optical backplane applications.

Index

Absorption, 133–34
    See also Electroabsorption modulator
Absorption spectrum, 276–77
Accelerated-aging tests, 96–97
Acousto-optic cellular array, 132–33
Acousto-optic modulator, 211–18
Active fiber backplane network, 90–96
Active processor, 130–31
Active splitter/active combiner, 342, 344
Active splitter/combiner network, 340–44
Aging tests, 96–97
Alignment, 13–14, 82
Analog network, 110
APD. *See* Avalanche photodiode
ASAC. *See* Active splitter/active combiner
AT&T MAC, 85, 87
AT&T Optiflex, 83
Avalanche detector, 240
Avalanche photodiode, 238

Backplane technology, 69–75
    capabilities of, 84–96
    environmental issues, 96–98
    fiber-embedded substrate, 77–84
    interconnect applications, 296–97
    intershelf communication, 75–76
    intrashelf communication, 69–75
    system requirements, 76–77
Banyan network, 127, 334–37
Baseline network, 335, 337
Basis-set approach, 53–54
Beam combination, 113–15
Beam propagation method, 204, 208
Benes network, 338–42, 347
BER. *See* Bit error rate
Bidirection switch, 324

Binary optics, 117, 119–22, 128
Bistable device, 131
Bit error rate, 5, 228, 230, 239, 296
Bit rate, 14
BIU. *See* bus interface unit
Blocking architecture. *See* Duobanyan network;
    Multistage interconnection network;
    n-Stage network
Board interconnection, 12, 234–35
Bottleneck, 6, 9–12, 322
BPM. *See* Beam propagation method
Bragg diffraction, 212, 214–18, 220–21, 265–66, 282, 284
Bragg modulator, 211–13
Break-even line length, 24–25, 29, 34–35
Buffer cladding, 160
Bus interconnection, 115–16
Bus interface unit, 228–29
Bus topology, 91, 100
Butt-coupled connector, 85–89

Capacitance, 16, 18, 22
Capacitance-limited break-even line, 25–26, 30–31
CCPD. *See* Charge coupled photo detector
Central processing unit, 4, 231–32
CGH. *See* Computer-generated hologram
Channel light leakage, 314–15
Channel light recycling, 315–16
Channel waveguide, 143, 160–66, 260–64
Charge coupled photo detector, 261
Chemical resistance testing, 97–98
Chip-level interconnect, 10–11
    electrical, 17–26
    fanout, 26–29
    optical, 15–17, 20–26, 31–35
    scaling, 29–31

Chip-to-chip interconnection, 236
Chirped Bragg gating lens, 190–93
Circuit-based switching network, 129
CMOS. *See* Complementary metal oxide semiconductor
Compatibility, 142–43
Complementary metal oxide semiconductor, 299–300
Complexity model, 45–47
Computer-generated hologram, 13, 50, 52, 54, 56
Connection density analysis
   introduction, 42–43
   lower boundaries, 43–50
   technology comparison, 55–60
   upper boundaries, 50–54
Connectivity, 3–4, 80–82
Cost, 14
Cotton-Mouton effect, 220
Coupled-mode theory, 281
Coupling, 5
   cryogenic signal, 299, 301
   evanescent field, 80–81, 85–86, 88–89, 93–94
   fiber-detector, 246–47, 249–50
   fiber-waveguide, 180–85
   fused-fiber, 92–93, 96
   longitudinal, 178, 180
   optical waveguide, 177–85, 233–34
   polymer circuits, 282
   polymetric, 93–94
   prism, 170
   single-mode, 182–83, 186
   transversal, 178–79, 185
CPU. *See* Central processing unit
Crossbar interconnection, 49–50
Crossbar network, 328–31, 347–51
Cross-linked hydrogen bond, 269
Crossover loss, 89–90
Crossover network, 123, 125
Crosstalk, 14, 266–68
Cryogenically generated signals, 299–316
Curved channel waveguide array, 260–64
Cutoff modulator, 205

Data rate, 14–15, 23, 41
DCG. *See* Dichromate gelatin
Deflection, 132–33
Delay, 34–35 232–2335
Demultiplexer, 193, 234, 265–66
Detector. *See* Integrated optoelectronic detector
Diagnostic tools, 135–37
Dichromate gelatin, 268
Diffraction, 120–21, 133, 212, 214–18, 220–21, 265–67, 281–84

Diffusion, 186
Digital routing network, 111
Directional coupler switch/modulator, 201–2
Dispersion curve, 153, 155–56, 163–64
Double-pass basis-set architecture, 54
Double-pass HOE architecture, 50–52
Driving gate, 17–18, 20
Dual-sided communication, 111–12
Duobanyan network, 336–38, 347

ECL. *See* Emitter-coupled logic
Edge emitting LED, 178
Effective index method, 162–64, 261
Effective refractive index, 145, 148
Efficiency, 31
Eigenequation, 151, 153, 158
Eigenfunction, 159–61, 184
Eigenvalues, 161, 184
Electrical interconnection, 4
   models, 37–42, 46–48
   performance, 70–76
   problem areas, 5–7
Electroabsorption modulator, 207–11
Electrode structure, 198, 200, 202
Electromagnetic pulse, 295
Electron, 13
Electro-optical harness, 83–84
Electro-optical transmitter, 143
Electro-optic effect, 171–72, 268
Electro-optic polymer, 268–72
Electro-optic waveguide devices
   directional coupler, 201–2
   electroabsorption, 207–11
   index distributed, 202–7
   modulator, 279–81
   phase modulator, 195–201
ELED. *See* Edge emitting LED
Embedding, 298
Emitter-coupled logic, 280
EMP. *See* Electromagnetic pulse
Energy dissipation, 22
Energy-limited line length, 31
Energy-versus-time plot, 22–23
Environmental issues, 96–98
EO. *See* Electro-optical effect;
         Electro-optic waveguide devices;
         Electro-optic polymer
Epitaxial growth, 237, 244
ERFC. *See* Error function
Error function, 172
ESA. *See* Excited state absorption
Evanescent field coupling, 80–81, 85–86, 88–89, 93–94

Excited state absorption, 274

Fabrication, 167–77
Fanout, 26–29, 95
Fanout density, 281–84
Faraday effect, 220
FDMA. *See* Frequency division multiple access
FET. *See* Field-effect transistor
Fiber amplifier, 95
Fiber-array connector, 85–89
Fiber coupler technology, 96
Fiber-detector coupling, 246–47, 249–50
Fiber-embedded substrate, 77–84
Fiber optics, 4–5, 70–76
Fiber packaging, 89–90
Fiber-to-the-home, 92
Fiber-waveguide coupling, 180–85
Field coupling, 80–81
Field-effect transistor, 232, 240
Flammability testing, 97–98
Flexible substrate, 78–82
Fluorescence lifetime, 274
Fluorescent spectrum, 277
Four-channel demultiplexer, 265–66
Four-channel laser array, 244, 246
Fourier plane filtering architecture, 52
Four-port switch. *See* Benes network; Crossbar network; Duobanyan network; Multistage network; *n*-Stage network
Free-space routing, 103–9
  architectures, 110–17
  components, 110–17
  networks, 117–28
  power supplies, 135–37
  processor issues, 128–35
  spatial light modulators, 128–35
  system diagnostic tools, 135–37
Frequency division multiple access, 92
Fresnel lenses, 187–91
FTTH. *See* Fiber-to-the-home
Fujikara MT connector, 85
Full width at half-maximum, 202, 207
Fused-fiber coupler, 92–93
FWHM. *See* Full width at half-maximum

GaAs fabrication, 172–76
Gabor's theorem, 45
Gate(s), 17, 20
  count, 65–66
  power equation, 7–8
Gaussian lens formula, 313
Gelatin, 268, 272
Geodesic lens, 185, 187, 190

Geometric optical elements, 117, 119–20, 128
Graded-index Fresnel lens, 190–91
Graded-refractive index amplifier, 277–78
Graded-refractive index modulator, 279
Graded-refractive-index profile, 144, 152, 258–60, 275
Graded-refractive-index waveguide, 152–54, 158, 166–67
Gradient-thickness Fresnel lens, 189–91
Graphical method, 164–65
Grating, 281, 285
  coupler, 193–94
  lens, 187
  modulation, 188–89, 267
GRIN profile. *See* Graded-refractive index profile
Growth rate, 58
GRTH Fresnel lens. *See* Gradient-thickness Fresnel lens
Guided-wave network, 324–27

HB method. *See* Horizontal Bridgeman method
HBT. *See* Heterojunction bipolar transistor
Heterojunction bipolar transistor, 195
HOE. *See* Holographic optical element
Hologram. *See* Waveguide hologram
Holographic optical element, 42, 50–52, 117, 119–28
Horizontal Bridgeman method, 236
Hostile environments, 295–96
  backplane applications, 296–97
  cryogenically generated signals, 299–316
  embedding, 298
  MCM interconnections, 316–18
  smart skins, 297–98
Hybrid mode, 160, 162–163, 165

IC. *See* Integrated circuit
Index distributed switch/modulator, 202–7
Index modulation, 261–62, 264
Index tuning, 258
Indiffusion fabrication, 171–172
Indium tin oxide, 270
Information processing, 3–4
Input/output pin, 74–75
Integrated circuit, 4, 227
Integrated optics, 142
Integrated optoelectronic detector, 237–40
Integrated optoelectronic receiver, 237–40
Integrated optoelectronic transmitter, 240–44
Intensity modulator, 199–201
Interconnect bottleneck, 9–12
Interconnection density, 89–90
Interconnection hierarchy, 65–66

Interconnection technology, 90
Interference, 4
Intershelf communication, 75–76
Intrashelf communication, 69–75
Inverse Wentzel-Kramers Brillouin, 258–59
Ion-exchange fabrication, 168–71
I/O pin. *See* Input/output pin
ITO. *See* Indium tin oxide
IWKB. *See* Inverse Wentzel-Kramers Brillouin

JFET. *See* Junction field-effect transistor
Junction field-effect transistor, 237–38

La Grange multiplier, 19
LAN. *See* Local area network
Large guided-wave network, 324–27
Laser diode, 178–82
Laser optical power supplies, 135–37
Lasers. *See* Waveguide lasers
LD. *See* Laser diode
LEC. *See* Liquid-encapsulated Czochralsky
LED. *See* Light emitting diode
Lenses. *See* Waveguide lenses
Light emitting diode, 135, 178
Light modulator, 32–34
Linear channel waveguide array, 260–64
Liouville theorem, 178
Lippman hologram, 286
Lippmann grating, 285–86
Liquid crystal array, 132–33
Liquid-crystal cell modulator, 133–34
Liquid-encapsulated Czochralsky, 236
Liquid phase epitaxy, 237
Littrow grating, 285
Local area network, 228
Locality, 12
Longitudinal coupling, 178, 180
Lorentz-Lorenz formulation, 258
Lower boundaries, 43–50, 55–60
LPE. *See* Liquid phase epitaxy
Luneberg lens, 185, 190

Mach-Zehnder interferometer, 200, 205, 280–81, 301–10, 314–15
Magneto-optic modulator, 218–20
Magnetostatic backward volume wave, 218
Magnetostatic forward volume wave, 218
Magnetostatic surface wave, 218
Magnetostatic wave, 218, 220
Marcatili's five-region method, 261
Maxwell equations, 151, 159
MBE. *See* Molecular beam epitaxy
MCM. *See* Multichip module
MCT. *See* Mercury cadmium telluride

Mechanical testing, 97
Medium-scale integration, 227
Mercury Cadmium Telluride, 299
MESFET. *See* Metal-semiconductor field-effect transistor
Messages, 7
Metal organic chemical vapor deposition, 237, 240, 247
Metal-semiconductor field-effect transistor, 230, 232, 237–38, 240
Metal semiconductor metal detector, 240, 242, 247
Methyl nitroaniline, 270
Michalson-Tocci architecture, 320–21
Microdot technique, 81
Microelectronic system, 9–12
Miller capacitance, 18
MIN. *See* Multistage interconnection network
Misalignment, 313
Mixing region ratio, 94
MM. *See* Multimode fiber
MNA. *See* Methyl nitroaniline
MOCVD. *See* Metal organic chemical vapor deposition
Modal field distribution. *See* Eigenfunction
Modal index, 155–57
Modal indices. *See* Eigenvalues
Mode converter, 206–9
Mode structures, 160
    See also Eigenfunction
MODFET, 237
Modulation
    acousto-optic, 211–18
    cutoff, 205
    electro-optic polymer, 271, 279–83
    grating, 188–89, 267
    index, 261–2, 264
    intensity, 199-201
    light, 32–34
    liquid crystal cell, 133–34
    magneto-optic, 218–20
    phase, 195–201
    signal transmission and, 32–34
    spatial light, 128–35, 329
Molecular beam epitaxy, 237
Monochromatic plane wave, 144
Monolithic receiver, 242–43
MQW. *See* Multiquantum well
MSBVW. *See* Magnetostatic backward volume waves
MSFVW. *See* Magnetostatic forward volume waves
MSI. *See* Medium-scale integration

MSM detector. *See* Metal semiconductor metal detector
MSSW. *See* Magnetostatic surface waves
MSW. *See* Magnetostatic waves
Multichip module, 9, 65, 261
    high-bandwidth interconnection, 316–18
    interconnection, 11–12, 35–42
Multifiber connector, 80
Multimode channel waveguide, 261
Multimode fiber, 78
Multimode waveguide, 148–51, 170
Multiplexer, 234, 244
Multiquantum well, 195, 210
Multiquantum well switch, 325, 351
Multistage interconnection network, 115–18, 332–36
Multiwire technology, 78
MUX. *See* Multiplexer
MZI. *See* Mach-Zehnder interferometer

Networks
    analysis of, 326–28, 344–47
    banyan, 127, 334–37
    benes, 338–42, 347
    circuit-based switching, 129
    crossbar, 328–31, 347–51
    digital routing, 111
    duobanyan, 336–38, 347
    large-guided wave, 324–27
    multistage, 115–18, 332–36
    n-stage, 331–32, 347
    parallel computing and, 319–20
    passive fiber backplane, 90–96
    passive optical, 92
    shared link, 51–52
    single-stage, 328
    splitter/combiner, 340–44, 347
    system operation, 351–53
    3-D integrated architecture, 320–24
    3-D integrated crossbar, 347–51
Network topology, 90–96
Neumann, John Von, 6
Nexus area, 45
Node capacitance, 239
Nonlinear all optical switch, 287–88
Nonreturn to zero, 244
Normalized effective index, 163
Normalized modal index, 155
Normalized thickness, 155, 176
NRZ. *See* Nonreturn to zero
N-stage network, 331–34, 347

OD. *See* Optical density
OEIC. *See* Optoelectronic integrated circuit

Omega network, 335, 337
1D-ODL. *See* One-dimensional optical data link
One-dimensional optical data link, 67
ONU. *See* Optical network unit
Optical density, 275
Optical interconnection networks. *See* Networks
Optical interconnection, 4–5, 13–14, 35–37, 46–47
    See also Optoelectronic interconnection
Optical link efficiency, 16
Optical link parameters, 31–32
Optical network unit, 92
Optical power, 135–37, 230–31, 233
Optical waveguide coupler, 177–85
Optical waveguide lens, 185–93
Optical waveguides, 77–84
    active guided wave devices, 193–94
        acousto-optic, 211–18
        electro-optic, 195–211
        magneto-optic, 218–20
    channel, 160–66
    embedded, 173–74
    fabrication, 167–77
    introduction, 143–44
    planar medium, 166–67
    properties, 144–51
    ridged, 173–74
    slab, 151–60
Optical wiring, 78–80
Optiflex, 83
Optoelectronic integrated circuit, 173, 227
Optoelectronic interconnection, 227–28
    design, 228–36
    detectors, 237–40
    electronics, 237
    materials, 236–37
    packaging, 244–251
    processing, 236–37
    receivers, 237–40
    transmitters, 240–44
Outdiffusion, 172

Packaging, 89–90, 244–51
    density, 261–262
Parallel optical interconnection, 67
Parallel processing, 6
Passive/active bus applications, 71–73
Passive device array, 129–30
Passive fiber backplane network, 90–96
Passive nodes, 129
Passive optical bus, 308, 311
Passive optical network, 92
Passive optical node, 317
Passive splitter active combiner, 342, 344

Passive splitter/combiner network, 340–44
PCB. *See* Printed circuit board
PE. *See* Processing element
Performance, 14
    backplane technology, 84–98
    fiber optics versus electrical, 66, 73–76
Phase modulators, 195–201
Photon, 13
PIH technology. *See* Pin-in-hole technology
p-i-n detector, 240–41
Pin-in-hole technology, 12
Pin spacing, 12
Planar fabrication, 187
Planar integration, 111–12
Planar medium, 166–67
Plank's constant, 16
Plastic cladded fiber, 78
Plastic optical fiber, 94
PMF. *See* Polarization maintaining fiber
PMSW. *See* Polymer microstructure waveguide
POF. *See* Plastic optical fiber
Point-to-point interconnection, 71–72, 75, 91, 100, 228
Polarization, 133, 151–60, 163
Polarization maintaining fiber, 298
Polishing, 82
Polymer-based fabrication, 176–77
Polymer-based photonic circuit, 255–56
    curved channel array, 260–64
    electro-optic modulator, 279–81
    fanout density, 281–84
    further applications, 287–90
    hologram formation, 264–68
    linear channel array, 260–64
    rare-earth amplifier, 272–78
    reliability test, 284–87
    waveguide formation, 256–60
    X^2 polymer, 268–72
Polymer microstructure waveguide, 256–60
Polymetric mixing rod coupler, 93–94
Polysilicon lines, 19–20
PON. *See* Passive optical network
Position sensors, 289–90
Power dissipation, 14, 41
Power-speed tradeoff, 14–15, 22–23
    chip-level interconnects, 15–35
    MCM-level interconnects, 35–42
Power supplies. *See* Laser optical power supplies
Printed circuit board, 4, 65
    backplane requirements, 76–77
    fiber-embedded substrate, 77–84
    intershelf communication, 75–76

    intrashelf communication, 69–75
    PCB-PCB interconnection, 65–68
Prism coupling, 170
Processing element, 42, 45–50
Processor issues, 128–35
Proton-exchange fabrication, 173
PSAC. *See* Passive splitter active combiner

Radiation hardness, 295
Radii of curvature, 262
Raman-Nath diffraction, 212
Rare-earth ion doped amplifier, 272–78
RC time constant. *See* Resistor-capacitor time constant
Receiver. *See* Integrated optoelectronic receiver; Monolithic receiver
Refractive index profile, 258–59
REI. *See* Rare-earth ion doped amplifier
Reliability, 14, 96–98, 284–87
Rent's rule, 6, 66
Resistor-capacitor delay, 28, 57
Resistor-capacitor-limited line, 25, 29–30
Resistor-capacitor time constant, 6, 57
Ridged waveguide, 184, 187
Rigid substrate, 78–82
Ring topology, 91, 100, 228–29
Rise time, 14, 17, 19, 21–24, 26, 28, 30, 32–33, 35–36
ROC. *See* Radii of curvature

SAW. *See* Surface acoustic wave
Scalability, 6
Scalable coherent interface, 73
Scaling, 29–31
Schneider architecture, 322
Schottky detector, 239
SCI. *See* Scalable coherent interface
SCSI. *See* Small computer system interface
SEED. *See* Self-electro-optic-effect device; Symmetric-SEED
Self-electro-optic-effect device, 322-24
    multiquantum well, 133–34
Serial processor, 6
Shared link network, 51–52
Shuffle interconnection, 124
Signal-to-noise ratio, 296, 300, 305
Single-mode coupling, 182–83, 186
Single-mode fiber, 78
Single-mode propagation, 158, 161–66
Single-mode waveguide, 148–51, 177, 179, 182, 184, 187, 261, 263
Single-pole-single-throw, 303
Single-sided communication, 111–13

Single-stage network, 328
Skew performance, 73–74, 76
Slab waveguide, 143, 151–60
SLM. *See* Spatial light modulator
Small computer system interface, 73
Smart skins, 297–98
SM fiber. *See* Single-mode fiber
S/N. *See* Signal-to-noise ratio
Snell's law, 146–47
Space-variant interconnects, 53
Spatial light modulator, 128–35, 329
SPICE program, 19
Splitter/combiner network, 340–44, 347
Splitter function, 92–93
SPST. *See* Single-pole-single-throw
S-SEED. *See* Symmetric SEED
Star topology, 91, 100, 297
Step-index channel waveguide, 165
Substrate technology, 77–78
    flexible, 82–84
    rigid, 78–82
Surface acoustic wave, 211, 214–16
Switches, 324–26, 345–46, 351
    See also Four-part switch; Three-port switch
Switching energy, 14–22, 24, 26, 33–38, 131, 202–3
Symmetric SEED, 123, 126–28, 134, 320–21, 323, 351–52
Systolic array, 288–89

Tapping power fraction, 305, 309–10
TDMA. *See* Time division multiple access
TE. *See* Transverse electric
Telephony over PON, 92
Thompson's model, 47
3-D integrated crossbar network, 347–51
3-D integrated network, 320–24
Three-dimensional waveguide.
    See Channel waveguide
Three-port switch. *See* Splitter/combiner network
Threshold limited break-even line, 24–25, 31
Threshold-limited region, 26
Threshold power, 31
Ti indiffusion, 171–72
Time division multiple access, 92, 296
TIR. *See* Total internal reflection
TM. *See* Transverse magnetic
Tocci architecture. *See* Michalson-Tocci architecture
Total internal reflection, 143, 146, 148, 204–5
TPF. *See* Tapping power fraction
TPON. *See* Telephony over PON
Transmission line, 38–39
Transmitter. *See* Integrated
    optoelectronic transmitter

Transparency, 142
Transversal coupling, 178–79, 185
Transverse electric, 279, 281, 283
    mode conversion, 206–7, 217
    polarization, 151, 155–60, 163
Transverse magnetic mode, 279
    polarization, 151, 155–56, 159
    mode conversion, 206–7, 217
Two-mode waveguide, 204

Ullman's model, 47

Very large scale integrated circuit, 10
Very large scale integrated design, 5–6
    connection density, 43–54
    cost, 55–60
Very large scale integrated optoelectronic computer
    connection density, 43–54
    cost, 55–60
VLSI design. *See* Very large scale integrated design
VLSIO computer. *See* Very large scale integrated optoelectronic computer
Von Neumann bottleneck, 6

Wafer stack, 349
Waveguide coupler, 177–85
Waveguide hologram, 264–68, 275–76, 286
Waveguide laser, 273, 275–76
Waveguide lens, 185–93
Waveguides. *See* Optical waveguides
Wavelength division demultiplexer, 265, 267–68
Wavelength division multiple access, 92, 289
Wave propagation, 141
WDDM. *See* Wavelength division demultiplexer
WDMA. *See* Wavelength division multiple access
Wentzel-Kramers Brillouin. *See* Inverse Wentzel-Kramers Brillouin
χ^2 *electro-optic polymer*, 268–72

YIG waveguide, 218, 221

Zigzag wave, 149–50, 152

The Artech House Optoelectronics Library

Brian Culshaw, Alan Rogers, and Henry Taylor, *Series Editors*

Acousto-Optic Signal Processing: Fundamentals and Applications, Pankaj Das

Amorphous and Microcrystalline Semiconductor Devices, Optoelectronic Devices, Jerzy Kanicki, editor

Electro-Optical Systems Performance Modeling, Gary Waldman and John Wootton

The Fiber-Optic Gyroscope, Hervé Lefèvre

Field Theory of Acousto-Optic Signal Processing Devices, Craig Scott

High-Power Optically Activated Solid-State Switches, Arye Rosen and Fred Zutavern, editors

Highly Coherent Semiconductor Lasers, Motoichi Ohtsu

Germanate Glasses: Structure, Spectroscopy, and Properties, Alfred Margaryan and Michael A. Piliavin

Introduction to Electro-Optical Imaging and Tracking Systems, Khalil Seyrafi and S. A. Hovanessian

Introduction to Glass Integrated Optics, S. Iraj Najafi

Optical Control of Microwave Devices, Rainee N. Simons

Optical Document Security, Rudolf L. van Renesse

Optical Fiber Amplifiers: Design and System Applications, Anders Bjarklev

Optical Fiber Sensors, Volume I: Principles and Components, John Dakin and Brian Culshaw, editors

Optical Fiber Sensors, Volume II: Systems and Applicatons, John Dakin and Brian Culshaw, editors

Optical Interconnection: Foundations and Applications, Christopher Tocci and H. John Caulfield

Optical Network Theory, Yitzhak Weissman

Optical Transmission for the Subscriber Loop, Norio Kashima

Principles of Modern Optical Systems, Volumes I and II, I. Andonovic and D. Uttamchandani, editors

Reliability and Degradation of LEDs and Semiconductor Lasers, Mitsuo Fukuda

Semiconductors for Solar Cells, Hans Joachim Möller

Single-Mode Optical Fiber Measurements: Characterization and Sensing, Giovanni Cancellieri

For further information on these and other Artech House titles, contact:

Artech House
685 Canton Street
Norwood, MA 01602
617-769-9750
Fax: 617-762-9230
Telex: 951-659
email: artech@world.std.com

Artech House
Portland House, Stag Place
London SW1E 5XA England
+44 (0) 71-973-8077
Fax: +44 (0) 71-630-0166
Telex: 951-659